1989

COMPUTABILITY AND LOGIC

Mathematics and its Applications

Series Editor: G. M. BELL, Professor of Mathematics, King's College London (KQC), University of London

Statistics and Operational Research

Editor: B. W. CONOLLY, Professor of Operational Research, Queen Mary College, University of London

Mathematics and its applications are now awe-inspiring in their scope, variety and depth. Not only is there rapid growth in pure mathematics and its applications to the traditional fields of the physical sciences, engineering and statistics, but new fields of application are emerging in biology, ecology and social organisation. The user of mathematics must assimilate subtle new techniques and also learn to handle the great power of the computer efficiently and economically.

The need for clear, concise and authoritative texts is thus greater than ever and our series will endeavour to supply this need. It aims to be comprehensive and yet flexible. Works surveying recent research will introduce new areas and up-to-date mathematical methods. Undergraduate texts on established topics will stimulate student interest by including applications relevant at the present day. The series will also include selected volumes of lecture notes which will enable certain important topics to be presented earlier than would otherwise be possible.

In all these ways it is hoped to render a valuable service to those who learn, teach, develop and use mathematics.

Mathematics and its Applications

Series Editor: G. M. BELL, Professor of Mathematics, King's College (KQC), University of London

Artmann, B.	**The Concept of Number**
Balcerzyk, S. & Joszefiak, T.	**Commutative Rings**
Balcerzyk, S. & Joszefiak, T.	**Noetherian and Krull Rings**
Baldock, G.R. & Bridgeman, T.	**Mathematical Theory of Wave Motion**
Ball, M.A.	**Mathematics in the Social and Life Sciences: Theories, Models and Methods**
de Barra, G.	**Measure Theory and Integration**
Bell, G.M. and Lavis, D.A.	**Co-operative Phenomena in Lattice Models Vols. I & II**
Berkshire, F.H.	**Mountain and Lee Waves**
Berry, J.S., Burghes, D.N., Huntley, I.D., James, D.J.G. & Moscardini, A.O.	**Mathematical Modelling Courses**
Berry, J.S., Burghes, D.N., Huntley, I.D., James, D.J.G. & Moscardini, A.O.	**Mathematical Methodology, Models and Micros**
Berry, J.S., Burghes, D.N., Huntley, I.D., James, D.J.G. & Moscardini, A.O.	**Teaching and Applying Mathematical Modelling**
Burghes, D.N. & Borrie, M.	**Modelling with Differential Equations**
Burghes, D.N. & Downs, A.M.	**Modern Introduction to Classical Mechanics and Control**
Burghes, D.N. & Graham, A.	**Introduction to Control Theory, including Optimal Control**
Burghes, D.N., Huntley, I. & McDonald, J.	**Applying Mathematics**
Burghes, D.N. & Wood, A.D.	**Mathematical Models in the Social, Management and Life Sciences**
Butkovskiy, A.G.	**Green's Functions and Transfer Functions Handbook**
Butkovskiy, A.G.	**Structural Theory of Distributed Systems**
Cao, Z-Q., Kim, K.H. & Roush, F.W.	**Incline Algebra and Applications**
Chorlton, F.	**Textbook of Dynamics, 2nd Edition**
Chorlton, F.	**Vector and Tensor Methods**
Cohen, D.E.	**Computability and Logic**
Crapper, G.D.	**Introduction to Water Waves**
Cross, M. & Moscardini, A.O.	**Learning the Art of Mathematical Modelling**
Cullen, M.R.	**Linear Models in Biology**
Dunning-Davies, J.	**Mathematical Methods for Mathematicians, Physical Scientists and Engineers**
Eason, G., Coles, C.W. & Gettinby, G.	**Mathematics and Statistics for the Bio-sciences**
Exton, H.	**Handbook of Hypergeometric Integrals**
Exton, H.	**Multiple Hypergeometric Functions and Applications**
Exton, H.	***q*-Hypergeometric Functions and Applications**

Series continued at back of book

COMPUTABILITY AND LOGIC

DANIEL E. COHEN, B.A., M.A., D.Phil.
Reader in Pure Mathematics
Queen Mary College, University of London

ELLIS HORWOOD LIMITED
Publishers · Chichester

Halsted Press: a division of
JOHN WILEY & SONS
New York · Chichester · Brisbane · Toronto

First published in 1987 by
ELLIS HORWOOD LIMITED
Market Cross House, Cooper Street,
Chichester, West Sussex, PO19 1EB, England
The publisher's colophon is reproduced from James Gillison's drawing of the ancient Market Cross, Chichester.

Distributors:
Australia and New Zealand:
JACARANDA WILEY LIMITED
GPO Box 859, Brisbane, Queensland 4001, Australia
Canada:
JOHN WILEY & SONS CANADA LIMITED
22 Worcester Road, Rexdale, Ontario, Canada
Europe and Africa:
JOHN WILEY & SONS LIMITED
Baffins Lane, Chichester, West Sussex, England
North and South America and the rest of the world:
Halsted Press: a division of
JOHN WILEY & SONS
605 Third Avenue, New York, NY 10158, USA

British Library Cataloguing in Publication Data
Cohen, Daniel E.
Computability and logic. — (Mathematics and its applications).
1. Machine theory
I. Title II. Series
519.4 QA267

Library of Congress Card No. 87–4088

ISBN 0–7458–0034–3 (Ellis Horwood Limited)
ISBN 0–470–20847–3 (Halsted Press)

Phototypeset in Times by Ellis Horwood Limited
Printed in Great Britain by Unwin Bros., Woking

Table of Contents

Preface . . . 9

Notation . . . 13

Part I COMPUTABILITY

1 Epimenides, Gödel, Russell, and Cantor
- 1.1 Epimenides . . . 17
- 1.2 Gödel . . . 18
- 1.3 Russell . . . 18
- 1.4 Cantor . . . 19

2 Informal theory of computable functions
- 2.1 Functions . . . 20
- 2.2 Strings . . . 22
- 2.3 Computable functions, listable sets and decidable sets . . . 23
- 2.4 Universal functions and undecidable sets . . . 29
- 2.5 Rice's theorem . . . 32

3 Primitive recursive functions
- 3.1 Primitive recursion . . . 36
- 3.2 Bounded quantifiers and minimisation . . . 41
- 3.3 Examples using bounded minimisation . . . 43
- 3.4 Extensions of primitive recursion . . . 45
- 3.5 Functions of one variable . . . 47
- 3.6 Some functions which are not primitive recursive . . . 51
- 3.7 Justifying definitions by primitive recursion . . . 57

4 Partial recursive functions
- 4.1 Recursive and partial recursive functions . . . 60
- 4.2 Recursive and recursively enumerable sets . . . 63

5 Abacus machines
5.1 Abacus machines 67
5.2 Computing by abacus machines 69
5.3 Partial recursive functions 73
5.4 Register programs 76

6 Turing machines
6.1 Turing machines 80
6.2 Computing by Turing machines 88

7 Modular machines
7.1 Turing machines 95
7.2 Modular machines 96
7.3 Partial recursive functions and modular machines 98
7.4 Kleene's Normal Form Theorem 99

8 Church's Thesis and Gödel numberings
8.1 Church's Thesis 103
8.2 Gödel numberings 107

9 Hilbert's tenth problem
9.1 Diophantine sets and functions 111
9.2 Coding computations 115
9.3 Removal of the relation $\leqslant$ 118
9.4 Exponentiation 119
9.5 Gödel's sequencing function and min-computable functions . 123
9.6 Universal diophantine predicates and Kleene's Normal Form Theorem 125
9.7 The four squares theorem 127

10 Indexings and the recursion theorem
10.1 Pairings 129
10.2 Indexings 130
10.3 The recursion theorem and its applications 132
10.4 Indexings of r.e. sets 137
10.5 The diophantine indexing 139

Part II LOGIC
11 Propositional logic
11.1 Background 145

11.2 The language of propositional logic 147
11.3 Truth 155
11.4 Proof 158
11.5 Soundness 167
11.6 Adequacy 168
11.7 Equivalence 170
11.8 Substitution 172

12 Predicate logic
12.1 Languages of first-order predicate logic 174
12.2 Truth 182
12.3 Proof 185
12.4 Soundness 189
12.5 Adequacy 191
12.6 Equality 197
12.7 Compactness and the Lowenheim–Skolem theorems 199
12.8 Equivalence 201

13 Undecidability and incompleteness
13.1 Some decidable theories 203
13.2 Expressible sets and representable functions 206
13.3 The main theorems 212
13.4 Further results 215

14 The natural numbers under addition
14.1 The order relation on **Q** 223
14.2 The natural numbers under addition 225

Notes 230

Further reading 234

Index of special symbols 237

Index 239

To the memory of my parents,
Amy and Leonard Cohen

Preface

The theory of computable functions could be described as the pure mathematics of computer science.

Mathematicians are finding more and more that they need to use computers (not only for their number-crunching power, which is how mathematicians found them helpful originally). Courses involving the practicalities of computing are becoming common in mathematics degrees. I feel strongly that a course explaining the nature and behaviour of computable functions should be an option in any well-designed mathematics degree.

The Incompleteness and Undecidability Theorems of mathematical logic are high points in our understanding of the scope and limits of mathematical reasoning, and their proofs should also be available in a course which is an option for a mathematics degree. The Incompleteness Theorem is one which many students will have vaguely heard of; it says (roughly) that for any reasonable set of axioms for the natural numbers **N** there will be a sentence which cannot be proved from these axioms but is nonetheless true in **N**. The Undecidability Theorem says that there is no mechanical process for obtaining all provable sentences. It could be regarded as saying that mathematics is essentially a creative process which cannot be mechanised (even theoretically, let alone practically). As such, it could be regarded as justifying the existence of mathematicians, which is surely something any student of mathematics should be aware of!

This book is a text for a course dealing with these matters. It is written by a pure mathematician, and is intended for mathematicians and theoretical computer scientists. Because I wrote it for mathematicians, I have not assumed any knowledge of programming. Most books on computability are intended primarily for computer scientists, and so assume a detailed knowledge of programming. Occasionally I mention some aspects of programming. This is just to help the reader who has some slight knowledge (no more, usually, than BASIC for a microcomputer) and can be ignored by other readers.

The amount of mathematical knowledge needed as a prerequisite for this book is slight. The algebra of sets, the principle of induction, and a precise definition of a function would be enough. As often happens with material which does not have many prerequisites, the level of mathematical maturity needed is quite high, and the intended readers are final-year undergraduates or beginning postgraduates.

There are several novelties in the treatment in this book, notably in the types of machines used and the proofs and uses of results about diophantine properties. In particular, the discussion of diophantine properties is carried through in a way which permits it to be used for the proofs of major theorems without needing a detailed discussion of Turing machines and modular machines.

The first chapter introduces various aspects of self-reference, a theme which plays a major part in the key results. Chapter 2 is about the informal theory of computable functions; it turns out that most of the results we encounter later can be proved using only a very intuitive notion of computability.

Chapters 3 and 4 formalise the theory, discussing primitive recursive and partial recursive functions.

The next three chapters look at various machines. Chapter 5 contains the first innovation, the use of abacus machines. These are just a variant of what are often called unlimited register machines. Because they are nicely structured, they are particularly easy to deal with. Chapter 6 considers Turing machines. Chapter 7 contains the second innovation, the use of modular machines. These are somewhat peculiar objects, but because they simultaneously act as a numerical coding of Turing machines and as machines in their own right, they provide very straightforward proofs of some major results (such as Kleene's Normal Formal Theorem). We finally discover that the sets of functions computable by each of these three types of machines coincide with the partial recursive functions.

Having proved this, in Chapter 8 we review the evidence for Church's Thesis which claims that the intuitively computable functions are exactly the partial recursive functions.

Chapter 9 is about Hilbert's Tenth Problem which asks if we can decide whether or not a polynomial with integer coefficients has integer solutions. Here the recent simplification due to Jones and Matijasevič is used. Kleene's Normal Form Theorem is given another proof using the results of this chapter, and the major Incompleteness and Undecidability Theorems of later chapters are also given several proofs, both conventional ones and also ones using the results of this chapter. This is the first time this approach has been used in a text at this level.

Chapter 10 is more technical. It covers aspects of enumerating the computable functions, including the important *s–m–n* and Recursion Theorems.

The remaining chapters are about logic. Chapter 11 is about propositional logic, and Chapter 12 about the deeper predicate logic. These chapters form a detailed introduction to mathematical logic. Chapter 13 contains the main Incompleteness and Undecidability Theorems, and Chapter 14 shows that, whereas number theory is undecidable, the theory of the natural numbers using addition (but not multiplication) is decidable.

This text had its origins in a course I have taught for some years at Queen Mary College, London University. This was a one-semester (33 hours) course on computability and logic given to final-year undergraduates who had already had a course in logic. The course consisted of Chapters 1 to 8 (omitting parts of Chapter 3), together with the recursive aspects of Chapter 12 and the first half of Chapter 13.

There are various ways of using this book as a text for students who have not had a previous course in logic. It could be used as a two-semester course in computability and logic, or the first part of the book could be used as a one-semester course in computability. It should also be possible to use the book as a one-semester course leading up to the Incompleteness and Undecidability Theorems by using the following approach. Begin with Chapters 1 to 5, followed by Chapter 9 (possibly only showing that r.e. sets are exponential diophantine, rather than the stronger result that they are diophantine). Then, instead of considering an arbitrary language of predicate logic and an arbitrary structure for the language, take only the language of number theory (possibly with an extra function symbol corresponding to exponentiation) and the structure **N**, and define truth only in **N**. Take the notion of proof to be the axiomatic one, and take for granted that this has the properties we would like. (I feel that if one wants to show the relevant properties of proof, then the natural deduction approach is better.) Finally, the Undecidability and Incompleteness Theorems can be proved, as shown, using the results of Chapter 9 (with the language extended by a symbol for exponentiation, and corresponding axioms, if desired).

Notation

The symbol ■ denotes the end of a proof.

'If and only if' is abbreviated to 'iff'.

The symbols **N**, **Z**, and **Q** stand for the natural numbers (that is, 0, 1, 2, and so on), the integers (positive, negative, or zero), and the rational numbers, respectively.

$\mathbf{N}^k$ is the set of k-tuples $(x_1,\ldots,x_k)$ of elements of **N**. Bold-face letters, such as **x**, **y**, and **z**, stand for tuples of elements of **N**; whether they are k-tuples, n-tuples and so on will depend on the situation in which they are used.

Standard notations of set theory, such as $\cup$, $\subset$, and so on, are used without further comment.

Standard logical symbols, such as $\wedge$ and $\exists$, are defined the first time they are used, but the reader is expected to be familiar with them.

Part I
COMPUTABILITY

1

Epimenides, Gödel, Russell, and Cantor

Some important themes of the book are given their simplest (but rather imprecise) forms in this chapter.

1.1 EPIMENIDES

The classical Greek writer Epimenides wrote, 'Cretans always lie'. This may seem like an example of ancient racial prejudice. But Epimenides was himself a Cretan. This doesn't prevent his remark from being a bit of racial self-hatred. However, it also leads to interesting logical and philosophical problems.

Is his statement true? If so, Cretans always lie. Since he was himself a Cretan, this statement, which was assumed true, must be a lie. We get a contradiction, and his statement cannot be true.

It then follows that this statement is a lie, and, hence, that Cretans do not always lie; we know, though, that one particular Cretan has lied on one particular occasion.

A related but stronger form of this paradox is the statement, 'This statement is false'. Is it true? If so, by what it says it must be false. Similarly, if it is false, since it asserts that it is false, it is true. So the statement cannot be either true or false. Presumably we have to decide that the statement is meaningless, though there is no obvious way of seeing this just from the form of the statement.

An even more troublesome form is to take two statements written on opposite sides of a piece of paper. One statement reads, 'The statement on the other side of this paper is true', while the other reads, 'The statement on the other side of this paper is false'. We get a very similar paradox to the previous one, but if we consider each statement in isolation there is no reason why it should not be true or false (for instance, there would be no problem if one of the statements was replaced by, 'The statement on the other side of this paper contains six words').

1.2 GÖDEL

Consider a statement closely related to the ones in the last section, namely, 'This sentence is unprovable'.

Can it be provable? If so, it is true. Hence, by what it says, it is unprovable. We get a contradiction, and so we conclude that the statement is in fact unprovable. It must also be true, since it says it is unprovable.

The alert reader may well feel that the matter does not stop here. After all, we have just shown that the statement is true. Isn't this a proof, and don't we run into the same problem as before?

The answer is that we must be more careful about what is meant by a proof. We should specify precisely what a proof is, using some formal system S (whose nature does not concern us as yet). Then the statement should be changed to read, 'This statement is not provable by means of the formal system S'. For the same reasons as before, the statement is not provable in S but is nonetheless true. This argument may be regarded as a proof of the statement; but if it is so regarded, it is a proof outside the system S, and no paradox occurs.

This is one major part of Gödel's Incompleteness Theorem. However, there has been a point glossed over in the above analysis. If we are looking at a formal system S, we must be looking only at statements formed in a language precisely specified as part of S. Our statement is not in that language. The other major part of Gödel's proof is his observation that numbers could be associated to statements. It is then possible to translate statements about statements into statements about numbers. Thus we can look at the statement, 'The statement in the system S whose number is n is unprovable in the formal system S'. This is a statement about the number n. Now it can be shown that there is a number k such that the meaning of the statement whose number is k turns out to be 'The statement whose number is k is unprovable in the system S'. So we get a true but unprovable statement, as required. The details, of course, are technical, and will be given in Chapter 13.

1.3 RUSSELL

In the early development of set theory it was thought that any property should define a set. It turns out that this assumption leads to a paradox, as was shown by Russell. Because of this, in formal set theory we have to be very careful with our definitions.

Russell considered the property that held of a set iff it was not a member of itself. Most sets are not members of themselves. For instance, the set of all cows is certainly not a cow. But if we consider the set of all thoughts that have ever been thought, this might be considered as a thought, and so might be regarded as a member of itself.

Suppose that every property defined a set. Following the above, we could define the set A to be $\{x; x \notin x\}$. So, by definition, for any x, we have $x \in A$ iff $x \notin x$. In particular, this holds for A. So we find that $A \notin A$ iff $A \notin A$, which is a contradiction. Hence we cannot have such a set, and not all properties define sets.

Russell based his paradox on some related results of Cantor.

1.4 CANTOR

Cantor was the founder of the theory of infinite sets. One of his results is that, if $P(X)$ denotes the set of all subsets of X, there can be no function from X onto $P(X)$.

For let f be a function from X to $P(X)$. For any x in X, fx is a subset of X which may or may not contain x. Now any property defines a set of all those elements of X which have the property. (Cantor assumed this. In formal set theory it is taken as an axiom. Unlike the closely related assumption that every property defines a set, it does not appear to lead to any paradoxes.)

Thus we have the set A defined as $\{x \in X; x \notin fx\}$. Hence, for x in X, $x \in A$ iff $x \notin fx$. It follows that A cannot be fa for any a. For, if so, taking x to be a, we would get the contradiction $a \in fa$ iff $a \notin fa$.

Cantor also showed that the set of all real numbers is uncountable. One way of showing this is to take a countable set A of reals strictly between 0 and 1 and to exhibit a real number not in A and strictly between 0 and 1.

So take such a set A, whose elements are $\alpha_1, \alpha_2, \ldots$. Each real between 0 and 1 can be uniquely written as an infinite decimal (recurring 0 being permitted, but not recurring 9). We write α_1 as $0{\cdot}a_{11}a_{12}a_{13}\ldots$, α_2 as $0{\cdot}a_{21}a_{22}a_{23}\ldots$, and, generally, α_n as $0{\cdot}a_{n1}a_{n2}a_{n3}\ldots$. Now define β to be $0{\cdot}b_1b_2b_3\ldots$, where b_n is 1 if $a_{nn}\neq 1$ and $b_n=2$ if $a_{nn}=1$. Then β is strictly between 0 and 1, and β cannot be in A since its nth decimal place differs from the nth place of α_n, and every element of A is α_n for some n.

The arguments used in both these results are very close to one another and to the arguments of Russell's Paradox, and, to some extent, to Gödel's method. If we regarded the elements a_{mn} as being arranged in an infinite square the elements a_{nn} would appear along the diagonal. So this argument, and other related arguments, is referred to as Cantor's Diagonal Argument. We shall see that it is at the heart of many of our negative results about computability.

2

Informal theory of computable functions

Most of the main results about computable functions can be proved with only an intuitive idea of what a computable function is. These results will be proved in the last three sections of this chapter (the first two sections contain some necessary introductory concepts). Later chapters will give various formal approaches to computable functions.

2.1 FUNCTIONS

The concept of function used in most branches of mathematics turns out to need modification when we consider computable functions, for reasons which will be made clear later in this chapter.

Definition Let X and Y be sets. A **partial function** f from X to Y is a function f from some subset A of X to Y. The set A is called the **domain** of f (abbreviated to dom f). If dom f is X itself we call f a **total function** from X to Y.

The word 'function' will always mean 'partial function'; if a function is total we will always state that it is total. In particular, the notation $f{:}X \to Y$ will mean that f is a partial function from X to Y. A total function from X to Y is often defined to be a subset S of $X \times Y$ such that for every x in X there is exactly one y in Y with (x, y) in S. Similarly, a partial function from X to Y can be regarded as a subset S of $X \times Y$ such that for every x in X there is at most one y in Y with (x, y) in S.

Examples For any sets X and Y, we have the **empty function** from X to Y. This is the function whose domain is the empty set; that is, the function is nowhere defined.

An example of a partial function f from **N** (**N** is the set $\{0, 1, \ldots\}$ of natural numbers) to **N** is given by $fx = y$ iff $x = y^2$; its domain is the set of squares.

We define $\mathbf{N}^k$ to be the set of k-tuples of elements of $\mathbf{N}$ (more generally, A^k is the set of k-tuples of elements of A, for any A). We use bold-face symbols, such as $\mathbf{x}$ and $\mathbf{y}$ to denote tuples, leaving it to the context to tell whether they are k-tuples, n-tuples, or what.

For any subset A of $\mathbf{N}^k$ we define the **characteristic function** of A, denoted by χ_A, to be the total function from $\mathbf{N}^k$ to $\mathbf{N}$ given by $\chi_A\mathbf{x} = 0$ if $\mathbf{x}$ is in A, and $\chi_A\mathbf{x} = 1$ if $\mathbf{x}$ is not in A. The **partial characteristic function of A**, denoted by χ_{Ap}, is the partial function from $\mathbf{N}^k$ to $\mathbf{N}$ given by $\chi_{Ap}\mathbf{x} = 0$ for $\mathbf{x}$ in A, and $\chi_{Ap}\mathbf{x}$ not defined if $\mathbf{x}$ is not in A. (It is more usual to define $\chi_A\mathbf{x}$ to be 1 for $\mathbf{x}$ in A and 0 for $\mathbf{x}$ not in A. However, the current definition seems to lead to simpler formulas, and the only important property is that the value of the characteristic function on $\mathbf{x}$ tells us whether or not $\mathbf{x}$ is in A.)

Let $f{:}X \rightarrow Y$ be a function, and let A and B be subsets of X and Y respectively. We define fA and $f^{-1}B$ just as for total functions. That is, fA is $\{y;\, y = fx \text{ for some } x \in A\}$ and $f^{-1}B$ is $\{x;\, fx \in B\}$. Notice, though, that fx need not be defined for all x in A, and that $fA = f(A \cap \operatorname{dom} f)$.

Let $f{:}X \rightarrow Y$ and $g{:}Y \rightarrow Z$ be functions. Their **composite** gf is the function from X to Z defined by $gfx = z$ iff there is some y in Y with $fx = y$ and $gy = z$. This is exactly the same definition as for total functions, but we have to take into account the fact that f and g are partial. Thus the domain of gf is $f^{-1}(\operatorname{dom} g)$.

Let f and g be functions from a set X to $\mathbf{N}$. Then the functions $f + g$ and $f.g$ from X to $\mathbf{N}$ are defined by $(f + g)x = fx + gx$ and $(f.g)x = fx.gx$ respectively. Since our functions are partial, both $f + g$ and $f.g$ have as domains the set $\operatorname{dom} f \cap \operatorname{dom} g$. Notice that when f is the total function from X to $\mathbf{N}$ which is constantly 0, the function $f.g$ is not total, but is the function whose domain is $\operatorname{dom} g$ and whose value is always 0. It can be shown that an attempt to define the product to be total in this case would not be compatible with the distributive law for addition and multiplication of functions.

The functions f and g from X to Y are regarded as equal if, for every x in X, either both fx and gx are defined and $gx = fx$ or neither fx nor gx are defined. This is the same as saying that $f = g$ if $\operatorname{dom} f = \operatorname{dom} g$ and f and g have the same value at any element of their common domain. If $gx = fx$ for every x in $\operatorname{dom} f$, we call g an **extension** of f and f a **restriction** of g.

The set $\mathbf{N}^2$ is bijective with $\mathbf{N}$. We shall need to give an explicit bijection. Write the elements of $\mathbf{N}^2$ in a table

$$\begin{array}{l}
(0,0)\ (0,1)\ (0,2)\ (0,3)\ \ldots \\
(1,0)\ (1,1)\ (1,2)\ (1,3)\ \ldots \\
(2,0)\ (2,1)\ (2,2)\ \ldots \\
(3,0)\ (3,1)\ \ldots \\
\ldots
\end{array}$$

Now write down the elements of this table by moving along the diagonals which go from north-east to south-west, that is, write them in the sequence

$(0,0),(0,1),(1,0),(0,2),(1,1),(2,0),(0,3),(1,2),(2,1),(3,0),(0,4),\ldots$. Since there are $k + 1$ pairs $(r.s)$ with $r + s = k$, we see that the pair (m,n) occurs in the position $1 + \ldots (m + n) + m$. Hence we have a bijection $J:\mathbf{N}^2 \to \mathbf{N}$ given by $J(m,n) = \frac{1}{2}(m + n)(m + n + 1) + m$.

We have two functions K and L from $\mathbf{N}$ to $\mathbf{N}$ such that $J^{-1}r = (Kr, Lr)$. They are given by the following formulas. Find s so that $\frac{1}{2}s(s + 1) \leqslant r < \frac{1}{2}(s + 1)(s + 2)$. Let m be $r - \frac{1}{2}s(s + 1)$. Then $m \leqslant s$, and $Kr = m$, and $Lr = s - m$.

We can use J to obtain bijections $J_k:\mathbf{N}^k \to \mathbf{N}$ for all k. We define J_1 to be the identity, and we define J_3 by $J_3(x_1,x_2,x_3) = J(x_1,J(x_2,x_3))$. If J_k has been defined, then J_{k+1} is defined by $J_{k+1}(x_1,\ldots,x_{k+1}) = J(x_1,J_k(x_2,\ldots,x_{k+1})$. The inverse of J_k has its components composites of K and L. For instance $J_3^{-1}r = (Kr, KLr, LLr)$.

2.2 STRINGS

Let A be a non-empty set, which we refer to as the **alphabet**. A **string** on A is just a finite sequence of elements of A. The set of all strings on A is denoted by A^+.

We write a string on A without commas between the elements. For instance, $a_2a_1a_4a_4$ is a string on the alphabet $\{a_1,a_2,a_3,a_4\}$, while 'dog' and 'dogma' are strings on the usual alphabet of twenty-six letters. Similarly 'CATS and dogs, not cats and DOGS!' is a string on an alphabet which contains both lower-case and upper-case letters together with various punctuation marks and the space. Also 137028 is a string on the alphabet consisting of the ten digits $\{0, 1, \ldots, 9\}$.

The string $a_1a_2\ldots a_n$ is said to have **length** n. If $1 \leqslant r < s \leqslant n$ the string $a_{r+1}\ldots a_s$ is called a **segment** of $a_1\ldots a_n$; it is a **proper segment** unless $r = 0$ and $s = n$, and is an **initial segment** if $r = 0$. For instance, 'dog' has length 3 and 137028 has length 6. The latter has among its segments 70 (of length 2) and 7028 (of length 4), while 13 and 1370 are initial segments; however, 1302 is not a segment of 137028.

Let $a_1\ldots a_n$ and $b_1\ldots b_m$ be strings. Their **concatenation** (also called their **join** or **product**) is the string $a_1\ldots a_nb_1\ldots b_m$ of length $m + n$. For instance, 137028 is the concatenation of 13 and 7028, while the concatenation of 'dogma' and 'tic' is 'dogmatic'.

It is easy to see that the set of strings on a finite or countably infinite alphabet is itself countably infinite. We will need for later use to obtain an explicit one–one mapping from this set into $\mathbf{N}$.

The usual n-ary representation does not provide a one–one map from the strings on $\{0,1,\ldots,n - 1\}$ into $\mathbf{N}$. For instance, the strings 1101, 01101, and 001101 all correspond to the same integer. However, for any positive n (including 1), there is a one–one map from $\{1,\ldots,n\}^+$ onto the set of positive integers. This map sends the string $a_0\ldots a_m$ to the integer $\Sigma a_i n^{m-i}$. If the set A is finite, there is a one–one map from A onto $\{1,\ldots,n\}$ for some n, and this map obviously extends to a one–one map from A^+ onto

$\{1,\ldots,n\}^+$. Combining this with the previous map, we obtain a one–one map from A^+ onto the set of positive integers.

We can map $\mathbf{N}^+$ into $\mathbf{N}$ by sending the string $a(1)\ldots a(k)$ to the integer $2^k\Pi p_i^{a(i)}$, where p_i is the ith prime (thus $p_0 = 2, p_1 = 3, p_2 = 5$, and so on). Uniqueness of prime factorisation ensures that this map is one–one. The initial factor 2^k is needed so that (for example) the strings 134, 1340, and 13400 map to different integers.

If A is countably infinite, there is a one–one map from A onto $\mathbf{N}$. This will extend to a one–one map from A^+ onto $\mathbf{N}^+$. Combining this with the previous map we obtain a one–one map from A^+ into $\mathbf{N}$. There are many other ways of finding a one–one map from A^+ into $\mathbf{N}$. One possibility is to regard the elements of A as being the symbols a, a', a'', a''', and so on. Then every string on A can be regarded as a string on the alphabet $\{a, '\}$. Since we have already shown how to obtain a one–one map from the set of these strings into $\mathbf{N}$, this gives a one–one map from A^+ into $\mathbf{N}$.

Exercise 2.1 Show, in detail, that the map sending $a_0\ldots a_m$ *into* $\Sigma a_i n^{m-i}$ is a one–one map from $[1,\ldots,n\}^+$ onto the set of positive integers.

2.3 COMPUTABLE FUNCTIONS, LISTABLE SETS AND DECIDABLE SETS

What is a computable function? In later chapters we shall give several apparently different answers to this question, which will all turn out to produce the same set of functions. In this chapter we shall be somewhat vague. It is enough to regard a computable function as being a function computed by some kind of a program, without needing to know exactly what a program is. Readers will have to accept certain properties of programs, which I hope will be easy to believe; in particular, they will have to accept certain specific functions as being computable, and will have to accept that certain constructions lead from computable functions to computable functions. No knowledge of any programming language is required; however, those readers who do have such knowledge (BASIC, which is the language currently most used for home micros, is quite satisfactory) are advised to justify claimed results by constructing suitable programs.

Our programs, when given an input, will compute in discrete steps. Readers with no experience of programming will probably find this assumption acceptable, as presumably actual machines compute in this way (though each step may be quite complex). Readers with a knowledge of BASIC will see that BASIC programs compute in this way, since each line is numbered and each step consists of performing the action on some line. Readers with a detailed knowledge of programming may have more difficulty with this simple idea. Many programming languages contain instructions such as WHILE ... DO or REPEAT ... UNTIL, and at first sight these do not appear to act in discrete steps. But it is not difficult to replace such programs by longer programs in a simpler language which do act as required. An example of this is given in Chapter 5, when abacus machines are replaced by register programs.

Our programs may either halt after a finite number of steps, in which case they provide an output, or they may continue computing for ever. Actual computers often have a third possibility, that they may stop without giving an output because of some error (for instance, the program might at some stage call for division by a quantity which turns out to be zero). However, it is easy to modify a program so that any such stopping situation is replaced by a computation continuing for ever (in BASIC, for instance, by requiring any error to go to a line of form n: GOTO n). Our computers will have to be able to work with arbitrarily large natural numbers. Plainly computers in the physical world cannot have this property (in whatever way they store numbers, some numbers will be too large to be stored), but this is the only way in which our ideal computers are required to differ from real ones.

We shall always be looking at functions from $\mathbf{N}^k$ to $\mathbf{N}$ (or, more generally, functions from $\mathbf{N}^k$ to $\mathbf{N}^r$). A function $f{:}\mathbf{N}^k \to \mathbf{N}$ will be (informally or intuitively) computable if there is a program which, when given an input $\mathbf{x}$ in $\mathbf{N}^k$, halts with output $f\mathbf{x}$ if $\mathbf{x}$ is in the domain of f, and continues computing for ever if $\mathbf{x}$ is not in the domain of f. More generally, a function $f{:}\mathbf{N}^k \to \mathbf{N}^r$ is computable if its components, that is the functions $f_1, \ldots, f_r{:}\mathbf{N}^k \to \mathbf{N}$ such that $f\mathbf{x} = (f_1\mathbf{x}, \ldots, f_r\mathbf{x})$, are all computable. Since our programs may compute for ever, it is clear that the function defined by a given program may well be a partial function. We might hope that this occurs only when the program has been badly defined, and that we could ensure our functions are total by taking carefully chosen programs. We shall see later that this is not possible.

Examples The empty function is computable; it is easy to obtain a program which never halts, and such a program will compute the empty function.

The functions J, K, and L defined in section 2.1 are computable. Because a precise definition of computability has not been given, it is not possible to give proofs of this and other claimed results. I hope readers will feel that these results will be true for any sensible notion of a computer program. Those readers who have a knowledge of some programming language are urged to give proofs of the results.

The function $f{:}\mathbf{N} \to \mathbf{N}$ given by $fx = y$ iff $x = y^2$ is a computable function, which is plainly not total.

Let f and g be computable functions from $\mathbf{N}^k$ to $\mathbf{N}$. Then both $f + g$ and $f.g$ are computable.

Let $f{:}\mathbf{N}^k \to \mathbf{N}^r$ and $g{:}\mathbf{N}^r \to \mathbf{N}^s$ be computable. Then their composite $gf{:}\mathbf{N}^k \to \mathbf{N}^s$ is also computable. In particular, the functions J_k and J_k^{-1} are computable. Further, let $f{:}\mathbf{N}^k \to \mathbf{N}^r$ be a function. Then f is computable iff $J_r f J_k^{-1}$ is computable. Because of this we shall usually consider functions from $\mathbf{N}^k$ to $\mathbf{N}$ rather than the more general case of functions from $\mathbf{N}^k$ to $\mathbf{N}^r$; however, it turns out not to be convenient to look only at functions from $\mathbf{N}$ to $\mathbf{N}$.

Let $f:\mathbf{N}^k \to \mathbf{N}$ be a computable function, and let r be in $\mathbf{N}$. Define g by $g\mathbf{x} = f\mathbf{x}$ if $f\mathbf{x} \leqslant r$ and $g\mathbf{x}$ not defined if $f\mathbf{x} > r$ (and, of course, $g\mathbf{x}$ not defined if $f\mathbf{x}$ is not defined). Then g is also computable. It should be reasonably clear that, given a program computing f, we simply have to modify it to compare $f\mathbf{x}$ with r and continue computing for ever if $f\mathbf{x}$ is greater than r.

Let $f:\mathbf{N}^{k+1} \to \mathbf{N}$ be a total function. We say that the function $g:\mathbf{N}^k \to \mathbf{N}$ comes from f by **minimisation**, and write $g\mathbf{x} = \mu y(f(\mathbf{x}, y) = 0)$, when $g\mathbf{x} = y$ iff $f(\mathbf{x}, y) = 0$ and, for all $z < y$ we have $f(\mathbf{x}, z) \neq 0$. If f is computable so is g, since we need only ensure that the program for f is given the inputs $(\mathbf{x}, 0)$, $(\mathbf{x}, 1)$, $(\mathbf{x}, 2)$, and so on, until, if ever, some input $(\mathbf{x}, y)$ has $f(\mathbf{x}, y) = 0$. Notice that g may well be partial, since there may be no y with $f(\mathbf{x}, y) = 0$. More generally, given total functions f_1 and f_2 from $\mathbf{N}^{k+1}$ to $\mathbf{N}$, we could consider the function $\mu y(f_1(\mathbf{x}, y) > f_2(\mathbf{x}, y))$; still more generally, $\mu y P(\mathbf{x}, y)$ denotes the least y such that $P(\mathbf{x}, y)$ holds, where P is any property. We shall consider minimisation of partial functions later.

Let f_1, f_2, and f_3 be defined as follows. First, $f_1 n = 0$ if the nth digit in the decimal expansion of π is 7, otherwise $f_1 n = 1$. Next, $f_2 n = 0$ if there is a block of exactly n consecutive 7s in the decimal expansion of π, the digits on either side of this block being different from 7, and $f_2 n = 1$ otherwise. Finally, $f_3 n = 0$ if there is a block of at least n consecutive 7s in the decimal expansion of π, otherwise $f_3 n = 1$.

Then f_1 is computable. Any of the methods for finding the decimal expansion of π can be made into a program for computing f_1.

I do not know whether or not f_2 is computable. At the time of writing no theorems which let us evaluate f_2 are known to me.

The situation with f_3 is more interesting. I do not know how to compute f_3. Nevertheless, it can be shown that f_3 must be computable. For either there are arbitrarily long blocks of consecutive 7s in the decimal expansion of π or there are not (one or the other case must hold, even though we do not know which). If there are arbitrarily long blocks, then, by definition, $f_3 n = 0$ for all n (since there is a block of more than n consecutive 7s). If not, there is an integer k (though we do not know how to find k) such that the longest block of consecutive 7s in the decimal expansion of π has length k. It then follows that $f_3 n = 0$ for $n \leqslant k$ and $f_3 n = 1$ otherwise. No matter what the value of k is, such a function is computable.

Definition Let A be a subset of $\mathbf{N}^k$. If the characteristic function of A is computable, A is called **decidable**. If A is either empty or equals $f\mathbf{N}$ for some total computable $f:\mathbf{N} \to \mathbf{N}$, then A is called **listable**.

These definitions are made in the informal theory; the similar concepts in the formal theory are given different names. The names are chosen so as to suggest the properties they relate to. Thus a set is decidable if we can tell (by means of a program for its characteristic function) whether or not an element belongs to it. Similarly, a non-empty set is listable if we can compute a list (possibly with repetitions) of its elements in the form $f0, f1, f2, \ldots$. One might at first believe that such a list of the members of A would enable us to tell whether or not a given element belongs to A. But this is not so, and we shall later give an example of a set which is listable but not decidable. The problem is that the members of A do not appear in the list in any particular order. If, for instance, we have found that 1 is not among the first 179 802 members of the list, this may be because it is not in the list at all, but it may equally well happen that it occurs as the very next member of the list. All we can be sure of is that any member of A will appear in the list sooner or later.

Let A be a subset of $\mathbf{N}^k$. Then A is decidable (listable) iff $J_k A$ is decidable (listable). The proof of this is left to the reader. Because of this we usually look at subsets of $\mathbf{N}$ rather than at subsets of $\mathbf{N}^k$.

Examples The empty set and $\mathbf{N}$ are decidable. The set of even numbers is decidable, as are the set of squares and the set of prime numbers. Any finite set is decidable.

A method for computing a total function, whether given as a formal program or informally, is often called an **algorithm**, and a method for computing a partial function is called a **partial algorithm**. For instance, the standard method of long division is an algorithm for computing the quotient of two integers. There is also a well-known procedure for finding the greatest common divisor of two integers; this is known as the Euclidean algorithm. An algorithm need not be numerical. We could easily construct an algorithm for determining whether or not a string on a given alphabet reads the same backwards as forwards. There are also algorithms for finding one's way through a maze, and there are many other algorithms in graph theory which have great practical importance.

Proposition 2.1 *Let A be a subset of $\mathbf{N}$. Then the following are equivalent*:

(*a*) *A is listable,*
(*b*) *$A = f\mathbf{N}$ for some partial computable f,*
(*c*) *χ_{Ap} is computable,*
(*d*) *$A = \operatorname{dom} g$ for some partial computable g.*

Proof Let A be listable. If A is empty it is the domain of the empty function, which is computable. Otherwise, $A = f\mathbf{N}$, where f is total and computable. Define g by $gn = \mu x(fx = n)$. Then g is computable, since f is total and computable. Plainly gn is defined iff there is some x with $fx = n$. This means that $\operatorname{dom} g = f\mathbf{N}$, and so (d) holds.

Let (d) hold. Let 0 denote the total function from $\mathbf{N}$ to $\mathbf{N}$ which is always zero. Then 0 is a computable function, and so $0.g$ is computable. However, $0.g$ has the same domain as g, and is 0 on any element of its domain. This means that $0.g$ is χ_{Ap}, and so (c) holds.

Let (c) hold. Let $I{:}\mathbf{N} \to \mathbf{N}$ be the identity function, so that $In = n$ for all n.

Now $I + \chi_{Ap}$ is computable. It is defined on n iff $\chi_{Ap}n$ is defined; that is, its domain is A. Its value on any n in its domain is $n + 0$. It follows that its set of values is exactly A, and so (b) holds.

Now let (b) hold. If A is empty, it is listable, so we may assume A non-empty. (It is not possible to tell, from a program for f, whether or not A is empty. But we only need the fact that A is either empty or not empty, without needing to know which case holds.) Take any a_0 in A. (Again, we do not need to know how to find such an a_0 explicitly from a program for f. It is enough to know that there is some a_0 in A; this is immediate bacause we are assuming A is not empty.)

Let P be a program for f. Each program, by definition, proceeds in a number of steps. Define a function $F:\mathbf{N}^2 \rightarrow \mathbf{N}$ by

> $F(n,x) = fx$ if P on input x has stopped in at most n steps (in this case, by definition, the output of the computation is fx), $F(n,x) = a_0$ otherwise.

Plainly F is total. Also, for any n and x, $F(n,x) \in$ A, since $F(n,x)$ is either a_0 or fx, both of which are in A. Also any $a \in A$ is fx for some x, and there will be some m such that the program P on input x stops in m steps. Hence $F(m,x) = fx = a$, and $A = F\mathbf{N}^2$. Now F is computable, since to obtain $F(n,x)$ it is enough to run the program P on input x for n steps and observe whether or not it has stopped, and, if so, what the output is. The function FJ^{-1} is then a total computable function from **N** to **N** whose set of values is A, as required.■

As already remarked (this will be proved later), it is not possible to tell, from a progam for computing the partial function f, whether or not $f\mathbf{N}$ is empty. We have seen that if $f\mathbf{N}$ is not empty there is some total computable ϕ with $\phi\mathbf{N} = f\mathbf{N}$. We have not seen how to construct such a ϕ, but have only shown that such a ϕ exists. There is a partial algorithm which, given a program for f, will provide a program for ϕ provided $f\mathbf{N}$ is not empty. The construction above is the relevant algorithm, if we have previously given a partial algorithm to find, from a program P for f, some a_0 in $f\mathbf{N}$ provided $f\mathbf{N}$ is not empty.

We cannot proceed by running the program P on inputs 0, 1, and so on, until we get an answer. If $f0$ is not defined but $f1$ is defined, we would take an infinite number of steps running P on input 0, and would never reach the input 1. Instead, we begin by running P for 0 steps on input 0, then 1 step on input 0, then for 0 steps on input 1, and so on, so that we systematically look through all pairs (m,n) and run the program for m steps on input n. If f is somewhere defined we will ultimately find a pair (m,n) for which P halts in m steps on input n, as needed. This approach, of looking at one input for a number of steps, leaving it for another input, and then returning to the original input for more steps, is called **dovetailing**. It will be used frequently.

The systematic search through all pairs is obtained by going through the pairs $J^{-1}r$ for $r = 0,1,, \ldots$. The algorithm is to run the program P for Kr steps on the input Lr for $r = 0,1,2,\ldots$ until, if ever, we find an r such that P

stops in Kr steps on input Lr, and then take a_0 to be the resulting output. If $f\mathbf{N}$ is not empty, this procedure will stop on some suitable r, while if $f\mathbf{N}$ is empty this procedure will continue for ever.

Proposition 2.2 *Let A be a subset of $\mathbf{N}$. Then A is decidable iff both A and $\mathbf{N} - A$ are listable.*

Proof Let A be decidable. By definition χ_A is computable. By a previous example, the function f given by $fx = \chi_A x$ if $\chi_A x = 0$, fx undefined otherwise, is computable. But this function is just χ_{Ap}. Hence A is listable by Proposition 2.1. It is easy to check that $\mathbf{N} - A$ is also decidable, and so $\mathbf{N} - A$ will also be listable.

Now let A and $\mathbf{N} - A$ be listable. On a very intuitive level, to decide whether or not n is in A we take the lists of the members of A and $\mathbf{N} - A$ and look at them alternately until we see which list n is in. More precisely, we take total computable functions f and g from $\mathbf{N}$ to $\mathbf{N}$ with $A = f\mathbf{N}$ and $\mathbf{N} - A = g\mathbf{N}$ (this is not possible if either A or $\mathbf{N} - A$ is empty, but the result is obvious in this case). Define h by $h(2n) = fn$, $h(2n+1) = gn$. Plainly h is total and computable and $h\mathbf{N} = f\mathbf{N} \cup g\mathbf{N} = \mathbf{N}$.

It follows that the function ϕ given by $\phi x = \mu y(hy = x)$ is total and computable. However, if $hy = x$ then x is in A if y is even and x is not in A if y is odd. Thus $x \in A$ iff ϕx is even, and so χ_A is the remainder when ϕx is divided by 2. Hence χ_A is computable, as needed.■

Exercise 2.2 Show that $f{:}\mathbf{N}^r \to \mathbf{N}^k$ is computable iff $J_k f$ is computable iff $J_k f J_r^{-1}$ is computable.

Exercise 2.3 Show that a subset A of $\mathbf{N}^k$ is listable (decidable) iff $J_k A$ is listable (decidable).

Exercise 2.4 Let A and B be listable subsets of $\mathbf{N}$. Show that $A \cup B$, $A \cap B$, and $A \times B$ are listable.

Exercise 2.5 Let A be listable and let $f{:}\mathbf{N} \to \mathbf{N}$ be computable. Show that both fA and $f^{-1}A$ are listable. (You will have to use Proposition 2.1 in these two exercises.)

Exercise 2.6 Is the union of countably many listable sets listable?

Exercise 2.7 Let A and B be decidable subsets of $\mathbf{N}$, and let $f{:}\mathbf{N} \to \mathbf{N}$ be total computable. Show that $A \cup B$, $A \cap B$, $A \times B$, and $f^{-1}A$ are decidable. What can be said about fA?

Exercise 2.8 Show that the subset A of $\mathbf{N}$ is listable iff there is a decidable subset B of $\mathbf{N}^2$ such that $m \in A$ iff there is n with $(m, n) \in B$. Show, more generally, that A is listable if there is a subset C of $\mathbf{N}^{k+1}$ for some k such that

$m \in A$ iff $(m, \mathbf{n}) \in C$ for some $\mathbf{n} \in \mathbf{N}_k$.

Exercise 2.9 Let f and g be total computable functions from $\mathbf{N}$ to $\mathbf{N}$. Suppose that whenever $m \in f\mathbf{N}$ there is some x with $x \leqslant gm$ and $m = fx$. Give an intuitive argument to explain why $f\mathbf{N}$ is decidable. Can you make this argument more precise?

Exercise 2.10 Show that a subset of $\mathbf{N}$ is infinite and decidable iff it can be written as $f\mathbf{N}$ where f is total computable and $fm < fn$ for $m < n$. (Use the previous exercise.)

Exercise 2.11 Show that a non-empty subset of $\mathbf{N}$ is decidable iff it is $f\mathbf{N}$ where f is total computable and $fm \leqslant fn$ for $m < n$. (Let A be $f\mathbf{N}$ for such an f. Then A is either finite or infinite, though we do not know which. If A is finite there is nothing to prove. So suppose A infinite, and show that the hypotheses of Exercise 2.9 hold.)

Exercise 2.12 Give an intuitive argument to show that any infinite listable set can be written as $f\mathbf{N}$ where f is total computable and one–one.

Exercise 2.13 Give an intuitive argument to show that any infinite listable set contains an infinite decidable set. Can you make the arguments of the last two exercises more precise?

2.4 UNIVERSAL FUNCTIONS AND UNDECIDABLE SETS

Proposition 2.3 *There is no universal total computable function. That is, there is no total computable $T:\mathbf{N}^2 \to \mathbf{N}$ with the property that for any total computable $f:\mathbf{N} \to \mathbf{N}$ there is some n such that $fx = T(n,x)$ for all x.*

Proof This is our first use of Cantor's diagonal process, which is a key tool in the whole theory.

Suppose there is such a function T. Then the function $f:\mathbf{N} \to \mathbf{N}$ defined by $fx = T(x,x) + 1$ is a total computable function. By definition of T, there will be some k such that $fx = T(k,x)$ for all x. Now look at fk. By definition of f, $fk = T(k,k) + 1$. By definition of k, $fk = T(k,k)$. This contradiction shows that no such function T can exist. ■

Theorem 2.4 *There is a universal partial computable function. That is, there is a partial computable function $\Phi:\mathbf{N}^2 \to \mathbf{N}$ with the property that for any partial computable $f:\mathbf{N} \to \mathbf{N}$ there is some n such that $fx = \Phi(n,x)$ for all x (more precisely, such that if one of fx and $\Phi(n,x)$ is defined so is the other and they are equal).*

Proof A very slight knowledge of the nature of programs is needed to prove this very important theorem.

Whatever notion of program is taken, it is clear that a program is a string on some finite alphabet; this is because we have already noted that a countable alphabet can be replaced by a finite one, by replacing letters a_0, $a_1, a_2, \ldots$ by $a, a', a'', \ldots$. The typical alphabet will consist of the twenty-six upper-case (capital) and lower-case letters of the English alphabet, the numbers $0, \ldots, 9$, and various punctuation marks (including parentheses and spaces as symbols).

Plainly, not all strings on this alphabet will be programs, but it must be possible to tell whether or not a string is a program; the fact that a computer can take action when a string of symbols is typed into it should justify this claim. Also, two different programs must have different strings (else the computer would not know which program to use when a string corresponding to two programs was typed in).

We have already seen how to obtain a one–one map from the set of all strings into **N**, in such a way that we can tell whether or not the number n corresponds to a string. From what we have said in the previous paragraphs, we can tell whether or not n corresponds to a program for computing a function of one variable. If so, we can find the relevant program, and in this case we define $\Phi(n,x)$ to be the result of applying the program corresponding to n to the input x. If n does not correspond to such a program, then $\Phi(n,x)$ is undefined.

The function Φ is computable, since, given n, we can determine whether or not n corresponds to a suitable program, and, if so, can carry out this program on the input x.

Now let $f:\mathbf{N}\rightarrow\mathbf{N}$ be any computable function. There must be some program P for f. Then P will correspond to some number n, and then, by definition, $\Phi(n,x) = fx$ for every x.■

It follows immediately that there are only countably many computable functions from **N** to **N**. Since there are uncountably many functions from **N** to **N** there must be functions which are not computable. In particular, Corollary 1 below shows that the function g given by $gx = \Phi(x,x) + 1$ if $\Phi(x,x)$ is defined, $gx = 0$ otherwise, is not computable.

Plainly there is a similar theorem giving a universal function from $\mathbf{N}^{k+1}$ to **N** for the class of partial computable functions from $\mathbf{N}^k$ to **N**. This can be proved by the same method, or can be deduced using composition with J_k and its inverse, as previously discussed.

For a full understanding of this theorem we must see why the argument of Proposition 2.3 does not lead to a contradiction here.

So define f by $fx = \Phi(x, x) + 1$. Then f is partial computable. So there must be n such that $fx = \Phi(n,x)$, in the sense that if one of these is defined so is the other and they are equal. If $\Phi(n,n)$ were defined, we would get a contradiction, as before. There is no problem with the theorem itself. We simply find that, for this n, $\Phi(\mathrm{n},n)$ is not defined.

This argument can be extended to the following corollary.

Corollary 1 *Let* $fx = \Phi(x,x) + 1$. *Let g be any computable function extending f. Then g is not total.*

Proof There is some m with $gx = \Phi(m,x)$ for all x. Suppose gm were defined. Then $\Phi(m,m)$ would be defined and equal to gm. Then, by definition, fm would also be defined and equal to $\Phi(m,m) + 1$. As g extends f, we would then have $gm = fm$, which gives a contradiction.■

It is customary to denote the function sending x to $\Phi(n,x)$ by ϕ_n, and to refer to n as an **index** of this function. Thus any computable function from **N** to **N** has at least one index; we shall see later (and it is easy to see directly, by adding irrelevant steps to a program) that it has infinitely many indexes.

Corollary 2 *Let A be a set such that if* $n \in A$ *then* ϕ_n *is total and such that every total computable function from* **N** *to* **N** *has at least one index in A. Then A is not listable. In particular, the set* $\{n;\ \phi_n \text{ is total}\}$ *is not listable.*

Proof Suppose A were the range of the total computable function f. Consider the function T defined by $T(k,x) = \Phi(fk,x)$. Plainly, T is computable. Also, for any k, the function ϕ_{fk} is total, by definition of f. Hence T is total. Also any total computable function g has index fk for some k, and hence $gx = T(k,x)$ for all x. This contradicts Proposition 2.3.■

We might have hoped that partial functions could be eliminated, and that we could manage to consider total functions only. One hope would be that when we encountered a partial function we could extend its definition so as to obtain a total function. Corollary 1 shows that this is not possible if we want the extended function to remain computable. Another hope might be that we could identify which programs gave total functions, but Corollary 2 shows that this is also impossible.

We have seen that there is no obvious reason why a list of elements of a set should enable us to decide whether or not a given element is in the set. The next theorem proves that listability and decidability are different.

Theorem 2.5 *The set* $\{n;\ \Phi(n,n) \text{ is defined}\}$ *is listable but not decidable.*

Proof This set, which we will call K, is listable, because it it the domain of a partial computable function.

If K were decidable, the set $\mathbf{N} - K$ would also be listable, and so the partial characteristic function f of $\mathbf{N} - K$ would be computable. There would be a number m such that $fx = \Phi(m,x)$ for all x. Then $x \notin K$ iff fx is defined, and so $x \notin K$ iff $\Phi(m,x)$ is defined. In particular $m \notin K$ iff $\Phi(m,m)$ is defined. But, by definition of K, $m \in K$ iff $\Phi(m,m)$ is defined. This contradiction shows that K cannot be decidable.■

We now look at minimisation of partial functions. There are two definitions which we might consider using; however, only one of them leads from computable functions to computable functions, so this is the only one worth using.

Let $f:\mathbf{N}^{k+1}\rightarrow\mathbf{N}$ be a function. We say $g:\mathbf{N}^k\rightarrow\mathbf{N}$ comes from f by **minimisation** if g has the following definition:

> $g\mathbf{x} = y$ iff $f(\mathbf{x},y) = 0$ and, for all $z<y$, $f(\mathbf{x},z)$ is defined and non-zero.

Then g will be computable if f is computable. In order to compute g on input $\mathbf{x}$ we simply compute f successively on the inputs $(\mathbf{x},0)$, $(\mathbf{x},1)$, $(\mathbf{x},2)$, and so on, until, if ever, the computation halts with value zero; if this happens, the output of the computation is the value of the $(k+1)$ st input at this time. This computation will stop after a finite number of steps with output y if $f(\mathbf{x},0)$, $\ldots, f(\mathbf{x},y-1)$ are defined and non-zero and $f(\mathbf{x},y) = 0$. In this case $y = g\mathbf{x}$. The computation will continue for ever if $f(\mathbf{x}, z)$ is defined for all z but is never zero; in this case $\mathbf{x}$ is not in the domain of g. The remaining case is that there is some y such that $f(\mathbf{x},z)$ is defined and non-zero for all $z<y$, but $f(\mathbf{x},y)$ is not defined. In this case also, $g\mathbf{x}$ is not defined, and the computation will continue for ever, because the computation for f on the input $(\mathbf{x},y)$ continues for ever.

The alternative definition for minimisation is the following, which I call **pseudo-minimisation**:

> h comes from f by pseudo-minimisation if $h\mathbf{x} = y$ iff $f(\mathbf{x},y) = 0$ and, for all $z<y$, either $f(\mathbf{x},z)$ is not defined or it is defined and non-zero.

For instance, let $f:\mathbf{N}^2\rightarrow\mathbf{N}$ be defined by $f(x,y) = 0$ unless $y = 0$ and $x\notin K$ (where K is any set which is listable but not decidable), in which case $f(x,y)$ is not defined. It is easy to see that f is computable. The function obtained from f by minimisation is the partial characteristic function of K, which is computable. However the function obtained by pseudo-minimisation is the characteristic function of K, which is not computable. Because pseudo-minimisation does not preserve computability we shall never use it. However, I feel it is worth mentioning, both because it is at first sight a reasonable definition and because some authors do use what I call pseudo-minimisation as their definition of minimisation. As a result they claim that minimisation of partial functions does not preserve computability, and they have to look at minimisation only for total functions.

2.5 RICE'S THEOREM

The next lemma requires a more detailed knowledge of what a program is than we need elsewhere in this chapter. Consequently, only a sketch proof is given. The ideas of this proof apply for many programming languages. A formal proof is given in Proposition 7.10. An indication in a specific language is given in Chapter 5, after Theorem 5.11.

Lemma 2.6 *Let $F:\mathbf{N}^2 \to \mathbf{N}$ be a computable function. Then there is a total computable $g:\mathbf{N} \to \mathbf{N}$ such that, for every n, the function with index gn sends x to $F(n,x)$ for all x.*

Sketch proof Let P be a program for F. The program P must have two inputs, corresponding to the variables of F. Let n be fixed. We want to find a program for the function sending x to $F(n,x)$.

Such a program will have only one input position, in which x will appear at the start. We need to transfer x to the second input place leaving the first one empty, and then put n into the first input place, and then apply P. We can put n into the first input place by adding 1 to this position n times (or by some similar technique, depending on the programming language). If we denote the number corresponding (as in the proof of Theorem 2.4) to this program by gn, it should be reasonably clear that g is a computable function. By definition the program with index gn computes the function sending x to $F(n,x)$, required.■

Theorem 2.7 (Rice's Theorem) *Let A be a set of computable functions from $\mathbf{N}$ to $\mathbf{N}$. Then $\{n;\ \phi_n \in A\}$ is decidable iff A is either empty or consists of all computable functions.*

Proof The set $\{n;\phi_n \in A\}$ is decidable iff $\{n;\phi_n \notin A\}$ is decidable. It follows that we may assume the empty function ω is not in A, replacing A by its complement, if necessary. We will then show A must be empty. Suppose not, and let f be in A.

Let K be any set which is listable but not decidable, and let P be a program for the partial characteristic function of K. Thus P halts from input x iff $x \in K$. Define $F:\mathbf{N}^2 \to \mathbf{N}$ by $F(x,y) = fy$ if $x \in K$, $F(x,y)$ undefined otherwise. Then F is computable, since to compute it we need only apply the program P to the input x, and when this halts, if ever, proceed to compute fy.

By Lemma 2.6 there is a computable function s such that sn is an index of f if $n \in K$ and is an index of ω otherwise. Hence K is the counter-image of the decidable set $\{m;\phi_m \in A\}$ by the computable function s. This is a contradiction by Exercise 2.7, as K is not decidable.■

An index for a function is essentially the same as a program for computing it. Thus Rice's Theorem can be stated informally as 'It is impossible to decide anything about a function from its program'. In particular, we cannot decide whether or not a given program computes a given function. This has important practical implications, since it means that there is no systematic procedure which will tell whether or not an arbitrary program is correct. Much work is being done in computer science on methods which can be applied in practice to show given programs are correct; however, Rice's Theorem shows that no such method can be guaranteed to work on all programs.

Notice that Rice's Theorem refers to all possible indexes for functions in

A. There may well be a decidable set B such that $f \in A$ iff there is some $n \in B$ with $\phi_n = f$; this refers only to some indexes for each function in A. For instance, such a B obviously exists if A consists of a single function or of all constant functions.

We now give an extension of Rice's Theorem from which the former can be derived.

Theorem 2.8 (Rice–Shapiro Theorem) *Let A be a set of computable functions from* **N** *to* **N** *such that* $\{n;\phi_n \in A\}$ *is listable.*

Then the computable function f:**N**→**N** *is in* A *iff some finite restriction of* f *is in* A.

Remark Recall that g is a restriction of f if $fx = gx$ whenever gx is defined, and that g is finite if its domain is finite.

Proof Let K and P be as in Rice's Theorem.

Suppose first that f has a finite restriction g in A. Define F:**N**^{2}→**N** by

$$F(x,y) = fy \text{ if } x \in K,\ F(x,y) = gy \text{ otherwise.}$$

Then F is computable. Given (x,y), we first check whether or not y is in the domain of g, which is possible as this domain is finite. If so, we compute gy, which we know is also fy. If not, we run the program P on the input x, and when it halts, if ever, we then proceed to compute fy. (Notice that we cannot begin by determining whether or not x is in K, and then computing either fy or gy, since if x is not in K we cannot discover this in finite time.)

By Lemma 2.6, there will be a computable function s such that ϕ_{sn} is f for $n \in K$ and is g otherwise. Hence **N** $-$ K is the counter-image under s of the listable set $\{m;\phi_m \in A\}$ if f is not in A. Since **N** $-$ K is not listable, and the counter-image is listable by Exercise 2.5, we see that f must be in A.

Conversely, suppose f is in A. This time define F:**N**^{2}→**N** by

$F(x,y) = fy$ if the program P on input x has not halted within y steps,
$F(x,y)$ undefined otherwise.

Plainly F is computable, since, given (x,y), we can simply run the program P on input x for y steps and see whether or not it has halted, and, if not, proceed to compute fy. Also, if x is not in K, then $F(x,y) = fy$ for every y, since P never halts on input x. But if x is in K, there will be some m such that P on input x halts after exactly m steps. In this case $F(x,y)$ is defined iff $y < m$, so $F(x,y)$, as a function of y for such an x, is a finite restriction of f.

Thus we have a computable function s such that ϕ_{sn} is f if n is not in K and is a finite restriction of f if n is in K. It follows that unless one of these finite restrictions is in A, the set **N** $-$ K is the counter-image under s of the listable set $\{m;\phi_m \in A\}$, which is impossible.■

Corollary 1 *Let $\{n;\phi_n \in A\}$ be listable. Then any extension of a function in A is itself in A.*

Proof Let h extend f, with f in A. Then some finite restriction g of f is also in A. Since g is also a restriction of h, it follows that h is in A.■

Corollary 2 *Let $\{n;\phi_n \in A\}$ be listable. If the nowhere-defined function ω is in A then all computable functions from* **N** *to* **N** *are in A.*

Proof Plainly the function ω is a finite restriction of every function.■

Rice's Theorem follows immediately, since if $\{n;\phi_n \in A\}$ is decidable both it and its complement are listable, and one of the two must contain ω and so will consist of all computable functions.

We can now give some examples. From Corollary 2 it follows at once that the sets $\{n;\phi_n$ one–one$\}$, $\{n;\phi_n$ finite$\}$, and $\{n;\phi_n$ has finite range$\}$ are not listable. By the theorem itself, $\{n;\phi_n$ has infinite range$\}$ cannot be listable.

The theorem does not give a condition which ensures that $\{n;\phi_n \in A\}$ is listable. For examples of this we show explicitly that the relevant sets are listable. We shall let P_n be the program corresponding to the integer n if there is such a program; otherwise let P_n be a fixed program computing ω (that is, a program which never halts).

The set $\{n;\phi_n$ somewhere defined$\}$ is listable. For the set $\{(n,x,t); P_n$ on input x halts in at most t steps$\}$ is plainly decidable, and the set we want is a projection of this (precisely, it is the image of this set under the map from $\mathbf{N}^3$ to $\mathbf{N}$ which sends any element of $\mathbf{N}^3$ to its first component), and so is listable by Exercise 2.5. Similarly $\{n;\phi_n$ is not one–one$\}$ is listable, being a projection of the decidable set $\{(n,x,y,t); x \neq y$ and P_n halts in at most t steps on both the input x and the input y with the same output in both cases$\}$.

It is possible to give a necessary and sufficient condition for $\{n;\phi_n \in A\}$ to be listable. The details of the proof are rather technical, and so the proof will be left to Chapter 10.

Exercise 2.14 Show that, for any r, the set $\{n;\phi_n$ takes at least r values$\}$ is listable, and so is $\{n;\phi_n$ is defined for at least r numbers$\}$. What happens if 'at least' is replaced by 'exactly'? What happens if we consider the set of those pairs (r,n) satisfying the various conditions?

Exercise 2.15 Let A be listable. Show that $\{n;A$ meets the range of $\phi_n\}$ and $\{n;A$ meets the domain of $\phi_n\}$ are listable. What happens if we consider $\{n;A$ contains the range of $\phi_n\}$ and $\{n;A$ contains the domain of $\phi_n\}$? Does it make any difference if A is decidable?

3

Primitive recursive functions

In this chapter we look at an important class of intuitively computable functions. This class includes most functions we are likely to consider in practice.

3.1 PRIMITIVE RECURSION

Let $g:\mathbf{N}^k \to \mathbf{N}$ and $h:\mathbf{N}^{k+2} \to \mathbf{N}$ be functions. We say that $f:\mathbf{N}^{k+1} \to \mathbf{N}$ is **defined from** g **and** h by **primitive recursion** if f satisfies the conditions $f(\mathbf{x},0) = g\mathbf{x}$ and $f(\mathbf{x},y+1) = h(\mathbf{x},y,f(\mathbf{x},y))$ for all $\mathbf{x}$ and y. Since the functions may be partial, these equations must be interpreted as saying that if one side exists so does the other and they are equal. We allow the case $k = 0$, when g is just a number.

It is easy to see (by induction) that, given g and h, there is at most one such function f. It is intuitively reasonable that there exists such an f and that f is total if g and h are total. This can be proved, but the proof is more difficult than one would expect and will be left to the final section of this chapter. If g and h are intuitively computable then so is f. To compute $f(\mathbf{x},5)$, for instance, we first compute $f(\mathbf{x},0)(= g\mathbf{x})$, then compute $f(\mathbf{x},1)$ using h and the already computed value of $f(\mathbf{x},0)$, then compute $f(\mathbf{x},2)$ using h and the already computed value of $f(\mathbf{x},1)$ and so on.

The **zero function** $Z:\mathbf{N} \to \mathbf{N}$ is given by $Zx = 0$ for all x. The **successor function** $S:\mathbf{N} \to \mathbf{N}$ is given by $Sx = x + 1$ for all x. The **projection functions** $\pi_{ni}:\mathbf{N}^n \to \mathbf{N}$ (where $1 \leqslant i \leqslant n$) are given by $\pi_{ni}(x_1, \ldots, x_n) = x_i$ for all $x_1, \ldots, x_n$. The zero, successor and projection functions are called **initial functions**. Note that π_{11} is the identity function on $\mathbf{N}$.

A set C of total functions from $\mathbf{N}^n$ to $\mathbf{N}$ (for all n) is called **primitive recursively closed** if it satisfies the following three conditions:

(1) all initial functions are in C.

(2) C is closed under primitive recursion. That is, if f comes from g and h by primitive recursion, and g and h are in C, then f is also in C.

(3) C is closed under composition. That is, if f comes from g and $h_1, \ldots, h_r$ by composition, so that $f\mathbf{x} = g(h_1\mathbf{x}, \ldots, h_r\mathbf{x})$, where g and $h_1, \ldots, h_r$ are in C, then f is also in C.

A function f is called **primitive recursive** if there is a sequence of functions $f_1, \ldots, f_n$ with f_n being f, and such that for all $r \leqslant n$ either f_r is an initial function, or there are i and j less than r such that f_r comes from f_i and f_j by primitive recursion, or there are i less than r and $j(1), \ldots, j(k)$ all less than r such that f_r comes from f_i and $f_{j(1)}, \ldots, f_{j(k)}$ by composition. Such a sequence will be called a **defining sequence** for f. Note that (by induction on r) each f_r, and hence f itself, will then be total.

Lemma 3.1 *(i) The set of all primitive recursive functions is a primitive recursively closed set. (ii) Any primitive recursively closed set contains every primitive recursive function.*

Proof (i) If f is one of the initial functions then the sequence with one term f shows that f is primitive recursive.

Let f be obtained by primitive recursion from the primitive recursive function g and h. Then there are defining sequences $g_1, \ldots, g_m$ and $h_1, \ldots, h_n$ for g and h. Then the sequence $g_1, \ldots, g_m, h_1, \ldots, h_n, f$ is a defining sequence for f.

Now let f be defined by composition from g and $h_1, \ldots, h_k$, all of which are primitive recursive. Take defining sequences $g_1, \ldots, g_m$ for g and $h_{i1}, \ldots, h_{i,n(i)}$ for h_i. Then the sequence

$$g_1, \ldots, g_m, h_{11}, \ldots, h_{1,n(1)}, h_{21}, \ldots, h_{k,n(k)}, f$$

is a defining sequence for f.

(ii) Let C be a primitive recursively closed set and f a primitive recursive function with $f_1, \ldots, f_n$ as a defining sequence. It is easy to see, by induction on r, that $f_r \in C$ for all $r \leqslant n$. In particular f_n, which is f, is in C. ■

We now look at some examples of primitive recursive functions. We normally use Lemma 3.1(i) to show that a function is primitive recursive rather than finding a defining sequence.

(I) Any constant function is primitive recursive.

First look at constant functions with domain N. Z is constantly 0, SZ is constantly 1, SSZ is constantly 2, etc. If we want a constant function with domain $\mathbf{N}^n$ we need only compose one of these with the projection π_{n1}.

(II) Addition is primitive recursive.

For addition is defined by $x+0=x$, $x+(y+1)=(x+y)+1$. More formally, define $\alpha(x,y)$ to be $x+y$ and we have $\alpha(x,0)=x=\pi_{11}x$, $\alpha(x,y+1)=S\alpha(x,y)$. It follows that α comes by primitive recursion from π_{11} and $S\pi_{33}$, and so α is primitive recursive.

Further, we can define $\alpha_n(x_1, \ldots, x_n)$ to be $x_1 + \ldots + x_n$, and α_n will be primitive recursive for any n. This is proved by induction on n. For we have

$$\alpha_{n+1}(x_1, \ldots, x_n, 0) = \alpha_n(x_1, \ldots, x_n)$$

and

$$\alpha_{n+1}(x_1, \ldots, x_n, y+1) = \alpha_{n+1}(x_1, \ldots, x_n, y) + 1.$$

Hence α_{n+1} comes by primitive recursion from α_n and $S\pi_{n+2,n+2}$. The latter is primitive recursive (as the composition of initial functions) and so α_{n+1} will be primitive recursive if α_n is primitive recursive.

(III) Multiplcation is primitive recursive.

For multiplication is defined by $x.0=0$, $x.(y+1)=x.y+x$. More formally, define $\mu(x,y)$ to be $x.y$ and we have $\mu(x,0)=0=Zx$, $\mu(x,y+1)=\alpha(x,\mu(x,y))$. Hence μ is primitive recursive, since it comes by primitive recursion from the functions Z and β, where β is the composite of the primitive recursive function α with π_{31} and π_{33}.

As in (II), the product $x_1. \ldots .x_n$ is a primitive recursive function, for any n.

(IV) Exponentiation is primitive recursive.

Here we have $x^0=1$, $x^{y+1}=x^y.x$. This can be formalised, as in (II) and (III), but in future the informal expression will be given, leaving the precise formalisation as an exercise for the reader.

(V) $x!$ is primitive recursive, being given by $0! = 1$, $(x+1)! = x!.(x+1)$.

(VI) The **sign** of x, $\text{sg}x$, is defined by $\text{sg}0=0$, $\text{sg}x=1$ for $x \neq 0$. This is primitive recursive, since it comes by primitive recursion from 0 and the constant function 1.

(VII) The **cosign** of x, $\text{co}x$, is defined by $\text{co}0=1$, $\text{co}x=0$ for $x \neq 0$. This is primitive recursive, as for (VI).

(VIII) The **predecessor** of x, $\text{pd}x$, is defined by $\text{pd}0=0$, $\text{pd}(x+1)=x$. It is primitive recursive, since it comes by primitive recursion from 0 and the function π_{21}.

(IX) $x-y$ does not map into **N**, so we do not consider it. Instead we look at $x \dot{-} y$, which is defined by $x \dot{-} y = x-y$ if $x \geqslant y$, $x \dot{-} y = 0$ otherwise. This is primitive recursive, since we can easily see that $x \dot{-} 0 = x$,
$x \dot{-} (y+1) = \text{pd}\,(x \dot{-} y)$.

$|x-y|$ is also primitive recursive, since it equals $(x \dot{-} y)+(y \dot{-} x)$.

(X) For the remainder of this chapter, C will be some primitive recursively closed set of functions. We shall show that certain constructions applied to functions in C produce functions which are also in C.

Let $f:\mathbf{N}^k \to \mathbf{N}$ be in C. Let $g:\mathbf{N}^r \to \mathbf{N}$ be given by $g(x_1, \ldots, x_r) = f(y_1, \ldots, y_k)$, where each y_j is either x_i for some i or is a constant. Then g is also in C.

This is true because g is the composite of f with functions which are either projection functions or constant functions. By definition the projection functions are in C, and by Lemma 3.1(ii) the constant functions, being primitive recursive, are also in C.

Let $g:\mathbf{N}^k \to \mathbf{N}$ and $h:\mathbf{N}^{k+2} \to \mathbf{N}$ be in C, and let $f:\mathbf{N}^{k+1} \to \mathbf{N}$ be given by $f(0,x_1, \ldots, x_k) = g(x_1, \ldots, x_k), f(x_0+1,x_1, \ldots, x_k) = h(x_0, \ldots, x_k, f(x_0, \ldots, x_k))$. Then f is also in C. That is, we can make definitions by primitive recursion on any variable, not just the last one.

To see this we define h' by $h'(x_1, \ldots, x_k,y,z) = h(y,x_1, \ldots, x_k,z)$. By earlier remarks, h' is also in C, and hence the function f' defined by primitive recursion from g and h' will also be in C. However, it is easy to check (by induction on x_0) that $f(x_0,x_1, \ldots, x_k) = f'(x_1, \ldots, x_k,x_0)$. As $f' \in C$ we also have $f \in C$.

(XI) Let $f:\mathbf{N}^{k+1} \to \mathbf{N}$ be in C. Then $\Sigma^{r \leq y} f(\mathbf{x},r)$ and $\Pi^{r \leq y} f(\mathbf{x},r)$ are in C.

Define $g(\mathbf{x},y)$ to be $\Sigma^{r \leq y} f(\mathbf{x},r)$. Then we have $g(\mathbf{x},0) = f(\mathbf{x},0)$ and $g(\mathbf{x},y+1) = g(\mathbf{x},y) + f(\mathbf{x},y+1)$. If we define $h(\mathbf{x},y,z)$ to be $z + f(\mathbf{x},y+1)$ then h will be in C, since we are assuming f is in C and addition, being primitive recursive, is in C. Since g comes by primitive recursion from $f(\mathbf{x},0)$ and h, which are both in C, we see g is in C. A similar proof works for Π.

Similarly, if we take the sum (or the product) over all $z < y$ (instead of over all $z \leq y$) we still get a function in C, by a variant of the formula; here if $y = 0$ we must take the sum to be 0 and the product to be 1. More generally, we could take the sum or product from y_1 to y, instead of from 0 to y, and still get a function in C.

(XII) Definition by cases.

Let $f_1, \ldots, f_n, g_1, \ldots, g_n$ be functions in C from $\mathbf{N}^r$ to $\mathbf{N}$. Suppose that for every $\mathbf{x}$ there is exactly one i such that $g_i\mathbf{x} = 0$. Define $f:\mathbf{N}^r \to \mathbf{N}$ by $f\mathbf{x} = f_i\mathbf{x}$ if $g_i\mathbf{x} = 0$. Then $f \in C$.

Notice first that if $g_i\mathbf{x} = 0$ then $\mathrm{co}g_i\mathbf{x} = 1$ while for $j \neq i$ we have $\mathrm{co}g_j\mathbf{x} = 0$, since, by assumption, we then have $g_j\mathbf{x} \neq 0$. Hence f is given by $f\mathbf{x} = f_1\mathbf{x}.\mathrm{co}g_1\mathbf{x} + \ldots + f_n\mathbf{x}.\mathrm{co}g_n\mathbf{x}$, and so f is in C.

There is another version of the definition by cases. Suppose we are given functions $f_1, \ldots, f_{n+1}, g_1, \ldots, g_n$ in C, such that for every $\mathbf{x}$ there is at most one i with $g_i\mathbf{x} = 0$. If we define $f\mathbf{x}$ to be $f_i\mathbf{x}$ if $g_i\mathbf{x} = 0$ (for $i = 1, \ldots, n$) and $f\mathbf{x} = f_{n+1}\mathbf{x}$ if $g_i\mathbf{x} \neq 0$ for all i, then again $f \in C$. It is enough to define a function g_{n+1} in C which is zero iff all of $g_1, \ldots, g_n$ are non-zero; for then the previous version applies since $f\mathbf{x} = f_{n+1}\mathbf{x}$ if $g_{n+1}\mathbf{x} = 0$. We need only define $g_{n+1}\mathbf{x}$ to be $\mathrm{co}(g_1\mathbf{x}. \ldots .g_n\mathbf{x})$.

(XIII) The quotient and remainder when y is divided by x, which we denote by $\mathrm{quo}(x,y)$ and $\mathrm{rem}(x,y)$, are primitive recursive.

It is easy to see that, for $x \neq 0$, we have $\text{quo}(x,0) = 0$ and $\text{quo}(x,y+1)$ is either $\text{quo}(x,y)$ or $\text{quo}(x,y)+1$. Precisely, $\text{quo}(x,y+1) = \text{quo}(x,y)+1$ iff $y+1 = x.(\text{quo}(x,y)+1)$. We simply define $\text{quo}(0,y)$ by this formula, which gives $\text{quo}(0,y) = 0$ for all y. Later we will give a different approach to showing that quo is primitive recursive, and then we will use a different definition of $\text{quo}(0,y)$. We have no interest in what happens when x is 0, except that we must have some meaning given to $\text{quo}(0,y)$ as primitive recursive functions are total; so no problems arise because the different approaches define $\text{quo}(0,y)$ differently.

The above definition is like a definition by cases, but is not quite of that form. In fact it shows that quo is defined by primitive recursion from Z and a function $h:\mathbf{N}^3 \to \mathbf{N}$ given by $h(x,y,z) = z+1$ if $y+1 = x.(z+1)$ and $h(x,y,z) = z$ otherwise. We can write this as $h(x,y,z) = z+1$ if $|y+1-x.(z+1)| = 0$ and $h(x,y,z) = z$ otherwise. This makes h primitive recursive, by the second version of definition by cases. As quo comes from Z and h by primitive recursion, quo will be primitive recursive.

Finally, rem is primitive recursive as $\text{rem}(x,y) = y \dot{-} x.\text{quo}(x,y)$.

(XIV) The pairing functions $J_k:\mathbf{N}^k \to \mathbf{N}$ are primitive recursive.

Since J_k is obtained by a sequence of compositions from J and projections, it is enough to show J is primitive recursive. Now $J(m,n)$ is $\text{quo}(2,(m+n)(m+n+1))+m$, and so the result follows by (XIII).

(XV) We say that a function $f:\mathbf{N}^r \to \mathbf{N}^k$ is in C if its coordinate functions $f_1, \ldots, f_k$, which are defined by $f\mathbf{x} = (f_1\mathbf{x}, \ldots, f_k\mathbf{x})$, are in C.

With this notation the functions J_k^{-1} are primitive recursive. Since the coordinate functions are obtained from K and L by sequences of compositions, it is enough to show that K and L are primitive recursive. One proof of this is given as Exercise 3.4, and another proof is (II) of section 3.3.

(XVI) Let X be any set and let $f:X \to X$ be a function. We define the **iterate** of f to be the function $F:X \times \mathbf{N} \to X$ given by $F(x,0) = x$, $F(x,n+1) = f(F(x,n))$ for all n. We also refer to the function from $\mathbf{N}$ to X which sends n to $F(x_0,n)$ as **an iterate** of f, or more precisely as **the iterate of** f **starting at** x_0.

If $f:\mathbf{N}^k \to \mathbf{N}^k$ is in C then so is its iterate $F:\mathbf{N}^{k+1} \to \mathbf{N}^k$ (and hence so is the iterate starting at x_0, for any x_0).

First assume $k = 1$. Then F is obtained from π_{11} and h, where $h(x,y,z) = fz$, by primitive recursion. Since f is in C so is h, and then F is in C.

Now let k be arbitrary. Define $g:\mathbf{N} \to \mathbf{N}$ to be $J_k f J_k^{-1}$. Since J_k and its inverse are primitive recursive, and hence in C, and f is in C, we also have g in C. By what we have just shown, the iterate G of g is in C. But it is clear (by induction on y) that $F(\mathbf{x},y) = J_k^{-1}G(J_k\mathbf{x},y)$, and so F is in C.

The following theorem will be useful in Chapter 5; it enables us to replace primitive recursion by iteration, which is easier to look at.

Theorem 3.2 *Let C be a class of partial functions which contains the initial functions and is closed under composition and iteration. Then C is also closed under primitive recursion.*

Proof Let $f:\mathbf{N}^{k+1} \to \mathbf{N}$ be obtained by primitive recursion from g and h, where g and h are in C. Let $\phi:\mathbf{N}^{k+2}\to\mathbf{N}^{k+2}$ be defined by $\phi(\mathbf{x},y,z)=(\mathbf{x},y+1,h(\mathbf{x},y,z))$ and let Φ be the iterate of ϕ. By induction on y we can show that $\Phi(\mathbf{x},0,g\mathbf{x},y)=(\mathbf{x},y,f(\mathbf{x},y))$. Now ϕ is in C since h is, and so Φ will be in C. As g is also in C, we find that f is in C.■

Exercise 3.1 Give formal proofs that the functions defined in (III) to (IX) are primitive recursive.

Exercise 3.2 Let sqr$x$ be the largest integer whose square is $\leqslant x$ (that is, sqr$x$ is the integer part of the square root of x). Show, as in (XIII), that sqr is primitive recursive.

Exercise 3.3 Show that $s(s+1)/2\leqslant r$ iff $(2s+1)^2\leqslant 8r+1$. Deduce that the function sending r to the largest s with $s(s+1)/2\leqslant r$ is primitive recursive.

Exercise 3.4 Use the previous exercise to show that the functions K and L are primitive recursive.

3.2 BOUNDED QUANTIFIERS AND MINIMISATION

We say that a subset A of $\mathbf{N}^k$ is in C if its characteristic function is in C. We say that a property (or predicate) P of k variables is in C if the set $\{\mathbf{x};\mathbf{x}\text{ has }P\}$ which it defines is in C. It seems to me inconvenient to give a formal definition of a property (the words 'property' and 'predicate' will be used to mean the same thing). Examples of properties are 'equals', 'divides', 'prime', 'between', whose corresponding sets are, respectively, $\{(x,y)\in\mathbf{N}^2;\ x=y\}$, $\{(x.y)\in\mathbf{N}^2;\ x\text{ divides }y\}$, $\{x\in\mathbf{N};\ x\text{ is prime}\}$ and $\{(x,y,z)\in\mathbf{N}^3; x<y<z\text{ or }x>y>z\}$.

We can apply logical connectives to properties to obtain new properties. Thus, $P\vee Q$, the **disjunction** of P and Q, holds for $\mathbf{x}$ if either P or Q (or both) holds for $\mathbf{x}$. The **conjunction** $P\wedge Q$ of P and Q holds for $\mathbf{x}$ if both P and Q hold for $\mathbf{x}$. The **negation** $\neg P$ of P holds for $\mathbf{x}$ if P does not holds for $\mathbf{x}$. If P is a property of $k+1$ variables then the **universal quantification** $\forall yP$ and **existential quantification** $\exists yP$ are properties of k variables such that $\forall yP$ holds for $\mathbf{x}$ if P holds for $(\mathbf{x},y)$ for all y, while $\exists yP$ holds for $\mathbf{x}$ if there is some y such that P holds for $(\mathbf{x},y)$.

The predicates $=$ and $\leqslant$ are primitive recursive, since their defining sets have characteristic functions sg$|x-y|$ and sg$(x\dot{-}y)$. Other examples will have to wait till the next few results are proved.

(I) Suppose there is a function ϕ in C such that x $\in A$ iff $\phi x=0$. Then A is in C, since its characteristic function is just sgϕ.

(II) Let A and B be subsets of $\mathbf{N}^k$ which are in C. Then $\mathbf{N}^k-A$, $A\cup B$, and $A\cap B$ are in C. To see this let χ_A and χ_B be the characteristic functions of A and B. Then $\mathbf{N}^k-A$ has characteristic function $1\dot{-}\chi_A$ and $A\cup B$ has characteristic function $\chi_A.\chi_B$, while $(\chi_A+\chi_B)\mathbf{x}=0$ iff x $\in A\cap B$.

Suppose that $f:\mathbf{N}^k\to\mathbf{N}^r$ is a function in C, and that X is a subset of $\mathbf{N}^r$ in C. Then $f^{-1}A$ and $A\times X$ are in C, since the former has $\chi_A f$ as its characteristic function, while the function ϕ given by $\phi(\mathbf{a},\mathbf{x}) = \chi_A\mathbf{a} + \chi_X\mathbf{x}$ is zero exactly on $A \times X$.

In particular, if $f:\mathbf{N}^r \to \mathbf{N}$ is in C, then $\{(\mathbf{x},y); f\mathbf{x} = y\}$ and $\{(\mathbf{x},y); f\mathbf{x} \leqslant y\}$, and other similar sets, are in C.

(III) Let P and Q be properties of k variables which are in C. Then the properties $\neg P$, $P \vee Q$, and $P \wedge Q$ are in C. We need only observe that if P and Q define the sets A and B, then these properties define respectively the sets $\mathbf{N}^k - A$, $A \cup B$ and $A \cap B$.

In particular the predicates 'less than' and 'greater than' will be primitive recursive.

The properties $P \to Q$ and $P \leftrightarrow Q$ are also in C, since the former is the same as $(\neg P) \vee Q$, and the latter is just $(P \to Q) \wedge (Q \to P)$.

Also if $f:\mathbf{N}^r \to \mathbf{N}^k$ is in C, then $\{\mathbf{x}; f\mathbf{x} \text{ has } P\}$ is in C, since its corresponding set is $f^{-1}A$.

(IV) Let C be the class of intuitively computable functions. Then a set is in C iff it is decidable. It follows from Theorem 2.5 and Exercise 2.8 that there is a subset B of $\mathbf{N}^2$ which is in C and such that $\{x; \exists y \text{ with } (x,y) \text{ in } B\}$ is not in C. Similarly, when C_1 is the class of primitive recursive functions, Theorem 4.9 and Exercise 4.3 show that there is a subset B of $\mathbf{N}^2$ in C_1 and such that $\{x; \exists y$ with (x,y) in $B\}$ is not in C_1. However, problems do not arise if we put a bound on the quantification.

To be precise, let P be a property of $k+1$ variables. The property $(\exists z \leqslant y)P(\mathbf{x},z)$ is defined to be true iff there is some $z \leqslant y$ for which $(\mathbf{x},z)$ has the property P. Similarly the property $(\forall z \leqslant y)P(\mathbf{x},z)$ is true iff $(\mathbf{x},z)$ has the property P for all $z \leqslant y$. Both of these are properties of $k+1$ variables.

If P is in C, then both $(\exists z \leqslant y)P(\mathbf{x},z)$ and $(\forall z \leqslant y)P(\mathbf{x},z)$ are in C. For let χ be the characteristic function of P. If there is $u \leqslant y$ with $(\mathbf{x},u)$ having P, then $\chi(\mathbf{x},u) = 0$ and hence $\Pi^{z \leqslant y}\, \chi(\mathbf{x},z)$ is 0. If there is no such u, then $\chi(\mathbf{x},z) = 1$ for all $z \leqslant y$ and hence $\Pi^{z \leqslant y}\chi(\mathbf{x},z)$ is 1. Thus the function $\Pi^{z \leqslant y}\, \chi(\mathbf{x},z)$, which is in C, is the characteristic function of $(\exists z \leqslant y)P(\mathbf{x},z)$. The property $(\forall z \leqslant y)P(\mathbf{x},z)$ is in C, since it can be written as $\neg(\exists z \leqslant y)(\neg P(\mathbf{x},z))$.

The related predicates $(\exists z < y)P(\mathbf{x},z)$ and $(\forall z < y)P(\mathbf{x},z)$ are also in C, since the product of the $\chi(\mathbf{x},z)$ for $z < y$ is also a function (of $\mathbf{x}$ and y) in C.

If we combine these results with the last result in (III) we find that if, in addition, $f:\mathbf{N}^k \to \mathbf{N}$ is a function in C, then the properties $(\exists z \leqslant f\mathbf{x})(P(\mathbf{x},z))$ and $(\forall z \leqslant f\mathbf{x})(\mathrm{P}(\mathbf{x},z))$ are in C, as are the properties obtained by replacing $\leqslant$ by $<$.

(V) Similarly, minimisation of a predicate in C may take us outside C. In particular, if we take a primitive recursive function f whose minimisation g is not total then f will be in C for all C but g cannot be in C as it is not total.

The situation is more complicated if we take a function f whose minimisation g happens to be total. In that case, when C is the class of recursive functions defined in the next chapter, it is immediate from the definition that g is in C if f is in C. By contrast, when we take C to be the class of primitive recursive functions, the discussion of Ackermann's function in section 3.6 shows that f can be in C but g not in C.

We will now show that bounded minimisation remains in C.

Let P be a predicate of $k+1$ variables. We define the function $(\mu z \leqslant y)P(\mathbf{x},z)$ to be the least $z \leqslant y$ such that P holds for $(\mathbf{x},z)$. We will show that this function from $\mathbf{N}^{k+1}$ to $\mathbf{N}$ is in C if P is in C.

Suppose that P holds for $(\mathbf{x},r)$ but not for $(\mathbf{x},i)$ when $i<r$. Let χ be the characteristic function of P. Then $\chi(\mathbf{x},r)=0$ and $\chi(\mathbf{x},i)=1$ for $i<r$. Hence $\Pi^{i \leqslant z}\,\chi(\mathbf{x},i)$ is 1 for $z<r$ and 0 for $z \geqslant r$. If we take the sum of these products for all $z \leqslant y$ the resulting function $\Sigma^{z \leqslant y}\Pi^{i \leqslant z}\,\chi(\mathbf{x},i)$ is a function in C, and equals r which is by definition $(\mu z \leqslant y)P(\mathbf{x},z)$, as required.

Notice that if $P(\mathbf{x},i)$ is false for all $i \leqslant y$ then $(\mu z \leqslant y)P(\mathbf{x},z)$ equals $y+1$ according to this definition. Hence the value of $(\mu z \leqslant y)P(\mathbf{x},z)$ is not, as one might suppose, always at most y. However, this definition seems to me better than the alternative definition which defines $(\mu z \leqslant y)P(\mathbf{x},z)$ to be 0 if $P(\mathrm{x},i)$ is false for all $i \leqslant y$. Firstly, this alternative definition does not provide such a simple formula. Secondly, if $(\mu z \leqslant y)P(\mathbf{x},z)=0$ with the alternative definition we would not know whether or not $P(\mathbf{x},0)$ holds, which would be inconvenient.

Suppose, in addition, that we have a function $f{:}\mathbf{N}^k \to \mathbf{N}$ in C. Then, by composition, the function $(\mu z \leqslant f\mathbf{x})P(\mathbf{x},z)$ is in C.

We can also look at the largest $z \leqslant y$ such that $P(\mathbf{x},z)$ holds. This is denoted by $(\mu' z \leqslant y)P(\mathbf{x},z)$. Observe that, if there is any such z, we have $P(\mathbf{x},z)$ and, for any $w \leqslant y$, if $w>z$ then $\neg P(\mathbf{x},w)$; that is, $w \leqslant z$ or $\neg P(\mathbf{x},w)$. Also if $z_1<z$ we do not have $\neg P(\mathbf{x},w)$ for all $w>z_1$ since we have $P(\mathbf{x},z)$. Thus we can define $(\mu' z \leqslant y)P(\mathbf{x},z)$ to be $(\mu z \leqslant y)[P(\mathbf{x},z) \wedge (\forall w \leqslant y)\ (w \leqslant z \vee \neg P(\mathbf{x},w))]$. Hence $(\mu' z \leqslant y)P(\mathbf{x},z)$ is also in C if P is.

In applying (IV) and (V) two principles are useful. The first is that if there is only one z with a property it is certainly the least z with that property. The second is that we may want to know about the least z with a property (or about the existence of such a z); in order to get rid of the unboundedness of the minimisation (or quantification) we may be able to find a suitable bound. This is illustrated in the examples in the next section.

3.3 EXAMPLES USING BOUNDED MINIMISATION

(I) The square root function, defined by $\mathrm{sqr}n=y$ if $y^2 \leqslant n$ but $(y+1)^2>n$, is primitive recursive. One proof of this is given as Exercise 3.2. But we can now simply observe that $\mathrm{sqr}n=(\mu' y \leqslant n)(y^2 \leqslant n)$.

(II) The functions K and L (which are the components of J^{-1}) are primitive recursive.

By definition Kr is the unique m for which there is n with $J(m,n)=r$. Thus $Kr=\mu m \exists n(J(m,n)=r)$. This is not enough to show K primitive recursive, since we have unbounded minimisation and quantification. However we know that $m,n \leqslant J(m,n)$, and hence $Kr=(\mu m \leqslant r)(\exists n \leqslant r)\ (J(m,n)=r)$. Since J is primitive recursive, the predicate $J(m,n) = r$ is primitive recursive. As $r(=\pi_{31}(r,m,n))$ is a primitive recursive function of (r,m,n) it follows first that $(\exists n \leq r)(J(m,n) = r)$ is a primitive recursive predicate of r and m, and then that $(\mu m \leq r)(\exists n \leq r)(J(m,n) = r)$ is a primitive recursive function of r, as required. Similarly for L.

(III) The predicate 'divides' is primitive recursive. First we have that x divides y iff $\exists z(xz = y)$. If there is such a z, then $z \leq y$ for $x \neq 0$, while if $x = 0$ either there is no such z (if $y \neq 0$) or we can take z to be 0. Thus x divides y iff $(\exists z \leq y)(xz = y)$. Since multiplication is primitive recursive, the predicate $xz = y$ is primitive recursive, and as y is a primitive recursive function of (x,y) the predicate $(\exists z \leq y)(xz = y)$ is primitive recursive.

(IV) The functions quo and rem are primitive recursive.

Let $\text{quo}(x,y) = q$. Then, if $x \neq 0$, $xq \leq y$ and $x(q+1) > y$ and also $q \leq y$. Thus we define $\text{quo}(x,y)$ to be $(\mu' z \leq y)(xz \leq y)$. As mentioned earlier, this definition is not the same as the previous definition of quo when x is 0, but this does not matter, as we shall never be interested in this case.

Finally, we define $\text{rem}(x,y)$ to be $y \dot{-} x.\text{quo}(x,y)$, which is also primitive recursive.

(V) The predicates 'composite' and 'prime' are primitive recursive.

By definition x is composite iff it is the product of two integers greater than 1. Since an integer greater than 1 is $y+2$ for some $y \in \mathbf{N}$, we have x composite iff $\exists y \exists z((y+2)(z+2) = x)$. Since y and z will then be at most x, we find that x is composite iff $(\exists y \leq x)(\exists z \leq x)[(y+2)(z+2) = x]$, which is a primitive recursive predicate.

Also, as 0 and 1 are not regarded as being either prime or composite, we see that x is prime iff $(x \neq 0) \wedge (x \neq 1) \wedge (x$ not composite), which is a primitive recursive predicate.

(VI) Let p_n be the nth prime. Then p_n is a primitive recursive function of n.

Here we have $p_0 = 2, p_1 = 3, p_2 = 5$, etc. Thus p_n is the iterate starting at 2 of the function f such that fx is the least prime $> x$. Hence it is enough to show f is primitive recursive.

Now the predicate 'p prime and $p > x$' is primitive recursive, and f is obtained from this by minimisation. If we can find a primitive recursive g which is *a* bound for f, so that $fx = (\mu p \leq gx)$ (p prime and $p > x$) the result will follow.

We can take gx to be $x! + 1$, which is primitive recursive. For $x! + 1$ cannot be divisible by any of $2,3, \ldots, x$, and so there must be a prime $\leq gx$ and $> x$ since gx has at least one prime factor.

(VII) Any positive integer a can be writen as the product of prime powers. If we write a as $p_0^{\alpha(0)} p_1^{\alpha(1)} \ldots$ we define the nth **exponent** of a, written $\exp_n a$, to be α_n. We also define the **length** of a to be the largest n such that $\alpha_n \neq 0$. Then $\exp_n a$ is a primitive recursive function of n and a, while the length of a is a primitive recursive function of a.

Observe first that $p_n^{\alpha(n)}$ divides a, while $p_n^{\alpha(n)+1}$ does not divide a. Also $\alpha_n < a$, so $\alpha_n + 1 \leq a$. Thus we can define $\exp_n a$ to be $(\mu' y \leq a)(p_n^y$ divides $a)$. We choose this to be the definition of $\exp_n a$ even when a is 0.

Let the length of a be k. Thus $\exp_i a = 0$ for $i > k$. But we already know that $\exp_i a$ must be 0 for $i > a$. Hence as $\exp_k a$ is not 0, by definition, we see that the length of a is $(\mu' n \leq a)(\exp_n a \neq 0)$. This shows that length is primitive recursive since $\exp_n a$ is a primitive recursive function of n and a.

3.4 EXTENSIONS OF PRIMITIVE RECURSION

We can now look at simultaneous primitive recursion and other extensions of the concept of primitive recursion.

Let g_0, g_1 be functions from $\mathbf{N}^k$ to $\mathbf{N}$ and let h_0, h_1 be functions from $\mathbf{N}^{k+3}$ to $\mathbf{N}$. We can define functions $f_0, f_1 : \mathbf{N}^{k+1} \to \mathbf{N}$ by $f_0(\mathbf{x},0) = g_0\mathbf{x}$, $f_1(\mathbf{x},0) = g_1\mathbf{x}$, $f_0(\mathbf{x},y+1) = h_0(\mathbf{x},y,f_0(\mathbf{x},y),f_1(\mathbf{x},y))$, $f_1(\mathbf{x},y) = h_1(\mathbf{x},y,f_0(\mathbf{x},y),f_1(\mathbf{x},y))$. Then f_0 and f_1 are in C if g_0, g_1, h_0, and h_1 are in C.

To see this define f, g, and h by $f(\mathbf{x},y) = J(f_0(\mathbf{x},y),f_1(\mathbf{x},y))$, $g\mathbf{x} = J(g_0\mathbf{x},g_1\mathbf{x})$ and $h(\mathbf{x},y,z) = J(u,v)$ where $u = h_0(\mathbf{x},y,Kz,Lz)$ and $v = h_1(\mathbf{x},y,Kz,Lz)$. Then, since J, K, and L are primitive recursive, f_0 and f_1 will be in C if f is in C. However, the functions g and h are in C and it is easy to check that f comes from g and h by primitive recursion. Hence the result follows.

Now suppose we have $g_0, g_1 : \mathbf{N}^k \to \mathbf{N}$ and $h : \mathbf{N}^{k+3} \to \mathbf{N}$ in C. Suppose f is defined by $f(\mathbf{x},0) = g_0\mathbf{x}$, $f(\mathbf{x},1) = g_1\mathbf{x}$, $f(\mathbf{x},y+2) = h(\mathbf{x},y,f(\mathbf{x},y),f(\mathbf{x},y+1))$. Then f is in C.

Here we reduce to the previous case by putting $f_1(\mathbf{x},y) = f(\mathbf{x},y+1)$. Then f and f_1 satisfy the equations $f(\mathbf{x},0) = g_0\mathbf{x}$, $f_1(\mathbf{x},0) = g_1\mathbf{x}$, $f(\mathbf{x},y+1) = f_1(\mathbf{x},y)$ and $f_1(\mathbf{x},y+1) = h(\mathbf{x},y,f(\mathbf{x},y),f_1(\mathbf{x},y))$. This is a simultaneous recursion of the kind we have just considered, bearing in mind that the third equation could be written as $f(\mathbf{x},y+1) = \pi_{k+3,k+3}(\mathbf{x},y,f(\mathbf{x},y),f_1(\mathbf{x},y))$.

Plainly these results can be extended to any number of simultaneous primitive recursions.

Proofs by induction are often made using complete induction instead of ordinary induction; that is, by proving a property for $n+1$ assuming it not just for n but for all $r \leq n$. Similarly we could require a function to be defined so that $f(\mathbf{x},n+1)$ depended on the values of $f(\mathbf{x},r)$ for all $r \leq n$.

To make this precise we first define for any $f : \mathbf{N}^{k+1} \to \mathbf{N}$ another function $F : \mathbf{N}^{k+1} \to \mathbf{N}$ which is called the **history** of f. We define $F(\mathbf{x},y)$ to be $\Pi^{r \leq y} p_r^{f(\mathbf{x},r)}$. Then, for any $r \leq y$ we have $f(\mathbf{x},r) = \exp_r F(\mathbf{x},y)$, so that $F(\mathbf{x},y)$ gives us knowledge of $f(\mathbf{x},r)$ for all $r \leq y$, which explains the name 'history of f'.

Lemma 3.3 *Let F be the history of f. Then F is in C iff f is in C.*

Proof Since $f(\mathbf{x},r) = \exp_r F(\mathbf{x},r)$, and $\exp_r a$ is a primitive recursive function, we see that f is in C if F is.

Conversely let f be in C. Since the functions p_r and a^b are primitive recursive and f is in C, the function $p_r^{f(\mathbf{x},r)}$ will be in C, and hence F will also be in C. ■

Let $g : \mathbf{N}^k \to \mathbf{N}$ and $h : \mathbf{N}^{k+2} \to \mathbf{N}$ be functions. We say $f : \mathbf{N}^{k+1} \to \mathbf{N}$ is obtained from g and h by **course-of-values recursion** if $f(\mathbf{x},0) = g\mathbf{x}$ and

$f(\mathbf{x},y+1) = h(\mathbf{x},y,F(\mathbf{x},y))$, where F is the history of f. The use of F is the only way we can keep track of the values of $f(\mathbf{x},r)$ for arbitrary values of r.

Lemma 3.4 *Let f be obtained from g and h by course-of-values recursion. If g and h are in C then so is f.*

Proof It is enough to show F is in C. Now $F(\mathbf{x},0) = 2^{g\mathbf{x}}$, and this latter function is in C. Further $F(\mathbf{x},y+1) = F(\mathbf{x},y).(p_{y+1})^{f(\mathbf{x},y+1)}$. Thus we define H by $H(\mathbf{x},y,z) = z.(p_{y+1})^{h(\mathbf{x},y,z)}$, and we see that H is in C. Now F comes from $2^{g\mathbf{x}}$ and H by primitive recursion, so that F is in C, as required.■

As an example suppose that $f:\mathbf{N}\to\mathbf{N}$ has its value at x defined in terms of its values at $x/2$ and $x/3$, or, more precisely (as we require all variables to be integers), in terms of quo$(2,x)$ and quo$(3,x)$. For instance suppose we have $f0 = 2$, $fx = f\text{quo}(2,x) + f\text{quo}(3,x)$. We must express this as a course-of-values recursion. Since both quo$(2,x+1)$ and quo$(3,x+1)$ are at most x, and $fr = \exp_r Fs$ whenever $r \leqslant s$, we can define f by $f0 = 2$ and $f(x+1) = \exp_{\text{quo}(2,x+1)}Fx + \exp_{\text{quo}(3,x+1)}Fx$, which is a course-of-values recursion from a primitive recursive function. Hence f is primitive recursive.

We conclude this section with a complicated example. This is included partly to show how quite complex definitions can be fitted into the frameworks we have considered. In Chapter 12 we shall argue informally that a certain set which we construct is decidable; a formal proof of that result would be a slightly more complex version of the example below.

Let A be a subset of $\mathbf{N}$ and let $f,g:\mathbf{N}^2\to\mathbf{N}$ be functions such that both $f(m,n)$ and $g(m,n)$ are greater than $\max(m,n)$. Let B be the smallest subset of $\mathbf{N}$ which contains A and is such that $f(m,n) \in B$ if m and n are in B and $g(m,n) \in B$ if $m \in B$ (whatever n is). If A, f, and g are in C then B is in C.

The definition of B (as the smallest set having the stated properties) and the assumed properties of f and g tell us that $0 \in B$ iff $0 \in A$, while $n+1 \in B$ iff either $n+1 \in A$ or there are r and s, both at most n, with r and s in B and $f(r,s) = n+1$ or there are r and s, both at most n, with r in B and $g(r,s) = n+1$. It is enough to find a function ϕ in C such that $\phi x = 0$ iff $x \in B$. We define ϕ in terms of f,g, and the characteristic function χ of A.

Suppose we have already defined ϕx for $x \leqslant n$. Then $\phi(n+1) = 0$ iff either $\chi(n+1) = 0$, or for some $r,s \leqslant n$ we have $f(r,s) = n+1$ and $\phi r = 0 = \phi s$ or for some $r,s \leqslant n$ we have $g(r,s) = n+1$ and $\phi r = 0$. These last two conditions can be combined into the condition that, for some $r,s \leqslant n$, $(\phi r + \phi s + |n+1-f(r,s)|)(\phi r + |n+1-g(r,s)|) = 0$. Hence we can define $\phi(n+1)$ to be

$$\chi(n+1).\Pi^{r\leqslant n}\Pi^{s\leqslant n}(\phi r + \phi s + |n+1-f(r,s)|)(\phi r + |n+1-g(r,s)|),$$

and we will have, as needed, $\phi(n+1) = 0$ iff $n+1 \in B$.

Finally we have to express this definition of $\phi(n+1)$ in terms of Φ, the history of ϕ. Since, for $r \leqslant n$, we have $\phi r = \exp_r \Phi n$, the expression is

$$\begin{aligned}\phi(n+1) = \chi(n+1).\Pi^{r\leqslant n}\Pi^{s\leqslant n}(&\exp_r\Phi n + \exp_s\Phi n + \\ &|n+1-f(r,s)|)(\exp_r\Phi n + |n+1-g(r,s)|).\end{aligned}$$

Hence ϕ is obtained by course-of-values recursion from $\chi 0$ and the function h given by

$$h(x,y) = \chi(x+1).\Pi^{r \leq x}\Pi^{s \leq x}(\exp_r y + \exp_s y + |x+1-f(r,s)|) \times \\ \times (\exp_r y + |x+1-g(r,s)|).$$

The complicated function h is obtained by procedures we have already discussed (forming $\Pi^{r \leq x}$ and $\exp_r$, etc.) from the functions f,g, and χ. Since f,g, and χ are in C, h will also be in C. It follows that ϕ is in C, as claimed.

Exercise 3.5 Let g_0,g_1:$\mathbf{N} \to \mathbf{N}$ and h:$\mathbf{N}^4 \to \mathbf{N}$ be functions. Let f:$\mathbf{N}^2 \to \mathbf{N}$ be defined by the *double recursion* $f(m,0) = g_0 m$, $f(0,n) = g_1 n$ (we must have $g_0 0 = g_1 0$) and $f(m+1,n+1) = h(m,n,f(m+1,n),f(m,n+1))$. Show that if g_0, g_1, and h are in C then f is in C. (You may wish to obtain a formula in terms of g_0, g_1, and h for the function D defined by $Dr = \Pi^{m+n=r}\, p_m^{f(m,n)}$.)

Exercise 3.6 Let g_0, g_1, and h be as above, and let ϕ:$\mathbf{N}^2 \to \mathbf{N}$ be a function such that $\phi(m,n) \leq n$ for all m and n. Let f be defined by

$$f(m,0) = g_0 m,\ f(0,n) = g_1 n \text{ and} \\ f(m+1,n+1) = h(m,n,f(m+1,\phi(m,n)),f(m,n+1)).$$

Show that if g_0, g_1, h, and ϕ are in C then f is in C. (You may wish to show the history of f is defined by a double recursion as in the previous exercise.)

3.5 FUNCTIONS OF ONE VARIABLE

The definition of primitive recursive functions requires functions of an arbitrary number of variables to occur in the defining sequence of a primitive recursive function of one variable. The remainder of this section, which can be omitted on a first reading, shows how a primitive recursive function of one variable can be obtained by a new kind of defining sequence in which all the functions are of one variable.

We need to define certain special functions. The *excess over a square* function, denoted by exq, is defined by $\text{exq}n = n - (\text{sqr}n)^2$. The function J^*:$\mathbf{N}^2 \to \mathbf{N}$ is defined by $J^*(m,n) = ((m+n)^2 + m)^2 + n$. The function L^*:$\mathbf{N} \to \mathbf{N}$ is just exq, while K^*:$\mathbf{N} \to \mathbf{N}$ is given by $K^* n = \text{exq}(\text{sqr}n)$. Plainly $J^*(0,0) = K^* 0 = L^* 0 = 0$. Also

$$((m+n)^2 + m + 1)^2 = ((m+n)^2 + m)^2 + 2((m+n)^2 + m) + 1 > \\ > J^*(m,n).$$

Hence $\text{sqr}J^*(m,n) = (m+n)^2 + m$, so $L^* J^*(m,n) = n$. Also, for similar reasons, $\text{sqr}(\text{sqr}J(m,n)) = m+n$, and so $K^* J^*(m,n) = m$. Also, from the definition of L^* it is easy to check that if $L^*(x+1) \neq 0$ then $L^*(x+1) = L^* x + 1$ and $K^*(x+1) = K^* x$. This latter property will be used later, and is the reason we need to look at J^*, K^*, and L^* as well as J, K, and L.

Let J^*_1 be π_{11}, and define inductively $J^*_{n+1}:\mathbf{N}^{n+1}\to\mathbf{N}$ by $J^*_{n+1}(x,\mathbf{y}) = J^*(x,J^*_n\mathbf{y})$. Thus $J^*_2 = J^*$. Define $J^*_1{}^{-1}$ to be π_{11}, and inductively define $J^*_{n+1}{}^{-1}:\mathbf{N}\to\mathbf{N}^{n+1}$ by $J^*_{n+1}{}^{-1}x = (K^*x, J^*_n{}^{-1}L^*x)$. As J^*_n is not a bijection it does not have a two-sided inverse. However, the remarks already made show that $J^*_n{}^{-1}J^*_n\mathbf{x} = \mathbf{x}$ for all $\mathbf{x}$, and this is all we shall need.

Proposition 3.5 *Let C be a set of total functions which contains the initial functions and is closed under composition and under iteration of functions of one variable. If C contains either J,K,L or J*,K*,L* then C is primitive recursively closed.*

Proof We prove the result for J,K,L. The other case is just obtained by putting * in suitable places.

By Theorem 3.2 it is enough to show C is closed under iteration of all functions. So let $f:\mathbf{N}^n\to\mathbf{N}^n$ be a function in C whose iterate is F. Let $g:\mathbf{N}\to\mathbf{N}$ be $J_n f J_n{}^{-1}$. Since J,K,L are in C, g will also be in C. Let G be the iterate of g. By hypothesis G is in C.

Hence it is enough to show that $F(\mathbf{x},y) = J_n{}^{-1}G(J_n\mathbf{x},y)$. We prove this by induction on y. Now $J_n{}^{-1}G(J_n\mathbf{x},0) = J_n{}^{-1}J_n\mathbf{x} = \mathbf{x} = F(\mathbf{x},0)$, as needed. Suppose that the formula holds for y. Then

$$J_n{}^{-1}G(J_n\mathbf{x},y+1) = J_n{}^{-1}gG(J_n\mathbf{x},y) = J_n{}^{-1}J_n f J_n{}^{-1}G(J_n\mathbf{x},y) = = fF(\mathbf{x},y),$$

by the inductive assumption, and this equals $F(\mathbf{x},y+1)$ by definition.■

Recall that $\phi:\mathbf{N}\to\mathbf{N}$ comes from $f:\mathbf{N}\to\mathbf{N}$ by iteration starting at 0 if $\phi 0 = 0$ and $\phi(n+1) = f(\phi n)$ for all n; equivalently $\phi n = F(n,0)$ where $F:\mathbf{N}^2\to\mathbf{N}$ is the iterate of f. We shall call a set of total functions from $\mathbf{N}$ to $\mathbf{N}$ *good* if it satisfies the following three conditions: (i) C contains the successor function S and the function exq, which we refer to as the initial good functions, (ii) if f and g are in C so is their sum $f+g$ and their composite fg, (iii) if f is in C so is its iterate starting at 0. We shall call C *very good* if C satisfies (ii) and (iii) and (i′) the functions S, exq, sqr, quo$(2,n)$, and n^2 (which we refer to as the initial very good functions) are in C. In fact all good sets are very good. However, this fact is somewhat messy to prove, and the heart of the results we are going to look at can be obtained without using this property; it therefore seems to me simplest to allow both concepts.

Plainly both K^* and L^* are in any very good set. Also if f and g are in the very good set C, then so is the composite of J^* with f and g, because the squaring function is in C. Inductively, it follows that if $f_1, \ldots, f_n$ are in C then the composite of $J_n{}^*$ with $f_1, \ldots, f_n$ is in C.

Let C be good. The identity function π_{11} is in C, since it is the iterate starting at 0 of S. The zero function is in C since it is the iterate starting at 0 of sqr. It follows that any constant function is in C. Also, since C is closed under addition, for any $f \in C$ and any fixed $k \in N$ the function $k.f$ is in C. The function sg is in C, since it is the iterate starting at 0 of the function which is constantly 1. Now observe that $2+2\text{sg}x$ is 2 if $x=0$ and is 4 otherwise. Hence the function co is in C, since it is given by $\text{co}x = \text{exq}(2+2\text{sg}x)$.

Let a and b be in $\mathbf{N}$ with $a \geqslant b$. Now

$$(a+b)^2+3a+b+1=(a+b+1)^2+(a-b)<(a+b+2)^2.$$

Hence $a-b=\mathrm{exq}((a+b)^2+3a+b+1)$. It follows that if C is very good, and $f,g \in C$ with $fn \geqslant gn$ for all n, then $f \dot{-} g$ is in C, since $(f \dot{-} g)n = \mathrm{exq}((fn+gn)^2+3fn+gn+1)$. It also follows that for any f and g in C the product $f.g$ is in C since $f.g=((f+g)^2 \dot{-} f^2 \dot{-} g^2)/2$, and both squaring functions in C and dividing them by 2 lead to functions in C.

Lemma 3.6 *Let C be a very good set. Let f be in C and let F be the iterate of f. Let $g=FJ^{*-1}$. Then $g \in C$.*

Proof We have $g0=F(0,0)=0$. We will show first that there is a function $\phi{:}\mathbf{N}^2 \to \mathbf{N}$ such that ϕJ^{*-1} is in C and $g(x+1)=\phi(x,gx)$ for all x. Now $g(x+1)=F(K^*(x+1),L^*(x+1))$. If $L^*(x+1)=0$ this gives $g(x+1)=K^*(x+1)$. If $L^*(x+1) \neq 0$, from what was noted when K^* and L^* were defined, it follows that $g(x+1)=F(K^*x,L^*x+1)=fF(K^*x,L^*x)=fgx$. This is a definition by cases, and the formula for definition by cases gives

$$\begin{aligned}&g(x+1)=\phi(x,gx) \text{ where}\\ &\phi(u,v)=K^*(u+1).\mathrm{co}L^*(u+1)+fv.\mathrm{sg}L^*(u+1).\end{aligned}$$

Then, as required, the function ϕJ^{*-1} is in C, since the functions sg, co, K^* and L^* are in any very good set, and very good sets are closed under multiplication.

Now define hx to be $J^*(x,gx)$. As $gx=L^*hx$, g will be in C if h is. But $h(x+1)=J^*(x+1,g(x+1))=J^*(x+1,\phi(x,gx))$. Since $x=K^*hx$, this can be written as $h(x+1)=\theta hx$, where $\theta u=J^*(K^*u+1,\phi J^{*-1}u)$. Since ϕJ^{*-1} is in C, so is θ. Finally $h0=J^*(0,0)=0$. Hence h is the iterate of θ starting at 0, and by hypothesis h is in C since θ is. ■

Lemma 3.7 *Let C be very good, and let $C'=\{fJ^*_n$; all f in C, all $n\}$; in particular $(n=1)$, $C' \supset C$. Then C' is primitively recursively closed.*

Proof We have already seen that C contains the zero function and π_{11}. Also C contains all the components of $J^*{}_n{}^{-1}$, since these are repeated composites of K^* and L^*. Hence C' contains all the components of $J^*{}_n{}^{-1}J^*{}_n$; that is, C' contains all the projections π_{ni}.

Let $f,g_1,\ldots,g_n$ be in C. Then the composite of fJ^*_n with $g_1J^*_k,\ldots,g_nJ^*_k$ is fhJ^*_k, where h is the composite of J^*_n with $g_1,\ldots,g_n$. We have already seen that the latter is in C. Hence C' is closed under composition.

Now let f be in C. By Lemma 3.6 and the definition of C' we have $F \in C'$, where F is the iterate of f. By Lemma 3.5 this shows C' is primitive recursively closed. ■

Lemma 3.8 *Let C_0 be the set of functions f for which there is a sequence $f_1,\ldots,f_n$ with $f_n=f$ and, for all $r \leqslant n$, either f_r is an initial good function or there* *is $i<r$ with f_r the iterate of f_i starting at 0 or there are $i,j<r$ with $f_r=f_i+f_j$ or $f_r=f_if_j$. Then C_0 is good.*

Let C_1 be the set of functions f for which there is a sequence $f_1,\ldots,f_n$ with $f_n=f$ and, for all $r\leq n$, either f_r is an initial very good function or there is $i<r$ with f_r the iterate of f_i starting at 0 or there are $i,j<r$ with $f_r=f_i+f_j$ or $f_r=f_if_j$. Then C_1 is very good.

Proof We prove the first case only; the second case is almost identical, and both parts are very similar to Lemma 3.1.

Plainly C_0 contains the initial good functions, for which there is a sequence of length one. Let f and g be in C_0, and suppose they are defined by sequences $f_1,\ldots,f_n$ and $g_1,\ldots,g_m$. Then $f_1,\ldots,f_n,g_1,\ldots,g_m,f+g$ is a suitable sequence for $f+g$, and similarly for fg, while, if ϕ is the iterate of f starting at 0, the sequence $f_1,\ldots,f_n,\phi$ is a suitable sequence for ϕ.■

The next proposition provides a characterisation of primitive recursive functions of one variable which does not use functions of more than one variable.

Proposition 3.9 *A function $f:\mathbf{N}^k\to\mathbf{N}$ is primitive recursive iff there is a sequence of functions of one variable $f_1,\ldots,f_n$ such that $f=f_nJ^*{}_k$, and each f_r is either an initial (very) good function or comes from two earlier functions by composition or addition or comes from one earlier function by iteration starting at 0. In particular $f:\mathbf{N}\to\mathbf{N}$ is primitive recursive iff there is such a sequence with $f=f_n$.*

Proof As $J^*{}_1$ is the identity, the one-variable case follows at once from the general case. We shall prove the result when very good initial functions are permitted. To show that we need only consider good initial functions we must show that any good set is very good, which will be proved in the next lemma.

By induction on r, if f can be represented in this form it is primitive recursive. To show that any primitive recursive function can be expressed in this form, it is enough, by Lemma 3.1, to show that the set of functions which can be expressed like this is primitive recursively closed. But this is immediate from Lemmas 3.7 and 3.8.■

Lemma 3.10 *Any good set is very good.*

Proof Let C be good. Suppose we show that the function ϕ given by $\phi n=n+2\text{sqr}n$ is in C. Then the squaring function is obtained from $\phi+1$ by iteration starting at 0, using the fact that $(n+1)^2=n^2+2n+1$.

Let f be defined by $fn=\text{co}n+2\text{co}(n-1)$. Then $f\in C$. Here we must regard $n-1$ as being given by the formula already considered in the paragraph before Lemma 3.6. This formula, which involves n^2, gives $n-1$ for $n\neq 0$, while its value is 2 for $n=0$. It follows that $f0=1, f1=2, fn=0$ for all other n.

Let g be the iterate of f starting at 0, so $g\in C$. Then the sequence of values of g is 0,1,2,0,1,2,0, . . . Let h be the iterate starting at 0 of the function $n+1+gn$, so that $h0=0$ and $h(n+1)=hn+1+ghn$. Then $h \in C$.

It is easy to check by induction that $h(2n)=3n$ and $h(2n+1)=3n+1$. Thus $h(2n)-2n=n=h(2n+1)-(2n+1)$; that is, $\text{quo}(2,x)=hx \dot{-} x$, which is in C. Further, $\text{sqr}n=\text{quo}(2,\phi n \dot{-} n)$ showing that sqr is in C.

Now $\phi(n+1)=n+1+2\text{sqr}(n+1)$. Hence we may write $\phi(n+1)=\phi n+1+2\alpha n$, where $\alpha n=0$ if $n+1$ is not a square and is 1 if $n+1$ is a square.

We shall show that $n+1$ is a square iff $\phi n+4$ is a square. Once this has been proved we will have the equation $\phi(n+1)=\phi n+1+2\text{co}(\text{exq}(\phi n+4))$. This shows that ϕ is obtained by iteration starting at 0 from the function $x+1+\text{co}(\text{exq}(x+4))$. This latter function is in C, and hence ϕ is in C.

If $n+1$ is a square, then $n=m^2+2m$ for some m, and hence $\phi n=m^2+2m+2m$, so $\phi n+4=(m+2)^2$. Conversely if $\phi n+4$ is a square we can write $\phi n=m^2+4m$ for some m. Now $m^2>n$ is not possible, since then $\text{sqr}n<m$ and we have $n+2\text{sqr}n<m^2+4m$. Also $(m+1)^2 \leqslant n$ is also impossible, since then $\text{sqr}n \geqslant m+1$, and $n+2\text{sqr}n \geqslant (m+1)^2+2(m+1)>m^2+4m$. It follows that $m=\text{sqr}n$, and then $n+2\text{sqr}n=m^2+4m$ gives $n=m^2+2m$, as needed.■

3.6 SOME FUNCTIONS WHICH ARE NOT PRIMITIVE RECURSIVE

3.6.1 Universal functions for primitive recursion

We have seen that many quite complicated functions are primitive recursive, and that a wide variety of constructions applied to primitive recursive functions still give us primitive recursive functions. The reader could be forgiven for believing that all intuitively computable total functions are primitive recursive; in fact this was thought to hold for some time in the early development of the subject. However, this is not true, and we shall give two examples of such functions. The first example will be theoretical, but the proof does define a function which could be explicitly calculated. The other example is of a function which is given by a nested recursion, which at first sight looks similar to simultaneous primitive recursion and other extensions of the original notion of primitive recursion.

The theorem below uses techniques similar to those in Proposition 2.3 and Theorem 2.4.

Theorem 3.11 *There is an intuitively computable function which is not primitive recursive.*

Proof Suppose we can find a set X of strings on some alphabet which satisfies the following conditions: (i) we can tell whether or not a string is in X, (ii) to any string in X corresponds (in some way to be made precise) a primitive recursive function of one variable, which can be computed when the string is known, (iii) every primitive recursive function of one variable corresponds in this way to some string in X.

We have a computable one–one mapping from the set of all strings into

N. Thus we can define, as in Theorem 2.4, a function $F:\mathbf{N}^2 \to \mathbf{N}$ by requiring that if the integer n is the image of a string in X then $F(n,x) = fx$ where f is the function corresponding to the string, and requiring $F(n,x)$ to be 0 if n is not the image of a string in X. Plainly F is total.

As the mapping from strings into **N** is computable, we can tell whether or not the integer n corresponds to a string; if so, we can find the string, and by (i) we can tell whether or not the string is in X. If the string is in X we can find the corresponding function f and can then compute fx, since f is primitive recursive. Thus F is computable.

However F cannot be primitive recursive. For if F were primitive recursive, then the function g defined by $gn = F(n,n) + 1$ would also be primitive recursive. Hence, by (iii), it would correspond to a string in X. If this string maps to k we would have $gx = F(k,x)$ for all x. As in Proposition 2.3, this is impossible. For, putting $x = k$, we get $gk = F(k,k)$, whereas the definition of g requires that $gk = F(k,k) + 1$.

We still have to find the set X of strings. One method is given in Chapter 5, where the set called the set of primitive abacus machines will satisfy our conditions. Another approach is given below, based on the defining sequence of a primitive recursive function. Take the alphabet to consist of the symbols Z,S, and $P(n,i)$ where $1 \leqslant i \leqslant n$, together with symbols $R(i,j)$ and $C(i,j(1), \ldots, j(s))$ where i,j, and $j(1), \ldots, j(s)$ are positive integers. Suppose the function f is given by the defining sequence $f_1, \ldots, f_n$. Corresponding to this we can obtain a string of length n whose rth symbol is Z,S, or $P(n,i)$ if f_r is the zero or successor function or the projection π_{ni}, and is $R(i,j)$ if f_r comes by primitive recursion from f_i and f_j, while it is $C(i,j(1), \ldots, j(s))$ if f_r comes by composition from f_i and $f_{j(1)}, \ldots, f_{j(s)}$. (Note that one defining sequence may correspond to more than one string, since, for instance, f_r could come by primitive recursion from f_i and f_j and also by composition from other functions in the sequence.)

Plainly, not all strings correspond to primitive recursive functions; for instance the string consisting of one symbol $R(2,3)$ does not correspond to a function. Suppose we have a string and that the first r symbols correspond to a defining sequence $f_1, \ldots, f_r$. We must check whether there is a function f_{r+1} such that the first $r+1$ symbols of the string correspond to the sequence $f_1, \ldots, f_r, f_{r+1}$. Obviously this is possible if the $(r+1)$st symbol is Z or S or $P(n,i)$. If this symbol is $R(i,j)$ then we must have i and j at most r; if this happens then there is a suitable function iff, for some k, f_i is a function of k variables and f_j is a function of $k+2$ variables, and we then take f_{r+1} to be the function of $k+1$ variables defined from these by primitive recursion. Finally if this symbol is $C(i,j(1), \ldots, j(s))$ we require $i,j(1), \ldots, j(s)$ all to be at most r; if this happens we further require f_i to be a function of s variables, and, for some k, each of $f_{j(1)}, \ldots, f_{j(s)}$ to be functions of k variables, and then f_{r+1} is the function of k variables obtained from these by composition. Thus we can tell whether or not a string corresponds to a primitive recursive function, and, if so, whether the function is a function of one variable, as required.■

3.6.2 Ackermann's function

We now look at **Ackermann's function** $A:\mathbf{N}^2 \to \mathbf{N}$ which is defined by $A(0,y) = y+1$, $A(x+1,0) = A(x,1)$ and $A(x+1,y+1) = A(x,A(x+1,y))$. (This is a simpler variant of the function originally considered by Ackermann.) We shall show that A is not primitive recursive, but is intuitively computable. At first sight this *nested recursion* used to define A is not very different from the double recursion looked at in Exercise 3.6. However the fact that $A(x+1,y+1)$ is defined from the value of A taken at arguments depending on A rather than from the values of a previously given function at arguments depending on A makes the difference. We shall show that A can be obtained from primitive recursive functions by composition and minimisation.

If we define $a_m:\mathbf{N} \to \mathbf{N}$ by $a_m n = A(m,n)$ it is clear that a_{m+1} is obtained from a_m by iteration, and hence each a_m is primitive recursive. However this does not let us conclude that A itself is primitive recursive.

It is easy to find what a_m is for $m = 1$, 2, or 3. By induction on y we find that $A(1,y) = y+2$, $A(2,y) = 2y+3$ and $A(3,y) = 2^{y+3} - 3$.

Hence $A(4,0) = A(3,1) = 5$, $A(4,1) = A(3,A(4,0)) = A(3,5) = 2^8 - 3 = 253$, $A(4,2) = A(3,A(4,1)) = A(3,253) = 2^{256} - 3$. Also $A(5,0) = A(4,1) = 253$, $A(5,1) = A(4,A(5,0)) = A(4,253)$. We see that $A(x,y)$ is extremely large even for small values of x and y. We will show that A is not primitive recursive by showing that it increases faster than any primitive recursive function.

Most proofs of properties of A are proofs by double induction. This is formalised in the lemma below. (Readers who know about well-oprdered sets should observe that it amounts to a proof that $\mathbf{N}^2$ is well-ordered, and that proof by double induction is a special case of proof by transfinite induction.)

Lemma 3.12 *Let P be a property of pairs of integers. Suppose that, for all m and n, the pair (m,n) has P provided that (m',n') has P for all pairs for which either m' < m (and arbitrary n') or m' = m and n' < n. Then all pairs have P.*

Proof Let $S = \{(m,n);(m,n)$ does not have $P\}$. We want to show S is empty. Suppose not. Let T be $\{m;\exists n$ with $(m,n) \in S\}$. Then T is non-empty, and we define m_0 to be the smallest member of T. By definition of T, $\{n;(m_0,n) \in S\}$ is not empty, and we let n_0 be the smallest element of this set.

By definition of S and m_0 the pair (m',n') has P if $m' < m_0$, whatever n' is. By definition of n_0 the pair (m_0,n') has P for all $n' < n_0$. So, by the assumption about P, the pair (m_0,n_0) has P. But (m_0,n_0) is in S, by definition, and so does not have P. This contradiction shows that S is empty.■

Proposition 3.13 *A can be obtained from primitive recursive functions by composition and minimisation.*

Proof We begin by defining a function f from the set of all finite sequences of members of **N** to itself. Let α be the sequence $a_1, \ldots, a_k$. If $k=1$ we let $f\alpha$ be α. If $k \neq 1$, the definition of $f\alpha$ will depend on the values of a_{k-1} and a_k. If $k \neq 1$ and $a_{k-1}=0$ then $f\alpha$ is $a_1, \ldots, a_{k-2}, a_k+1$. If $k \neq 1$, $a_{k-1} \neq 0$ and $a_k=0$, then $f\alpha$ is $a_1, \ldots, a_{k-2}, a_{k-1}-1, 1$. Finally if $k \neq 1$ and $a_{k-1} \neq 0 \neq a_k$ then $f\alpha$ is $a_1, \ldots, a_{k-2}, a_{k-1}-1, a_{k-1}, a_k-1$. Let F be the iterate of f.

The apparently strange definition of f is made so as to parallel the definition of A. For instance, $A(5,4)=A(4,A(5,3))=A(4,A(4,A(5,2)))=A(4,A(4,A(4,A(5,1))))=\ldots$, and $f(5,4)=(4,5,3)$, $f(4,5,3)=(4,4,5,2)$, $f(4,4,5,2)=(4,4,4,5,1)$.

Because $f\alpha$ depends only on the last two entries of α, we see that if $f\alpha=\beta$ then, provided that α has length greater than 1, $f(\gamma\alpha)=\gamma\beta$ for any sequence γ. Iterating, we see that if $F(\alpha,r)=\beta$ then $F(\gamma\alpha,r)=\gamma\beta$ provided that $F(\alpha,r-1)$ has length greater than 1.

We now show, by double induction on m and n, that for all m and n there is r such that $F((m,n),r)$ has length 1, and that, for any such r, if $F((m,n),r)=u$, then $u=A(m,n)$.

Suppose that p is the smallest integer such that $F((m+1,n),p)$ has length 1, and that $F((m+1,n),p)=u$, where $u=A(m+1,n)$. Suppose that q is the smallest integer such that $F((m,u),q)$ has length 1, and that $F((m,u),q)=v$, where $v=A(m,u)$. Then $f(m+1,n+1)=(m,m+1,n)$. By previous remarks it follows that p is the smallest integer such that $F((m+1,n+1),p+1)$, which equals $F((m,m+1,n),p)$, has length 2, and that $F((m+1,n+1),p+1)=(m,u)$. It then follows, setting $r=p+q+1$, that r is the smallest integer such that $F((m+1,n+1),r)$ has length 1, and that $F((m+1,n+1),r)=F(m,u)=v$. Further $v=A(m,u)=A(m,A(m+1,n))=A(m+1,n+1)$. This is the main step in the double induction. We also have to look at the case $(0,n)$, which is immediate, and the case $(m+1,0)$, which is similar to the case discussed but simpler.

We now have to define a map from the set of sequences into **N**. Let α be $a(1), \ldots, a(k)$. We define the *code* of α to be $2^k \Pi^{i \leqslant k} p_i^{a(i)}$. In particular the code of m,n is $2^2 3^m 5^n$. The set of codes is primitive recursive, since x is a code iff $\exp_0 x \neq 0$ and length $x \leqslant \exp_0 x$.

We define a function $g:\mathbf{N} \to \mathbf{N}$ by $gx=x$ if x is not a code, and $g(\text{code}\alpha)=\text{code}(f\alpha)$. Let G be the iterate of g. Then g is given by a complicated definition by cases, and we find that g, and hence G, is primitive recursive. Further the previous discussion of F, and the relation between G and F, shows that $A(m,n)=\exp_1 G(2^2 3^m 5^n,r)$, where $r=\mu t(\exp_0 G(2^2 3^m 5^n,t)=1)$, as required.■

The details of this proof are of interest, as well as the result itself. They show how constructions of a more complicated kind than we have previously considered can be coded up to give computations in **N** involving primitive recursion, composition and minimisation only. An alternative proof of the proposition will be given later, which proves slightly more.

We now proceed to develop a number of properties of A. The proofs are by double induction, and many details will be left to the reader.

(I) For all x and y, $A(x,y)>y$.

This is obvious for $x = 0$. Now $A(x+1,0) = A(x,1)$, and (inductively) $A(x,1) > 1$, so $A(x+1,0) > 1$. Further $A(x+1,y+1) = A(x,A(x+1,y)) > A(x+1,y)$ inductively, and (also inductively) $A(x+1,y) > y$, so that $A(x+1,y) \geqslant y+1$. Hence $A(x+1,y+1) > y+1$, completing the inductive proof.

(II) For all x and y, $A(x,y+1) > A(x,y)$.

This is obvious for $x = 0$. Also $A(x+1,y+1) = A(x,A(x+1,y)) > A(x+1,y)$, by (I).

(III) For all x, y_1, and y_2 with $y_1 < y_2$, we have $A(x,y_1) < A(x,y_2)$. This is easy (by ordinary induction on $y_2 - y_1$) from (II).

(IV) For all x and y, $A(x+1,y) \geqslant A(x,y+1)$.

This is proved by induction on y. For $y = 0$ we have $A(x+1,0) = A(x,1)$ by definition.

$A(x+1,y+1) = A(x,A(x+1,y))$. By the inductive assumption, $A(x+1,y) \geqslant A(x,y+1)$, and by (I) $A(x,y+1) \geqslant y+2$. Hence, by (III), $A(x,A(x+1,y)) \geqslant A(x,y+2)$, which is the required property for $y+1$.

(V) For all x and y, $A(x,y) > x$.

It follows from (IV), by induction on z, that $A(x+z,y) \geqslant A(x,y+z)$ for all x,y and z. In particular $A(z,y) \geqslant A(0,y+z)$ for all y and z. As $A(0,y+z) = y+z+1$, we have $A(z,y) > y+z \geqslant z$, as required.

(VI) For all x_1, x_2, and y, if $x_1 < x_2$ then $A(x_1,y) < A(x_2,y)$.

This follows easily (by induction on $x_2 - x_1$) if we show that $A(x+1,y) > A(x,y)$ for all x and y. But, by (IV), $A(x+1,y) \geqslant A(x,y+1)$, and, by (II), $A(x,y+1) > A(x,y)$.

(VII) for all x and y, $A(x+2,y) > A(x,2y)$.

We prove this by induction on y. For $y = 0$ we require that $A(x+2,0) > A(x,0)$, which is true by (VI).

Now $A(x+2,y+1) = A(x+1,A(x+2,y))$. By the inductive assumption, $A(x+2,y) > A(x,2y)$, so by (III) $A(x+2,y+1) > A(x+1,A(x,2y))$ and this latter is greater than $A(x,A(x,2y)+1)$ by (IV). From (I) we have $A(x,2y)+1 > 2y+1$, so that $A(x,2y)+1 \geqslant 2(y+1)$. Using (III) again, we see that $A(x,A(x,2y)+1) \geqslant A(x,2(y+1))$, which gives what we want for $y+1$.

We shall say that a total function $f{:}\mathbf{N}^k \to \mathbf{N}$ is *within level r* if, for all $\mathbf{x}, f\mathbf{x} \leqslant A(r,X)$ where $X = \max x_i$. Because $A(4,y)$ is so large, it seems likely that any function which occurs in practice will differ at only finitely many places from a function within level 4.

Proposition 3.14 *Let f be primitive recursive. Then f is within level r for some r.*

Proof We shall show that the set of functions which are within level r for some r is a primitive recursively closed set, which will give the result by Lemma 3.1.

Since $A(0,y) = y+1$, the initial functions are all within level 0.

Let f come by composition from $g{:}\mathbf{N}^m \to \mathbf{N}$ and $h_1, \ldots, h_m{:}\mathbf{N}^k \to \mathbf{N}$. Let g

be within level r and let h_i be within level s_i for $i \leqslant m$. Let $s = \max(r, s_1, \ldots, s_m) + 3$. We shall show f is within level s.

Write $y_i = h_i\mathbf{x}$, $Y = \max y_i$. Since $s \geqslant 2$ we have, by (IV), $A(s,X) \geqslant A(s-1, X+1) = A(s-2, A(s-1,X))$. As $s-1 > s_i$ we have $A(s-1,X) > A(s_i,X) > y_i$, since h_i is within level s_i. By (III) it follows that $A(s,X) > A(s-2,Y)$. As $s-2 \geqslant r$, by (VI) it follows that $A(s-2,Y) \geqslant A(r,Y) \geqslant g\mathbf{y}$, by definition of r. Since $f\mathbf{x} = g\mathbf{y}$, f is within level s.

Now let f come from g and h by primitive recursion. Let g be within level r and let h be within level s. Let $p = \max(r,s) + 3$. We shall show f is within level p.

We first show that for any $\mathbf{x}$ and any $y \in \mathbf{N}$ we have $A(p-2, X+y) > f(\mathbf{x},y)$. This is proved by induction on y.

When $y = 0$, $f(\mathbf{x},0) = g\mathbf{x} \leqslant A(r,X)$. As $r < p-2$, we have $A(r,X) < A(p-2, X+0)$ by (VI), as required.

Now suppose the result is true for y and prove it for $y+1$. Write $z = f(\mathbf{x},y)$, so that $f(\mathbf{x},y+1) = h(\mathbf{x},y,z)$. Our inductive assumption is that $A(p-2,X+y) > z$. Also we know that $A(p-2,X+y) > X+y \geqslant \max(X,y)$. Hence, by (III), $A(p-2,X+y+1) = A(p-3, A(p-2,X+y)) > A(p-3, \max(X,y,z))$. By (VI) this latter is at least $A(s, \max(X,y,z))$ which is greater than $h(\mathbf{x},y,z)$ by definition of s. This completes the inductive step.

Finally $A(p, \max(x_1, \ldots, x_k, y)) = A(p, \max(X,y)) \geqslant A(p-2, 2\max(X,y))$, by (VII), $\geqslant A(p-2, X+y)$ (since $2\max(X,y) > X+y$), and we have just proved this latter is at least $f(\mathbf{x},y)$, as needed.■

Proposition 3.15 *A is not primitive recursive.*

Proof We use the diagonal argument as usual. Suppose A were primitive recursive. Then the function $f:\mathbf{N} \to \mathbf{N}$ defined by $fn = A(n,n) + 1$ would also be primitive recursive. By the previous lemma there would be some r such that $fn \leqslant A(r,n)$ for all n. In particular we would have $fr \leqslant A(r,r)$, which contradicts the definition of A.■

We have seen that A can be obtained from primitive recursive functions by minimisation and composition. The next result shows that A can be obtained by minimising a primitive recursive function. It follows that unbounded minimisation can lead from a primitive recursive function to one which is not primitive recursive.

Proposition 3.16 *The set* $\{(x,y,z); z = A(x,y)\}$ *is primitive recursive.*

Corollary *Let* χ *be the characteristic function of* $\{(x,y,z); z = A(x,y)\}$. *Then* χ *is primitive recursive and* $A(x,y) = \mu z(\chi(x,y,z) = 0)$.

Proof The corollary is immediate from the proposition.

We use the notation of Proposition 3.13. Let $\phi(x,y)$ be the least t such that $F((x,y),t)$ has length1. The proof in Proposition 3.13 that there is such a t gives the following properties for ϕ:$\phi(0,y) = 1$, $\phi(x+1,0) = 1 + \phi(x,1)$ and $\phi(x+1,y+1) = 1 + \phi(x+1,y) + \phi(x,u)$ where $u = A(x+1,y)$ and so $A(x,u) = A(x+1,y+1)$.

We shall show inductively that $\phi(x,y) \leqslant (A(x,y)+1)^x$ for all x and y. This

is obvious for $x=0$. Suppose that $\phi(x,1) \leqslant (A(x,1)+1)^x$. As $A(x,1)=A(x+1,0)$ we have
$\phi(x+1,0) \leqslant 1+(A(x+1,0)+1)^x \leqslant (A(x+1,0)+1)^{x+1}$, as required.

Now suppose that $\phi(x+1,y) \leqslant (A(x+1,y)+1)^{x+1}$ and that $\phi(x,u) \leqslant (A(x,u)+1)^x$, where $u=A(x+1,y)$. We know that $A(x+1,y+1) > A(x+1,y)$, so that $A(x+1,y+1) \geqslant A(x+1,y)+1$. It follows that

$$\begin{aligned}&(A(x+1,y+1)+1)^{x+1}-(A(x+1,y+1)+1)^x\\&=(A(x+1,y+1)+1)^x A(x+1,y+1) \geqslant\\&(A(x+1,y)+1+1)^x(A(x+1,y)+1) \geqslant (A(x+1,y)+1)^{x+1}+1.\end{aligned}$$

Since $A(x+1,y+1)=A(x,u)$, this gives

$$\begin{aligned}&(A(x+1,y+1)+1)^{x+1} \geqslant \phi(x,u)+\phi(x+1,y)+1\\&=\phi(x+1,y+1),\end{aligned}$$

completing the inductive step.

We know that $G(2^2 3^x 5^y,t)=2.3^{A(x,y)}$ if $t \geqslant \phi(x,y)$, while $\exp_0 G(2^2 3^x 5^y,t) \neq 1$ for $t < \phi(x,y)$. It follows that if $z \neq A(x,y)$, then $G(2^2 3^x 5^y,t) \neq 2.3^z$ for all t, and, in particular, that $G(2^2 3^x 5^y,(z+1)^x) \neq 2.3^z$. But if $z=A(x,y)$ then $(z+1)^x \geqslant \phi(x,y)$ and so $G(2^2 3^x 5^y,(z+1)^x)=2.3^z$. Thus $\{(x,y,z); z=A(x,y)\}$ is defined by the primitive recursive property $G(2^2 3^x 5^y,(z+1)^x)=2.3^x$.■

Exercise 3.7 Show that a function f can be written as $\mu t(v(x,t)=0)$ for some primitive recursive function v iff the graph of f, that is, $\{(x,y); fx=y\}$ is a primitive recursive set.

Exercise 3.8 Let f have primitive recursive graph. Show that if f takes on only the values 0 and 1 then f is primitive recursive.

Exercise 3.9 Let F be the function of Theorem 3.11. Show that the function co$F$ cannot be primitive recursive. Deduce that there is an intuitively computable function which cannot be written as $\mu t(v(x,t)=0)$ for any primitive recursive function v.

3.7 JUSTIFYING DEFINITIONS BY PRIMITIVE RECURSION

Let $g:\mathbf{N}^k \to \mathbf{N}$ and $h:\mathbf{N}^{k+2} \to \mathbf{N}$ be functions. We now show that there is a function defined from g and h by primitive recursion, and that this function is total if g and h are total. The proof may seem very complicated. I have seen alleged proofs simpler than the one given here (and the other proof given in the exercises), but they were wrong. The trouble is that, while the existence of such a function seems intuitively obvious, detailed justification is needed, and it is quite common to take for granted part of what needs proof.

A subset A of $\mathbf{N}^{k+2}$ will be called *admissible* if it satisfies the following two conditions: (i) if $g\mathbf{x}$ is defined then $(\mathbf{x},0,g\mathbf{x}) \in A$, (ii) if $(\mathbf{x},y,z) \in A$ and $h(\mathbf{x},y,z) \in A$ then $(\mathbf{x},y+1,h(\mathbf{x},y,z)) \in A$. Plainly $\mathbf{N}^{k+2}$ is admissible.

Let A_α be a collection of admissible sets. It is easy to check that $\cap A_\alpha$ is admissible. In particular let F be the intersection of all admissible sets. Then F is admissible. Also (since F is a subset of every admissible set) no proper subset of F is admissible.

Lemma 3.17 *If* $g\mathbf{x}$ *is defined then* $(\mathbf{x},0,z) \in F$ *iff* $z = g\mathbf{x}$. *If* $g\mathbf{x}$ *is not defined then there is no* z *with* $(\mathbf{x},0,z) \in F$.

Suppose either that $g\mathbf{x}$ is defined and $z \neq g\mathbf{x}$ or that $g\mathbf{x}$ is not defined and z is arbitrary. In either case we see that $F - \{(\mathbf{x},0,z)\}$, the set obtained by removing $(\mathbf{x},0,z)$ from F, is still admissible. Hence this set cannot be a proper subset of F, and so $(\mathbf{x},0,z) \notin F$.■

Lemma 3.18 (*i*) *For some* $\mathbf{x}$ *and* y, *suppose that* $(\mathbf{x},y,z) \in F$ *and that* $(\mathbf{x},y,w) \notin F$ *for* $w \neq z$. *If* $h(\mathbf{x},y,z)$ *is defined then* $(\mathbf{x},y+1,w) \in F$ *iff* $w = h(\mathbf{x},y,z)$. *If* $h(\mathbf{x},y,z)$ *is not defined then there is no* w *with* $(\mathbf{x},y+1,w) \in F$.

(*ii*) *If there is no* z *with* $(\mathbf{x},y,z) \in F$ *then there is no* w *with* $(\mathbf{x},y+1,w) \in F$.

Proof Suppose either that there is no z with $(\mathbf{x},y,z) \in F$ or that there is exactly one such z and, in the latter case, that if $h(\mathbf{x},y,z)$ is defined then $w \neq h(\mathbf{x},y,z)$; in all other cases w can be arbitrary. Then we can easily see that $F - \{(\mathbf{x},y+1,w)\}$ is still admissible. It follows, since no proper subset of F is admissible, that $(\mathbf{x},y+1,w) \notin F$.■

From the two previous lemmas it follows, by induction on y, that for every $\mathbf{x}$ and y there is at most one z with $(\mathbf{x},y,z) \in F$. Hence we can define a function $f{:}\mathbf{N}^{k+1} \to \mathbf{N}$ by $f(\mathbf{x},y) = z$ iff $(\mathbf{x},y,z) \in F$. By Lemma 3.17, if $g\mathbf{x}$ is not defined then $f(\mathbf{x},0)$ is not defined, while if $g\mathbf{x}$ is defined then, as F is admissible, we have $f(\mathbf{x},0) = g\mathbf{x}$. Also if $f(\mathbf{x},y) = z$ and $h(\mathbf{x},y,z)$ is defined then, by Lemma 3.18, we have $f(\mathbf{x},y+1) = h(\mathbf{x},y,z)$. If either $f(\mathbf{x},y)$ is not defined or $f(\mathbf{x},y) = z$ and $h(\mathbf{x},y,z)$ is not defined, then by Lemma 3.18 $f(\mathbf{x},y+1)$ is not defined. Hence f is defined by primitive recursion from g and h.

Lemma 3.19 *Let* f *be obtained by primitive recursion from* g *and* h. *If both* g *and* h *are total then* f *is total.*

Proof It is easy to check, by induction on y, that for every $\mathbf{x}$ and y there is some z with $(\mathbf{x},y,z) \in F$. By definition of f, this means that $f(\mathbf{x},y)$ is defined for every $\mathbf{x}$ and y.■

In the exercises an alternative proof is given that definition by primitive recursion is meaningful.

Exercise 3.10 Show that there is at most one function defined by primitive recursion from g and h.

Exercise 3.11 Show that the intersection of admissible sets is admissible.

Exercise 3.12 Fill in the omitted details in the proofs of the lemmas above.

A function $\phi{:}\mathbf{N}^{k+1} \to \mathbf{N}$ will be called *allowable* if it satisfies the following three conditions: (i) given $\mathbf{x}$, either $\phi(\mathbf{x},y)$ is defined for all y or it is not defined for any y or there is some n such that $\phi(\mathbf{x},y)$ is defined iff $y \leqslant n$, (ii) if $\phi(\mathbf{x},0)$ is defined then $g\mathbf{x}$ is defined and $\phi(\mathbf{x}0) = g\mathbf{x}$, (ii) if both $\phi(\mathbf{x},y)$ and

$\phi(\mathbf{x},y+1)$ are defined then $\phi(\mathbf{x},y+1) = h(\mathbf{x},y,\phi(\mathbf{x}y))$. Allowable functions exist; for instance, the function such that $\phi(\mathbf{x},y)$ is defined only if $y = 0$ and $\phi(\mathbf{x},0)$ is defined iff $g\mathbf{x}$ is defined, and then $\phi(\mathbf{x},0) = g\mathbf{x}$, is allowable. Notice that an allowable function is not necessarily defined by primitive recursion from g and h, since we do not require $\phi(\mathbf{x},y+1)$ to be defined whenever $\phi(\mathbf{x},y)$ and $h(\mathbf{x},y,\phi(\mathbf{x},y))$ are defined.

Exercise 3.13 Show, by induction on y, that if ϕ and ψ are allowable functions and both $\phi(\mathbf{x},y)$ and $\psi(\mathbf{x},y)$ are defined then $\phi(\mathbf{x},y) = \psi(\mathbf{x},y)$.

It follows that we can define a function f by requiring $f(\mathbf{x},y)$ to be defined iff there is some allowable function defined on $(\mathbf{x},y)$ and then requiring $f(\mathbf{x},y)$ to be $\phi(\mathbf{x},y)$ for any allowable function ϕ defined on $(\mathbf{x},y)$. The exercise shows that we get the same value for $f(\mathbf{x},y)$ whatever allowable function we choose.

Exercise 3.14 Show that f is allowable.

Exercise 3.15 Let ϕ be an allowable function such that $\phi(\mathbf{x},y)$ and $h(\mathbf{x},y,\phi(\mathbf{x},y))$ are both defined but $\phi(\mathbf{x},y+1)$ is not defined. Define a function ψ to be equal to ϕ wherever ϕ is defined, and with $\psi(\mathbf{x},y+1)$ defined to be $h(\mathbf{x},y,\phi(\mathbf{x},y))$ and with ψ not defined anywhere else. Show that ψ is allowable.

Exercise 3.16 Use the previous exercise to show that if $f(\mathbf{x},y)$ and $h(\mathbf{x},y,f(\mathbf{x},y))$ are both defined then $f(\mathbf{x},y+1)$ is defined and equals $h(\mathbf{x},y,f(\mathbf{x},y))$.

It follows from the last exercise that f is defined from g and h by primitive recursion.

Exercise 3.17 Let g and h be total. Using Exercise 3.15, show that if there is an allowable function defined for some $\mathbf{x}$ on all y with $y \leqslant n$ then there is also an allowable function defined for that $\mathbf{x}$ on all $y \leqslant n+1$. Deduce that f is total.

Exercise 3.18 We took for granted in section 3.6 that the equations for Ackermann's function defined a unique function, which is total. Prove this. (A proof by double induction can be given along similar lines to the current section. Simpler is to show by induction on m that for each m there is a unique function α_m such that $\alpha_m n$ satisfies the defining condition for $A(m,n)$, and that each function α_m is total.)

4

Partial recursive functions

In this chapter we formalise the theory of Chapter 2, with a class of functions called partial recursive functions taking the place of intuitively computable functions. In later chapters we shall see the evidence suggesting that the intuitively computable functions are exactly the partial recursive functions.

4.1 RECURSIVE AND PARTIAL RECURSIVE FUNCTIONS

We alreay know what is meant by minimisation of a function. We need also a related concept.

Definition The function $g:\mathbf{N}^k \to \mathbf{N}$ comes from $f:\mathbf{N}^{k+1} \to \mathbf{N}$ by **regular minimisation** if

(a) g comes from f by minimisation,
(b) f is total,
(c) g is total.

Plainly (c) can be replaced by (c′) $\forall \mathbf{x} \exists y (f(\mathbf{x},y)=0)$.

A set C of functions is said to be **closed under (regular) minimisation** if $g \in C$ whenever $f \in C$ and g comes from f by (regular) minimisation; we already know the meanings of 'initial function', 'closed under composition' and 'closed under primitive recursion'.

Definition The function f is called **partial recursive** if there is a sequence of functions. $f_1, \ldots, f_n$ such that $f_n = f$ and, for all $r \leqslant n$, either f_r is an initial function, or there are $i, j(1), \ldots, j(k)$, all less than r, with f_r coming from f_i and $f_{j(1)}, \ldots, f_{j(k)}$ by composition, or there are i and j less than r with f_r coming from f_i and f_j by primitive recursion, or there is i less than r with f_r coming from f_i by minimisation. If, in addition, whenever the last condition applies the minimisation is regular, then f is called **recursive.**

We know that composition, primitive recursion, and minimisation when applied to intuitively computable functions produce intuitively computable functions. Hence partial recursive functions are intuitively computable.

The next lemma has a proof almost identical to the proof of Lemma 3.1; the details will be left to the reader.

Lemma 4.1 *The set of partial recursive functions contains the initial functions, and is closed under composition, primitive recursion, and minimisation. Also, any set containing the initial functions and closed under composition, primitive recursion, and minimisation will contain every partial recursive function.*

The set of recursive functions contains the initial functions, and is closed under composition, primitive recursion, and regular minimisation. Also, any set of functions containing the initial functions and closed under composition, primitive recursion, and regular minimisation will contain every recursive function. ■

This lemma tells us that the set of partial recursive functions (recursive functions) is the smallest set containing the initial functions, and closed under composition, primitive recursion, and minimisation (regular minimisation). There are many situations where we want the smallest set with certain properties, and begin by defining a set constructively by procedures similar to the above definitions, and then showing, as in the above lemma and Lemma 3.1, that it has the required properties.

Obviously, the set of recursive functions is primitive recursively closed. In fact, the main reason for considering primitive recursively closed sets was to obtain results which are valid both for the set of primitive recursive functions and for the set of recursive functions.

Since Ackermann's function may, by Proposition 3.13, be written as $A(m,n)=\phi(m,n,s)$ where $s=\mu t(\psi(m,n,t)=0)$, for some primitive recursive functions ϕ and ψ, (and Proposition 3.16 gives an even simpler form) we see that it is recursive.

Plainly, every recursive function is both partial recursive and total. The converse is true, as will be shown in Corollary 1 to Theorem 4.2 below. But the converse is not, as might be thought, obvious. Suppose we take some recursive function f:$\mathbf{N}\rightarrow\mathbf{N}$ and some partial recursive g:$\mathbf{N}\rightarrow\mathbf{N}$ whose domain contains $f\mathbf{N}$. Then the composite gf is both partial recursive and total; but there is no immediately obvious reason why gf should be recursive.

Theorems 4.2 and 4.3 below are crucial to the development of the theory. However, they will not be proved until Chapter 7 (with an alternative proof in Chapter 9), although they will be used immediately. They are stated for functions of one variable, but may be extended to functions of any number of variables, using the bijection J_k:$\mathbf{N}^k\rightarrow\mathbf{N}$. Direct proofs are possible, but the details seem to me to be both complicated and not particularly natural.

Theorem 4.2 *Let f:$\mathbf{N} \to \mathbf{N}$ be partial recursive. Then there are primitive recursive u:$\mathbf{N} \to \mathbf{N}$ and v:$\mathbf{N}^2 \to \mathbf{N}$ such that $fx = u(\mu t(v(x,t)=0))$.*

We have seen in Exercise 3.9 that there are recursive functions which cannot be written as $\mu t(v(x,t)=0)$ for any primitive recursive v. The function u (and the function U of Theorem 4.3) can be taken to be K. This is shown in section 9.6.

Corollary 1 *A function which is both partial recursive and total is recursive.*

Proof Let f be both partial recursive and total, and write f in the form given by the theorem. Now u and v, being primitive recursive, are recursive. Since f is total, for every x there must exist t such that $v(x,t)=0$. It follows that the one minimisation used is a regular minimisation, and so f is recursive.■

Corollary 2 *The set of partial recursive functions may be defined in either of the following two ways:*

(a) *it is the smallest set containing the initial functions, and closed under composition of partial functions, primitive recursion of partial functions, and minimisation of total functions;*
(b) *it is the smallest set containing the initial functions, and closed under composition of partial functions, primitive recursion of total functions, and minimisation of total functions.*

Proof That we can refer to the smallest set having the properties of (a) or (b) follows from the remarks made after Lemma 4.1. Evidently the set of partial recursive functions contains the set defined in (a), which in turn contains the set defined in (b). Evidently this latter set contains all primitive recursive functions, and will also contain all functions of the form $u(\mu t(v(x,t)=0))$ with u and v primitive recursive. Hence, by the theorem, the set defined in (b) contains all partial recursive functions.■

Some authors take (a) or (b) as the definition of partial recursive functions (or the related constructive definition). This is unsatisfactory for a number of reasons. If we are concerned with a set of partial functions, it is aesthetically unpleasant to restrict minimisation to total functions if this is not necessary. Also, as shown for intuitively computable functions in Corollary 2 to Theorem 2.4 (which extends to partial recursive functions, as indicated later), it is not possible to tell, from a definition of a function (by a suitable sequence of simpler functions, by an abacus machine — these are defined in the next chapter — or other methods), whether or not the function is total. The restriction to this case seems to arise from these authors considering pseudo-minimisation of partial functions (which does not preserve computability) instead of the correct notion of minimisation.

The next theorem extends Theorem 4.2, and is the analogue in the current theory of Theorem 2.4.

Theorem 4.3 (Kleene's Normal Form Theorem) *There exist primitive recursive functions* $U:\mathbf{N}\to\mathbf{N}$ *and* $V:\mathbf{N}^3\to\mathbf{N}$ *with the following property. To any partial recursive* $f:\mathbf{N}\to\mathbf{N}$ *there is* k *such that* $fx=U(\mu t(V(k,x,t)=0))$ *for all* x.

4.2 RECURSIVE AND RECURSIVELY ENUMERABLE SETS

Just as partial recursive functions are a formal version of the intuitive notion of computable functions, we have a similar formal version of the intuitive notions of decidable and listable sets.

Definition The set A is **recursive** if its characteristic function is a recursive function. The set A is **recursively enumerable** (usually abbreviated to **r.e.**) if either $A=\varnothing$ or $A=f\mathbf{N}$ for some recursive function $f:\mathbf{N}\to\mathbf{N}$.

Most of the results of Chapter 2 translate immediately to this situation, simply replacing 'decidable' by 'recursive', 'listable' by 'recursively enumerable', 'intuitively computable' by 'partial recursive', and 'computable total' by 'recursive'; we may also have to use Corollary 1 to Theorem 4.2 which tells us that a function which is both partial recursive and total is recursive. All these results will be stated, but the proof will only be given where it cannot be obtained by this direct translation.

Proposition 4.4 *Let* A *be a subset of* **N**. *Then the following properties of* A *are equivalent:*

(a) *A is r.e.,*
(b) *A is empty or $A=f\mathbf{N}$, where $f:\mathbf{N}\to\mathbf{N}$ is primitive recursive,*
(c) *$A=f\mathbf{N}$, where $f:\mathbf{N}\to\mathbf{N}$ is partial recursive,*
(d) *the partial characteristic function of A is partial recursive,*
(e) *A is the domain of some partial recursive function.*

Proof We show that (b) is true if (c) is true; we will use Theorem 4.2 and the ideas of Proposition 2.1. As for the other parts, if (b) is true, obviously (a) is true, and the remainder of the proof comes from Proposition 2.1 by the translation already mentioned.

Suppose (c) holds. By Theorem 4.2, there are primitive recursive functions $u:\mathbf{N}\to\mathbf{N}$ and $v:\mathbf{N}^2\to\mathbf{N}$ such that $fx=u(\mu t(v(x,t)=0))$. If $A=\varnothing$ then (b) holds. If $A\neq\varnothing$ take any $a_0\in A$. Define $F:\mathbf{N}^2\to\mathbf{N}$ by

$F(x,n)=u(\mu t \leqslant n(v(x,t)=0))$ if $\exists t \leqslant n(v(x,t)=0)$,
$F(x,n)=a_0$ otherwise.

Now F is primitive recursive, being given by a definition by cases from primitive recursive functions and predicates. Exactly as in Proposition 2.1, $F\mathbf{N}^2=A$, and A is the range of the primitive recursive function FJ^{-1}.■

The next proposition has the same proof as Proposition 2.2.

Proposition 4.5 *Let A be a subset of* **N**. *Then A is recursive iff both A and* **N**$-A$ *are r.e.*■

The informal versions of the next results were given as exercises in Chapter 2. As they are of importance later, a full proof is given now.

Proposition 4.6 *Let A and B be r.e. subsets of* **N**. *Then both* $A \cup B$ *and* $A \cap B$ *are r.e. If* $\alpha:\mathbf{N}\to\mathbf{N}$ *is partial recursive then the counter-image* $\alpha^{-1}A$ *of A under* α *and the image* αA *of A under* α *are r.e. Further, if A and* α *are recursive then* $\alpha^{-1}A$ *is recursive.*

Proof If either A or B is empty, there is nothing to prove. So we may take recursive functions f and g such that $A=f\mathbf{N}$ and $B=g\mathbf{N}$. Then $A\cup B=h\mathbf{N}$, where h is defined by $h(2n)=fn$, $h(2n+1)=gn$, or, equivalently,

$$hx=f(\text{quo}(2,x)) \text{ if } \text{rem}(2,x)=0,\ hx=g(\text{quo}(2,x)) \text{ otherwise.}$$

Then h is recursive, being given by a definition by cases, and so $A\cup B$ is r.e. by definition.

By Proposition 4.4, there are also partial recursive function ϕ and ψ from **N** to **N** such that ϕ has domain A and ψ has domain B. Then $\phi+\psi$ is partial recursive and its domain is $A\cap B$. By Proposition 4.4, this shows that $A\cap B$ is r.e.

$\alpha^{-1}A$ is the domain of $\phi\alpha$, and $\alpha A=\alpha f\mathbf{N}$, so they are both r.e. If α is total then $\mathbf{N}-\alpha^{-1}A=\alpha^{-1}(\mathbf{N}-A)$, so it is r.e. if $\mathbf{N}-A$ is r.e.■

Proposition 4.7 *Let A be a subset of* **N**. *Then A is recursive and infinite iff there is a strictly increasing function* $f:\mathbf{N}\to\mathbf{N}$ *such that* $A=f\mathbf{N}$.

Proof Suppose there is such a function f. Since f is strictly increasing, we see that $fn\geqslant n$ for all n. Plainly A is then infinite. Also we find that $a \in A$ iff $\exists n\leqslant a(fn=a)$. Since the quantification is bounded, this is a recursive property, and so A is recursive.

Now let A be recursive and infinite and let χ be its characteristic function. Define ϕ by $\phi x=\mu y(y>x$ and $\chi y=0)$. Since χ is recursive, ϕ is partial recursive. Since A is infinite ϕ is total. Hence ϕ is recursive.

Define f to be the iterate of ϕ starting at a_0, where a_0 is the smallest element of A. Then f is recursive, and f is strictly increasing, since $f(n+1)=\phi(fn)>fn$ by the definition of ϕ.

We still have to show $A=f\mathbf{N}$. Take any $a \in A$. Since the smallest element a_0 of A is f_0, we may assume $a>a_0$. Since f is strictly increasing there will be n

such that $fn<a$ and $f(n+1)\geqslant a$. Now $f(n+1)$ is, by definition, the smallest element of A which is greater than fn. Thus the conditions $fn<a\leqslant f(n+1)$ ensure that $a=f(n+1)$.■

If we want to use a definition by cases with partial recursive functions the methods used in Chapter 3 do not work, since, in general, the relevant sum is nowhere defined. However, a different technique permits definition by cases here.

Proposition 4.8 *For $i=0,1,\ldots, n-1$ let $f_i:\mathbf{N}\to\mathbf{N}$ be partial recursive functions and let A_i be r.e. subsets of* **N** *with $A_i\cap A_j$ empty for $i\neq j$. Let f be given by $fx=f_ix$ for $x \in A_i$ (and fx not defined if $x \notin \cup A_i$). Then f is partial recursive.*

Proof We know that domain f_i is r.e., and hence so is $A_i \cap$ domain f_i. Replacing A_i by this set, we may assume that $A_i \subset$ domain f_i. We may also assume A_i is not empty, since we can just ignore any empty A_i. Hence there are recursive functions ϕ_i such that $A_i=\phi_i\mathbf{N}$. The functions $f_i\phi_i$ will also be recursive, being partial recursive and total. Hence, using definition by cases for recursive functions, we have recursive functions ϕ and g given by $\phi x=\phi_i(\text{quo}(n,x))$ if $\text{rem}(n,x)=i$ and $gx=f_i\phi_i(\text{quo}(n,x))$ if $\text{rem}(n,x)=i$.

Plainly $\phi\mathbf{N}=\cup A_i$. Also, if $a \in A_i$ and $\phi x=a$ we must have $\text{rem}(n,x)=i$ and $a=\phi_i(\text{quo}(n,x))$; we would then also have $fa=f_ia=gx$. Hence $f=g\psi$, where ψ is defined by $\psi a=\mu x(\phi x=a)$. Since ψ is partial recursive, so is f.■

Kleene's Normal Form Theorem tells us that there is a universal partial recursive function from $\mathbf{N}^2$ to **N** (and tells us about the form of such a function). Consequently, results using the existence of a universal computable function may be proved in the current context. In particular we have the following theorem.

Theorem 4.9 *There is a set K which is r.e. but not recursive.*

Proof Take U and V as in Theorem 4.3. Define K to be the set of those x for which $U(\mu t(V(x,x,t)=0))$ is defined (so in fact $K=\{x;\exists t(V(x,x,t)=0)\}$). As in Theorem 2.5 the set K has the required properties.■

Rice's Theorem and the Rice–Shapiro Theorem may also be proved for partial recursive functions by a direct translation of the earlier proofs. However, the proof requires the analogue of Lemma 2.6, which has no obvious proof at this stage. The relevant lemma will be proved in Chapter 7, with an alternative proof in Chapter 10.

Exercise 4.1 Let A be an infinite r.e. set. Show that $A=f\mathbf{N}$ for some recursive one–one f.

Exercise 4.2 Let A be an infinite r.e. set. Show that A contains an infinite recursive set. (Compare these exercises with Exercises 2.12 and 2.13.)

Exercise 4.3 Let A be a subset of $\mathbf{N}$. Show that the following are equivalent; (i) A is r.e., (ii) for every k, there is a primitive recursive subset B of $\mathbf{N}^{k+1}$ such that $A=\{x;\exists \mathbf{y} \text{ with } (x,\mathbf{y}) \in B\}$, (iii) for some k there is a ecursive subset C of $\mathbf{N}^{k+1}$ such that $A=\{x;\exists \mathbf{y} \text{ with } (x,\mathbf{y}) \in C\}$.

Exercise 4.4 Using the previous exercise or otherwise, show that if A is r.e. then so is $\{n;\exists x\leqslant n \text{ with } x \in A\}$.

5

Abacus machines

The original abacus was a frame of wires, on each of which beads were free to move. By moving the beads from one side of the frame to the other, quite complex calculations could be performed, and the abacus (which some readers may recognise as the counting-frame of their childhood) was used in many parts of the world as a cash-register. A trained abacus operator could work extremely fast, and in the early days of electronic computers, a contest between an abacus operator and a computer operator was won by the abacus (mainly because feeding information into the computer was slow).

The machines we call abacus machines are given that name because their fundamental operations consist only of adding or subtracting 1 (corresponding to the move of a bead from one side to another). The first section contains the definition and some formal properties, and computations by these machines are defined and discussed in the second section. We then proceed to prove that partial recursive functions are abacus computable, and conversely that abacus computable functions are partial recursive. The final section looks at a related class of machines.

5.1 ABACUS MACHINES

Abacus machines will be defined as strings on a certain alphabet, and the reader should review the properties of strings in Chapter 2. The alphabet consists of infinitely many symbols a_k and s_k, the left parenthesis (and infinitely many right parentheses $)_k$; the subscript k in a_k, s_k and $)_k$ can be any positive integer.

We define inductively **simple abacus machines of depth** n and **abacus machines of depth** n as follows.

(1) a_k and s_k are the only simple abacus machines of depth 0,

(2) the abacus machines of depth n are the strings $M_1 \ldots M_r$, where each M_i is a simple abacus machine of depth $\leqslant n$, and some M_i has depth exactly n (we allow $r = 1$, so that any simple abacus machine is an abacus machine),
(3) the simple abacus machines of depth $n+1$ are the strings $(M)_k$ where M is an abacus machine of depth n.

Examples s_2a_1 is an abacus machine of depth 0, and $(s_2a_1)_1$ is a simple abacus machine of depth 1. Hence $(s_2a_1)_1s_4a_7$ is an abacus machine of depth 1, and $((s_2a_1)_1s_4a_7)_4$ and $((s_2a_1)_1s_4a_7)_3$ are simple abacus machines of depth 2. Also $((s_2a_1)_1s_4a_7)_4a_5$ and $((s_2a_1)_1s_4a_7)_4(s_5)_5a_6$ are abacus machines of depth 2. We see that the depth is the number of nestings of parenthesis pairs inside each other.

It is possible to tell whether or not a string is an abacus machine, and, if so, it is possible to find the simple abacus machines $M_1, \ldots, M_r$ of which it is the product. The next two lemmas show this. The reader may decide to take these results for granted, on the reasonable belief that if this were not so a different definition would have been chosen.

Lemma 5.1 (*i*) *Any abacus machine has the same number of left and right parentheses.*

(*ii*) *Any proper initial segment of a simple abacus machine has more left parentheses than right parentheses.*

Proof The proofs are by induction on the depth.

Plainly (i) holds for simple abacus machines of depth 0. If (i) holds for simple abacus machines of depth $\leqslant n$, it obviously holds for all abacus machines of depth $\leqslant n$. If (i) holds for all abacus machines of depth n, it plainly holds for all simple abacus machines of depth $n+1$, and so (i) holds for all abacus machines by induction.

Plainly (ii) holds for simple abacus machines of depth 0, since these do not have any proper initial segments at all. Suppose (ii) holds for all simple abacus machines of depth $\leqslant n$, and let M be a simple abacus machine of depth $n+1$. By definition M is $(M_1 \ldots M_r)_k$, where each M_i is a simple abacus machine of depth $\leqslant n$. Now the proper initial segments of M are just (and the strings $(M_1 \ldots M_{i-1}M'_i$ for some i, where M'_1 is an initial segment of M_i. By (i) each of $M_1, \ldots, M_{i-1}$ has the same number of left and right parentheses, while by (i) or the inductive assumption M'_i has at least as many left parentheses as right parentheses. Thus any proper initial segment of M has more left parentheses than right parentheses, and (ii) follows by induction. ∎

Lemma 5.2 (*i*) *We can tell whether or not a string is an abacus machine.*

(*ii*) *If a string S is an abacus machine, there is exactly one value of r and one sequence of simple abacus machines $M_1, \ldots, M_r$ such that S is $M_1 \ldots M_r$.*

(*iii*) *If a string S is a simple abacus machine there is exactly one k and one abacus machine M such that S is $(M)_k$.*

Proof First note that (iii) is obvious, since k is determined by the fact that $)_k$ is the last symbol of S, and M is obtained from S by deleting the first and last symbol of S.

Now suppose S is an abacus machine, so that S is $M_1 \ldots M_r$ for some simple abacus machines $M_1, \ldots, M_r$. By Lemma 5.1 M_1 has the same number of left and right parentheses, while any proper initial segment of M_1 has more left parentheses than right parentheses. Hence M_1 is the smallest initial segment of S for which the segment has the same number of left parentheses and right parentheses. If M_1 is not the whole of S, then we can write S as M_1S'. Then, for similar reasons, M_2 will be the smallest initial segment of S' that has the same number of left parentheses and right parentheses. This procedure can be continued, and determines $M_1, \ldots, M_r$ (and hence r itself), proving (ii).

Finally consider any string S. As we have just seen S cannot be an abacus machine unless it has an initial segment with the same number of left and right parentheses. So we determine whether or not there is such a segment, and if so let S_1 be the smallest such segment. If S is not S_1 write S as S_1S' for some string S'. Then S is an abacus machine iff S_1 is a simple abacus machine and S' (if it exists) is an abacus machine.

Now S_1 is a simple abacus machine if it is a_k or s_k; otherwise it is a simple abacus machine only if its first symbol is (and its last symbol is $)_k$ for some k and it is not just $()_k$. We can test whether or not this condition holds, and if so we can write S as $(S'')_k$ for some string S''. Then in this case S_1 is a simple abacus machine iff S'' is an abacus machine.

Thus the procedure to test whether or not a string is an abacus machine is the following one, which works inductively by reducing the string to one or more smaller strings, which in turn have to be tested, and so on. Given a string S, we first see if it has an initial segment with the same number of left parentheses and right parentheses; if not, S is not an abacus machine. If so, we find the smallest such string S_1. If S_1 is not the whole of S we write S as S_1S'. If S_1 is just a_k or s_k and S is S_1 then S is a simple abacus machine while if S is S_1S' then S will be an abacus machine iff S' is an abacus machine. Otherwise S will not be an abacus machine unless S_1 is $(S'')_k$ for some k and some string S'', and we can check whether or not this holds. If it does hold S will be an abacus machine iff S'' and S' (or just S'' if S is S_1) are abacus machines. ■

5.2 COMPUTING BY ABACUS MACHINES

As we have mentioned in Chapter 2, intuitively a computer may be regarded as having registers, each of which can contain an arbitrary natural number. Since there is no reason to place any fixed limit on the number of registers, it may be convenient to assume that there are an infinite number of registers, only a finite number of which are in use at any time. We now proceed to formalise this.

We look at the set Σ consisting of those infinite sequences of natural numbers which have only a finite number of non-zero terms. We use small Greek letters for members of Σ and the corresponding subscripted ordinary letters for the terms of the sequence. For instance we write $\xi = (x_1, x_2, \ldots)$ and $\eta = (y_1, y_2, \ldots)$. In line with the intuitive notions, we refer to x_i as the **entry in register** i of ξ, and say that **register** i **is empty** if $x_i = 0$.

We shall define, by induction on the depth, to each abacus machine M a function from Σ to Σ, which we shall also denote by M. The effect of M on ξ is denoted by ξM (and not by $M\xi$) for reasons which will soon be seen. We often refer to ξM as the **output** of M on the **input** ξ.

We begin by defining ξa_k to be η where $y_i = x_i$ for $i \neq k$ and $y_k = x_k + 1$. Similarly ξs_k is ζ where $z_i = x_i$ for $i \neq k$ and $z_k = x_k \dot{-} 1$.

Next we explain how to define the function on an abacus machine of depth n if we know how to define it on all simple abacus machines of depth $\leqslant n$. If M is $M_1 \ldots M_r$ we define ξM to be $\xi M_1 \ldots M_r$; that is, ξM is obtained by applying M_1 to ξ, then applying M_2 to the result, and so on. Because we have used the notation ξM, in this formula the machines $M_1, \ldots, M_r$ are applied in the order in which they are naturally read. Observe that Lemma 5.2(ii) is needed to ensure that ξM is uniquely given from M.

Finally we have to explain what $\xi(M)_k$ is, assuming that we already know how M acts. We define $\xi(M)_k$ as ξM^t (that is, $\xi M \ldots M$, where there are t copies of M) where t is chosen to be as small as possible such that the kth register of ξM^t is empty. Of course if ξM^t has its kth register non-empty for all t then $\xi(M)_k$ is not defined. Also if $x_k = 0$ then $t = 0$ and $\xi(M)_k$ is just ξ.

Examples Three particular machines will be of special interest to us and will be given individual names.

Clear$_k$ is the machine $(s_k)_k$. This machine keeps subtracting 1 from register k, until that register becomes empty. That is $\xi(s_k)_k$ is η, where $y_i = x_i$ for $i \neq k$ and $y_k = 0$, which we can refer to as **clearing** register k.

Descopy$_{p,q}$ is the machine $(s_p a_q)_p$. This repeatedly subtracts 1 from register p and adds 1 to register q, stopping when register p is empty. Thus ξDescopy$_{p,q}$ is η, where $y_i = x_i$ for $i \neq p,q$, and $y_p = 0$, $y_q = x_p + x_q$. We could have named this machine Add$_{p,q}$ for obvious reasons, but we will be more concerned with the fact that if $x_q = 0$ then $y_q = x_p$, i.e. the machine **copies** register p into the originally empty register q. However register p is emptied in the process, which we therefore call a **destructive copy.**

If we wish to make a copy without destroying the contents of register p, we must use a third register which is initially empty. Thus we define Copy$_{p,q,r}$ to be $(s_p a_q a_r)_p (s_r a_p)_r$. This machine will leave all registers except p,q,r unchanged. If $x_q = x_r = 0$, then after applying $(s_p a_q a_r)_p$ we get ζ where $z_q = z_r = x_p$ and $z_p = 0$. Applying $(s_r a_p)_r$ to this will give η where $y_q = z_q$ (since this does not change register q) and $y_r = 0$ and $y_p = z_r$. Thus $y_p = x_p$ and $y_q = x_p$ while $y_r = 0$ which is also the value of x_r.

As another example observe that $a_1 s_1$ defines the identity function, while $s_1 a_1$ does not.

We now say that a machine M **computes** a function $f . \mathbf{N}^n \to \mathbf{N}$ if M, on the input ξ which has x_i in register i for $i \leq n$ and all other registers empty has ξM defined iff $f\mathbf{x}$ is defined and then has $f\mathbf{x}$ in register 1 of ξM. Note that M defines a function from $\mathbf{N}^n$ to $\mathbf{N}$ for every n. We say that f is **abacus computable** if it is computable by some abacus machine.

For convenience, whenever we say that some η has $f\mathbf{x}$ as the contents of some register this will be taken to mean also that η exists iff $f\mathbf{x}$ is defined.

Since, by Lemma 5.2, we can tell whether or not a string is an abacus machine, Theorem 2.4 is valid if computable means abacus computable.

The next group of lemmas shows that, when considering abacus computable functions, we can put the input variables and the output value in any registers we like, and can also keep registers unchanged. We need to define the registers **used by** the machine M and prove an easy lemma first.

Definition (i) The machines a_k and s_k use only register k,

(ii) The machine $M_1 \ldots M_r$ uses those registers which are used by M_i for at least one i,

(iii) The machine $(M)_k$ uses register k and those registers used by M.

Notice, however, that unless the register k is already used by M, then the machine $(M)_k$ is uninteresting. For if register k is not used by M, then $\xi(M)_k$ is defined iff $x_k = 0$ and then $\xi(M)_k = \xi$.

The following lemma is almost obvious, if we consider what is meant by a machine using a register. The detailed proof, which would use induction on the depth, is left to the reader.

Lemma 5.3 *(i) Let $\xi M = \eta$. Then $y_i = x_i$ if register i is not used by M.*

(ii) Let $\xi M = \eta$, and let ξ' have $x'_i = x_i$ for all i such that register i is used by M. Then $\xi' M = \eta'$ where $y'_i = y_i$ for all i such that register i is used by M.

Lemma 5.4 *Let $f:\mathbf{N}^n \to \mathbf{N}$ be abacus computable. Then there is a machine which on the input with x_i in register i for $i \leq n$, remaining registers empty, has output with $f\mathbf{x}$ in register 1 and the remaining registers empty.*

Proof Let M compute f. Choose $m \geq n$ such that the registers used by M are among registers $1, \ldots, m$. By Lemma 5.3(i), on the given input the output will have at most registers $1, \ldots, m$ non-empty. Then the machine $M\text{Clear}_2 \ldots \text{Clear}_m$ will do what we want.■

Lemma 5.5 *Let $f:\mathbf{N}^n \to \mathbf{N}$ be abacus computable. Then there is a machine which on every input with x_i in register i for $i \leq n$ will have output with $f\mathbf{x}$ in register 1.*

Proof Take M and m as before. By Lemma 5.3(ii) M has output with $f\mathbf{x}$ in register 1 on every input with x_i in register i for $i \leq n$ and with registers

$n+1,\ldots,m$ empty. It follows that the machine we want, which has suitable output even when registers $n+1,\ldots,m$ are not empty for the input, will be $\text{Clear}_{n+1}\ldots\text{Clear}_m M$.■

Lemma 5.6 *Let $f:\mathbf{N}^n \to \mathbf{N}$ be a function such that there is a machine M which has output with $f\mathbf{x}$ in register k on the input with x_i in register i for $i \leqslant n$, remaining registers empty. Then f is abacus computable.*

Proof If $k=1$ there is nothing to prove. Otherwise $M\text{Clear}_1\text{Descopy}_{k,1}$ will transfer the value $f\mathbf{x}$, which occurs in register k of the output of M, into register 1 where it is needed.■

Lemma 5.7 *Let $i(1),\ldots,i(n)$ be distinct registers. Suppose there is a machine M which, on input having $x_1,\ldots,x_n$ in registers $i(1),\ldots,i(n)$ respectively, and remaining registers empty, has output with $f\mathbf{x}$ in register k for some k. Then f is abacus computable.*

Proof We first look for a machine which, on input with x_j in register j for $j \leqslant n$, remaining registers empty, has output with x_j in register $i(j)$ for $j \leqslant n$, remaining registers empty. We have to be careful since the set $\{i(1),\ldots,i(n)\}$ may meet the set $\{1,\ldots,n\}$. So we take $r(1),\ldots,r(n)$ distinct from all of $1,\ldots,n,i(1),\ldots,i(n)$. Then the machine K defined as $\text{Descopy}_{1,r(1)}\ldots\text{Descopy}_{n,r(n)}\text{Descopy}_{r(1),i(1)}\ldots\text{Descopy}_{r(n),i(n)}$ does this, and KM will have output with $f\mathbf{x}$ in register k when the input has x_j in register j for $j \leqslant n$, remaining registers empty. The result follows by Lemma 5.6.■

Lemma 5.8 *Let $f:\mathbf{N}^n \to \mathbf{N}$ be abacus computable. Let $i(1),\ldots,i(n)$ be distinct registers. Let $j(1),\ldots,j(p)$ be any registers (for any p) and k a register different from these. Then there is a machine which, on any input with $x_1,\ldots,x_n$ in registers $i(1),\ldots,i(n)$ has output $f\mathbf{x}$ in register k, and has the output values in register $j(1),\ldots,j(p)$ the same as the input values in these registers.*

Proof Take $m \geqslant i(1),\ldots,i(n),j(1),\ldots,j(p),k$. Let M be a machine which on any input with $x_1,\ldots,x_n$ in registers $1,\ldots,n$ has output with $f\mathbf{x}$ in register 1. This is possible by Lemma 5.5. Obtain M' from M by increasing each subscript by m (for instance, if $m=5$ and M is $(a_1 s_2)_2$ then M' will be $(a_6 s_7)_7$). Plainly M', on any input with $x_1,\ldots,x_n$ in registers $m+1,\ldots,m+n$, will have output with $f\mathbf{x}$ in register $m+1$ and with the first m registers having the same value on output as on input.

Let K be the machine $\text{Clear}_{m+1}\ldots\text{Clear}_{m+n}\text{Clear}_{m+n+1}$. Let L be the machine $\text{Copy}_{i(1),m+1,m+n+1}\ldots\text{Copy}_{i(n),m+n,m+n+1}$. Then KL, on any input with $x_1,\ldots,x_n$ in registers $i(1),\ldots,i(n)$, will put $x_1,\ldots,x_n$ into registers $m+1,\ldots,m+n$ and will leave registers $1,\ldots,m$ with the same values for the output as they had for the input. Thus $KLM'\text{Clear}_k\text{Descopy}_{m+1,k}$ is the machine we want.■

Lemma 5.9 *Let $f_1,\ldots,f_r:\mathbf{N}^n\to\mathbf{N}$ be abacus computable functions. Let $i(1),\ldots,i(n)$ be distinct registers. Let $j(1),\ldots,j(p)$ and $k(1),\ldots,k(r)$ be distinct registers (which need not be distinct from $i(1),\ldots,i(n)$). Then there is a machine, which on any input with $x_1,\ldots,x_n$ in registers $i(1),\ldots,i(n)$ has output with $f_q\mathbf{x}$ in register $k(q)$ for $q\leqslant r$ and has the output values in registers $j(1),\ldots,j(p)$ the same as their input values.*

Proof Take $m\geqslant i(1),\ldots,i(n),j(1),\ldots,j(p),k(1),\ldots,k(r)$. Let M_q be a machine which, on the given input, has output with $f_q\mathbf{x}$ in register $m+q$ and otherwise leaves registers $1,\ldots,m+r$ with the same values on output that they had on input. This is possible by Lemma 5.8. Then the machine $M_1\ldots M_r$, on the given input, has output with $f_q\mathbf{x}$ in register $m+q$ for all $q\leqslant r$ and with the first m registers having the same value for the output as they had for the input. Thus $M_1\ldots M_rK$ is the machine we want, where K is $\text{Clear}_{k(1)}\ldots\text{Clear}_{k(r)}\text{Descopy}_{m+1,k(1)}\ldots\text{Descopy}_{m+r,k(r)}$. ■

Exercise 5.1 Find abacus machines which compute n^2, 2^n, and other functions which you care to try. (There are systematic ways of constructing suitable machines given in the next section.)

5.3 RECURSIVE FUNCTIONS

Theorem 5.10 *Partial recursive functions are abacus computable.*

Proof By Theorem 3.2 it is enough to show that the set of abacus computable functions contains the initial functions and is closed under composition, iteration and minimisation.

Plainly Clear_1 computes the zero function and a_1 computes the successor function. Using Lemma 5.6, we see the machine a_1s_1 is enough to show all projection functions are abacus computable.

Let $f_1,\ldots,f_r:\mathbf{N}^n\to\mathbf{N}$ and $g:\mathbf{N}^r\to\mathbf{N}$ be abacus computable. Let M be a machine which on input x_i in register i for all $i\leqslant n$ has output with $f_q\mathbf{x}$ in register q for all $q\leqslant r$. Such an M exists by Lemma 5.9. Let M' be a machine which on any input with y_q in register q for $q\leqslant r$ has output with $g\mathbf{y}$ in register 1. Then MM' computes the composite of g with $f_1,\ldots,f_r$.

Let $f:\mathbf{N}^n\to\mathbf{N}^n$ be abacus computable; that is, the components $f_q:\mathbf{N}^n\to\mathbf{N}$ are all abacus computable. By Lemma 5.9 there is a machine M which, on any input with $x_1,\ldots,x_n$ in registers $1,\ldots,n$ will have output with $f_1\mathbf{x},\ldots,f_n\mathbf{x}$ in registers $1,\ldots,n$, and with register $n+1$ having the same value for the output that it had for the input. Then $(Ms_{n+1})_{n+1}$ computes the iterate of f.

Finally, let $f:\mathbf{N}^{n+1}\to\mathbf{N}$ be abacus computable. Let M be a machine which on any input with $x_1,\ldots,x_{n+1}$ in registers $1,\ldots,n+1$ has output with $f(x_1,\ldots,x_{n+1})$ in register $n+2$, and with the first $n+1$ registers having the same values for the output that they had for the input. Consider the machine $M(a_{n+1}M)_{n+2}$ on the input which has $x_1,\ldots,x_n$ in registers $1,\ldots,n$ and all the remaining registers empty. This will first compute $f(\mathbf{x},0)$, putting the result in register $n+2$; here $\mathbf{x}=(x_1,\ldots,x_n)$. If $f(\mathbf{x},0)=0$, the machine now stops with 0 in register $n+1$. Otherwise it first adds 1 to register $n+1$, then computes f on the result; that is, it computes $f(\mathbf{x},1)$. If $f(\mathbf{x},1)=0$ it stops and has 1 in register $n+1$. Otherwise it adds 1 to register $n+1$ and computes f of the result; that is, it computes $f(\mathbf{x},2)$, putting the result in register in $n+2$. If $f(\mathbf{x},2)=0$, it will stop and will have 2 in register $n+1$. We see that, on the given input, the output has $\mu y(f(\mathbf{x},y)=0)$ in register $n+1$. Thus this function is also abacus computable, as required.■

Conversely all abacus computable functions are partial recursive. A simple direct proof of this fact will now be given. Another proof of this is contained in Chapters 6 and 7, where it is also shown that these functions are exactly the functions computed by very different machines, and where Kleene's Normal Form Theorem is also proved.

Let M be an abacus machine, and let m be such that at most registers $1,\ldots,m$ are used by M. Then M defines a function, which we will still call M, from $\mathbf{N}^m$ to $\mathbf{N}^m$ as follows. We define $(x_1,\ldots,x_m)M$ to be $(y_1,\ldots,y_m)$ iff M on some input with x_i in register i for all $i \leqslant m$ has output with y_i in register i for all $i \leqslant m$. By Lemma 5.3, if this holds for one such input, then it holds for all such inputs.

Theorem 5.11 *The functions* $M{:}\mathbf{N}^m \to \mathbf{N}^m$ *are partial recursive.*

Proof We use induction on the depth of M. The result is clear if M is a_k or s_k.

If M is $M_1 \ldots M_r$, then the function M is the composite of the functions $M_1,\ldots,M_r$, and hence it is partial recursive if each function M_i is partial recursive.

Let M' be $(M)_k$. We have already remarked that if register k is not used by M, then the function M' is not defined on any input with $x_k \neq 0$, while it is the identity on any input with $x_k=0$. Thus M' is partial recursive in this case. Now suppose register k is used in M. Then $\mathbf{x}M'$ is just $\mathbf{x}M^s$, where s is defined to be $\mu t(\mathbf{x}M^t$ has 0 in register $k)$. Hence the function M' is obtained from the function M and a projection function by the use of iteration, composition and minimisation. It follows that M' is partial recursive if M is.■

Corollary *Let* $f{:}\mathbf{N}^n \to \mathbf{N}$ *be abacus computable. Then* f *is partial recursive.*

Proof Let f be computed by M. Take $m \geqslant n$ so that at most registers $1,\ldots,m$ are used in M. Then f is obtained by composition from the function sending $\mathbf{x}$ to $(x_1,\ldots,x_n,0\ldots,0)\in\mathbf{N}^m$, the function M and a projection function. Since M is partial recursive, so is f.■

Let the abacus machine M compute $F{:}\mathbf{N}^2 \to \mathbf{N}$. Then, for each x, the function sending y to $F(x,y)$ is computed by $\mathrm{Descopy}_{1,2}\, a_1{}^x M$. This provides a justification of Lemma 2.6.

We now show that a certain subset of the set of all abacus machines, which we call the set of **primitive abacus machines**, will compute exactly the set of primitive recursive functions. There may also be abacus machines not in this set which happen to compute primitive recursive functions.

Definition (i) The machines a_k and s_k are the only simple primitive abacus machines of depth 0.

(ii) An abacus machine of depth n is primitive iff it is $M_1 \ldots M_r$ where each M_i is a simple primitive abacus machine of depth $\leqslant n$.

(iii) The simple primitive abacus machines of depth $n+1$ are the machines $(Ms_k)_k$ where M is a primitive abacus machine of depth n and register k is not used in M.

Notice that all the machines Clear, Descopy and Copy (with any subscripts) are all primitive machines. Hence Lemmas 5.4 to 5.8 are valid if we restrict consideration to primitive machines only.

Theorem 5.12 *A function is primitive recursive iff it is computable by a primitive abacus machine.*

Proof If we look at the proof of Theorem 5.10 it is clear that the set of functions computable by primitive abacus machines contains the initial functions and is closed under composition and iteration. Hence it contains all primitive recursive functions.

Suppose M' is $(Ms_k)_k$ where register k is not used in M, and let m be such that at most registers $1,\ldots,m$ are used in M'. Take $\mathbf{x} \in \mathbf{N}^m$ and let $\mathbf{y}$ be the result of applying the function M x_k times to $\mathbf{x}$. Because register k is not used in M it is easy to see that $\mathbf{x}M'$ has register k empty but agrees with $\mathbf{y}$ in all other registers. Thus the function M' is primitive recursive if the function M is primitive recursive. The remainder of the proof follows by the arguments of Theorem 5.11 and its corollary.■

Notice that we can tell whether or not a string is a primitive abacus machine. This provides the justification needed for Theorem 3.11 where it was shown that there are intuitively computable total functions which are not primitive recursive.

Exercise 5.2 Let f come by primitive recursion from the abacus computable functions g and h. Find an abacus machine which computes f. (If you solve this exercise directly, compare your construction with Theorem 3.2. If you have difficulty, look at the proof of Theorem 3.2, and try to convert it into a construction.)

Exercise 5.3 Find abacus machines which compute some of the primitive recursive functions defined in Chapter 3. (One way is to use the results of this chapter, including the previous exercise, starting from a defining sequence for the function.)

5.4 REGISTER PROGRAMS

Abacus machines are one of a variety of machines which are sometimes called *unlimited register machines*, for obvious reasons. We now look at another class of unlimited register machines, which will be called **register programs.** (They are often called register machines, but I prefer to call them programs to emphasise the difference between their definition and that of abacus machines. It is true, though, that abacus machines can be regarded as just a different type of program.) Historically register programs (in one of a number of different versions) were introduced before abacus machines were invented. It seems to me that the approach using abacus machines leads to rather easier proofs than are found using register programs.

A register program is defined to consist of a number of **lines,** each line consisting of a **label** and an **instruction.** The labels are just the integers $1,\ldots,r$ for some r, different lines having different labels. Instead of referring to the line with label i, this line will simply be called line i. The instructions are of four kinds, namely the **add** and **subtract** instructions a_k and s_k, the **stop** instruction STOP and the **jump** instructions $J_k(i_1,i_2)$ where i_1 and i_2 are labels (which may be the same). We also require that the last line (that is, the line with label r) has STOP as its instruction.

A **configuration** of a program P is a pair (i,ξ) where i is a label and $\xi\in\Sigma$. We call (i,ξ) **terminal** if line i has STOP as its instruction. If line i has instruction a_k we say (i,ξ) **yields** $(i+1,\xi')$ where $x'_i=x_i$ for $i\neq k$, and $x'_k=x_k+1$; similarly if line i has instruction s_k then (i,ξ) yields $(i+1,\xi'')$ where $x''_i=x_i$ for $i\neq k$ and $x''_k=x_k\dot{-}1$. If line i has instruction $J_k(i_1,i_2)$ then (i,ξ) yields (i_1,ξ) if $x_k=0$ and yields (i_2,ξ) if $x_k\neq 0$.

The **computation of** P starting from ξ is the finite or infinite sequence (i_1,ξ_1), $(i_2,\xi_2)\ldots$where (i_1,ξ_1) is (i,ξ), (i_n,ξ_n) yields (i_{n+1},ξ_{n+1}) unless (i_n,ξ_n) is the last term of the sequence, and (i_n,ξ_n) is terminal if it is the last term of a finite sequence.

The function $P{:}\Sigma\to\Sigma$ is given by $\xi P=\eta$ if the computation of P starting from ξ is a finite sequence whose last term is (i,η) for some i.

We call two programs (or two machines, or a program and a machine) **equivalent** if they define the same function from Σ to Σ.

Readers who are familiar with programming will observe that register programs are badly structured because of the jump instructions, which can be written in the form IF $x_k=0$ GOTO i_1 ELSE GOTO i_2. By contrast, the abacus machine $(M)_k$ can be expressed as WHILE $x_k\neq 0$ DO M, and abacus machines can be regarded as well-structured programs. We will see that it is easy to find a register program equivalent to a given abacus machine. It is not possible to find an abacus machine equivalent to a given register program P, but for each n there is an abacus machine equivalent to P on the set of those ξ with $x_i=0$ for $i>n$. This is hard to prove directly (it amounts to finding a WHILE program corresponding to a program with GOTO); it will follow from results in Chapter 7.

We say that the function $f{:}\mathbf{N}^n\to\mathbf{N}$ is **computed by** the register program P if P started on $(1,\xi)$, where 1 is the label of the first line and ξ has x_i in register

i for $i \leq n$ and all other registers empty, ends in a configuration (k,η) for some k where η has $f\mathbf{x}$ as its first entry. Then Proposition 5.13 shows that partial recursive functions may be computed by register programs. Because of the bad structure of register programs, it is not as easy as in Theorem 5.11 to show that functions computed by register programs are partial recursive. This will be shown informally in Chapter 8. It also follows from the results on Turing machines and the fact that register programs can be simulated by Turing machines; this is briefly indicated in the next chapter.

Proposition 5.13 *To any abacus machine there is an equivalent register program whose only stop instruction is on the last line.*

Proof Plainly the programs $1{:}a_k$ 2:STOP and $1{:}s_k$ 2:STOP are equivalent to the machines a_k and s_k. Suppose P_1 with labels $1,\ldots,r$ and P_2 with labels $1,\ldots,s$ are equivalent to M_1 and M_2. Relabel the lines of P_2 as $r+1,\ldots,r+s$ (of course changing the jump instructions from $J_k(i_1,i_2)$ to $J_k(r+i_1,r+i_2)$). Replace line r (the only stop instruction of P_1) by $r{:}J_1(r+1,r+1)$, but leave the other lines of P_1 unchanged. Let P_1P_2 be the program with these $r+s$ lines. Then P_1P_2 is equivalent to M_1M_2.

For plainly the computation of P_1P_2 starting from ξ is the same as the computation of P_1 starting from ξ until, if ever, the latter reaches a pair (r,η). This happens iff ξM_1 is defined and equals η. At this point (r,η) yields $(r+1,\eta)$. Because of the way the lines of P_2 relate to those of P_1P_2 the computation of P_1P_2 now follows that of P_2, and ends with $(r+s,\zeta)$ iff the computation of P_2 starting with η ends with (s,ζ), as needed. The result for a product $M_1 \ldots M_n$ follows similarly.

Now suppose M is equivalent to P with labels $1,\ldots,r$. Let P' be obtained as follows. First increase all labels of P (including references to them in jump instructions) by 1, and delete the last line. Add new lines $r+1{:}J_k(r+2,2)$ and $r+2$:STOP and a new first line $1{:}J_k(r+2,2)$. Then P' is equivalent to $(M)_k$.

For suppose we start the computation of P' from ξ. If $x_k = 0$, we jump at once to line $r+2$, the STOP line. Otherwise we move to line 2, from which we follow the computation of P (with labels increased by 1) until, if ever, P reaches the terminal pair (r,η), at which point P' will be at $(r+1,\eta)$. If $y_k = 0$, we move to the STOP line. Otherwise we go back to line 2 and follow the computation of P again starting from η, and so on, as needed.■

Note that any program is equivalent to one whose only stop instruction is on the last line. For if the last line has label r we need only replace a line i:STOP with $i{:}J_1(r,r)$. When the new program reaches line i, it will immediately move to line r and will then stop.

Further, any program is equivalent to one in which the instruction s_k is used only when register k is non-empty; this will be used in Chapter 9. For let P be a program with labels $1,\ldots,r$ and let line i be $i{:}s_k$. First increase the labels $i+1,\ldots,r$ of P (and any reference to them in jump instructions) by 1.

The program P' is to consist of these lines $i+2,\ldots,r+1$, together with the lines $1,\ldots,i-1$ of P and the two lines $i{:}J_k(i+2,i+1)$ and $i+1{:}s_k$.

For P, the pair (i,ξ) yields $(i+1,\xi')$ where $x_i' = x_i$ for $i \neq k$ and $x_k' = x_k \dot{-} 1$; in particular $\xi' = \xi$ if $x_k = 0$. For P' the pair (i,ξ) yields $(i+2,\xi)$ if $x_k = 0$, and this is just $(i+2,\xi')$ in this case; if $x_k \neq 0$, then (i,ξ) yields $(i+1,\xi)$ which in turn yields $(i+2,\xi')$. As line $i+2$ of P' is just line $i+1$ of P (and the same for later lines) we see that the computation of P' from a given start is just a simple modification of the corresponding computation of P, as needed.

Finally we note that in P' line $i+1$ is only reached from line i (and not by a jump instruction from any other line), and by the definition of line i it can only be reached if $x_k \neq 0$, as required. We perform this change of program for each s_k instruction of P.

We can also replace our program by an equivalent one in which any line $j{:}J_i(j',j'')$ has j' and j'' different from j. This result will also be needed in Chapter 9. All we need to do is increase by 1 the labels on lines $j+1,\ldots,r$, and replace line j by $j{:}J_1(j+1,j+1)$ and add the extra line $j+1{:}J_i(j',j'')$.

It is also possible to use different varieties of jump instructions. The *unconditional jump* $J(i_1)$ has (i,ξ) yielding (i_1,ξ) for any ξ if line i is $i{:}J(i_1)$. This is the same as $i{:}J_1(i_1,i_1)$. The instructions $J_k(i_1)$ and $J_k'(i_1)$ jump when x_k is or is not 0 respectively, and otherwise move to the next line; that is, the line $i{:}J_k(i_1)$ is just $i{:}J_k(i_1,i+1)$ and the line $i{:}J_k'(i_1)$ is just $i{:}J_k(i+1,i_1)$. The exercises show how these instructions may be used instead of our instructions $J_k(i_1,i_2)$.

Exercise 5.4 Show that any program is equivalent to one whose jump instructions are all of the form $J_k(i)$ or $J_k'(i)$.

Exercise 5.5 Show that any program is equivalent to one whose jump instructions are all of the form $J_k(i)$ or $J(i)$. (In this and the next exercise you may want to use Exercise 5.4.)

Exercise 5.6 Show that any program is equivalent to one whose jump instructions are all of the form $J_k'(i)$ or $J(i)$.

Exercise 5.7 Show that, for any n, any program is equivalent on the set of those ξ with $x_i = 0$ for $i > n$ to a program whose jump instructions are all of form $J_k(i)$.

Exercise 5.8 Let P be a program whose only jump instructions are of form $J_k(i)$. Show that if ξ is a sequence for which the registers used by P contain large enough values then the computation of P starting from ξ never uses a jump instruction. Deduce that for any such ξ the output is obtained by adding (or subtracting) suitable constants to each register used by P.

Exercise 5.9 Show that there are programs which are not equivalent to a program whose only instructions are of the form $J_k(i)$.

Exercise 5.10 Show that any program is equivalent to a program in which the only instructions are of form $J'_k(i)$. (Use Exercise 5.6. Whenever we want to perform an instruction $J(i)$ we precede it by an instruction a_1, then use the instruction $J'_1(i)$ and follow it by an instruction s_1. The details are tricky, because we may have to put in other a_1 or s_1 instructions to allow for the fact that a line may be reached either by a jump instruction or from the preceding line.)

6

Turing machines

Abacus machines and register programs are sometimes referred to as random access machines. This means that all registers are equally easy to write to or read from. By contrast, if the contents of the registers are arranged linearly on a tape then to look at the seventeenth register it is first necessary to go through all the previous sixteen registers. Readers with experience of microcomputers are likely to have met this problem; data on tape is much more slowly accessed than data on disk.

The original computing machine defined by Turing is now known as a Turing machine; its information is contained on tape in the above way.

In the first section we shall define Turing machines and give examples of them. In the second section we shall show that abacus computable functions are computable by Turing machines.

6.1 TURING MACHINES

6.1.1 Turing

Turing gave the following justification that the class of machines now known as Turing machines captures the intuitive idea of computability. (He discussed computable real numbers rather than computable functions, but this distinction is unimportant. Note that he used the word 'computer' to mean a person making a computation, and not to refer to a machine; when he wrote computing machines as now known had not been built.)

Turing wrote:

> We may compare a man in the process of computing a real number to a machine which is only capable of a finite number of conditions $q_1, q_2, \ldots, q_R$ which will be called '*m*-configurations'. The machine is supplied with a 'tape' (the analogue of paper) running through it,

and divided into sections (called 'squares') each capable of bearing a 'symbol'. At any moment there is just one square, say the rth, bearing the symbol $S(r)$, which is 'in the machine'. We may call this square the 'scanned square'. The symbol on the scanned square may be called the 'scanned symbol'. The 'scanned symbol' is the only one of which the machine is, so to speak, 'directly aware'. However, by altering its m-configuration the machine can effectively remember some of the symbols which it has 'seen' (scanned) previously. The possible behaviour of the machine at any monent is determined by the m-configuration q_n and the scanned symbol $S(r)$. This pair q_n, $S(r)$ will be called the 'configuration': thus the configuration determines the possible behaviour of the machine. In some of the configurations in which the scanned square is blank (i.e. bears no symbol) the machine writes down a new symbol in the scanned square: in other configurations it erases the scanned symbol. The machine may also change the square which is being scanned, but only by shifting it one place to right or left. In addition to any of these operations the m-configuration may be changed. Some of the symbols written down will form the sequence of figures which is the decimal of the real number being computed. The others are just rough notes to 'assist the memory'. It is only these rough notes which are liable to erasure.

It is my contention that these operations include all those which are used in the computation of a number.

Later in his paper Turing wrote:

Computing is normally done by writing certain symbols on paper. We may suppose this paper is divided into squares like a child's arithmetic book. In elementary arithmetic the two-dimensional character of the paper is sometimes used. But such a use is always avoidable, and I think that it will be agreed that the two-dimensional character of paper is no essential of computation. I assume then that the computation is carried out on one-dimensional paper, i.e. on a tape divided into squares. I shall also suppose that the number of symbols which may be printed is finite. If we were to allow an infinity of symbols, then there would be symbols differing to an arbitrarily small extent. The effect of this restriction of the number of symbols is not very serious. It is always possible to use sequences of symbols in the place of single symbols. Thus an Arabic numeral such as 17 or 999999999999 is normally treated as a single symbol. Similarly in any European language words are treated as single symbols (Chinese, however, attempts to have an enumerable infinity of symbols). The differences from our point of view between the single and compound symbols is that the compound symbols, if they are too lengthy, cannot be observed at one glance. This is in accordance with experience. We cannot tell at a glance whether 9999999999999999 and 999999999999999 are the same.

The behaviour of the computer at any moment is determined by the symbols which he is observing, and his 'state of mind' at that moment. We may suppose there is a bound B to the number of symbols or squares which the computer can observe at one moment. If he wishes to observe more, he must use successive observations. We will also suppose that the number of states of mind which need to be taken into account is finite. The reasons for this are of the same character as those which restrict the number of symbols. If we admitted an infinity of states of mind, some of them will be 'arbitrarily close' and will be confused. Again, the restriction is not one which seriously affects computation, since the use of more complicated states of mind can be avoided by writing more symbols on the tape.

Let us imagine the operations performed by the computer to be split up into 'simple operations' which are so elementary that it is not easy to imagine them further divided. Every such operation consists of some change of the physical system consisting of the computer and his tape. We know the state of the system if we know the sequence of symbols on the tape, which of these are observed by the computer (possibly with a special order), and the state of mind of the computer. We may suppose that in a simple operation not more than one symbol is altered. Any other change can be split up into simple changes of this kind. The situation in regard to the squares whose symbols may be altered in this way is the same as in regard to the observed squares. We may, therefore, without loss of generality, assume that the squares whose symbols are changed are always 'observed' squares.

Besides these changes of symbols, the simple operations must include changes of distribution of observed squares. The new observed squares must be immediately recognisable by the computer. I think it is reasonable to suppose that they can only be squares whose distance from the closest of the immediately previously observed squares does not exceed a certain fixed amount. Let us say that each of the new observed squares is within L squares of an immediately previously observed square.

In connection with 'immediate recognisability', it may be thought that there are other kinds of square which are immediately recognisable. In particular, squares marked by special symbols might be taken as immediately recognisable. Now if these squares are marked only by single symbols there can only be a finite number of them, and we should not upset our theory by adjoining these marked squares to the observed squares. If, on the other hand, they are marked by a sequence of symbols, we cannot regard the process of recognition as a simple process. This is a fundamental point and should be illustrated. In most mathematical papers the equations and theorems are numbered. Normally the numbers do not go beyond (say) 1000. It is, therefore, possible to recognise a theorem at a glance by its number. But if the paper was very long, we might reach Theorem 157767733443477; then, further on in the paper, we might find '... hence (applying Theorem 157767733443477) we have ...'. In order to make sure which was the relevant theorem we should have to compare the two numbers figure by figure, possibly ticking the figures off in pencil to make sure of their not being counted twice. If in spite of this it is still thought that there are other 'immediately recognizable' squares, it does not upset my contention so long as these squares can be found by some process of which my type of machine is capable. This idea is developed in III below.

The simple operations must therefore include:

(*a*) Changes of the symbol on one of the observed squares.
(*b*) Changes of one of the squares observed to another square within L squares of one of the previously observed squares.

It may be that some of these changes necessarily involve a change of state of mind. The most general single operation must therefore be taken to be one of the following:

(*A*) A possible change (*a*) of symbol together with a possible change of state of mind.
(*B*) A possible change (*b*) of observed squares, together with a possible change of state of mind.

The operation actually performed is determined, as has been suggested, by the state of mind of the computer and the observed symbols. In particular, they determine that state of mind of the computer after the operation is carried out.

We may now construct a machine to do the work of the computer. To each state of mind of the computer corresponds an '*m*-configuration' of the machine. The machine scans B squares corresponding to the B squares observed by the computer. In any move the machine can change a symbol on a scanned square or can change any one of the scanned squares to another square distant not more than L squares from one of the other scanned squares. The move which is done, and the succeeding configuration, are determined by the scanned symbol and the *m*-configuration. The machines just described do not differ essentially from computing machines as defined [earlier], and corresponding to any machine of this type a computing machine can be constructed to compute the same sequence, that is to say the sequence computed by the computer.

In his section (III) (referred to above) Turing continued:

We suppose, as [previously], that the computation is carried out on a tape; but we avoid introducing the 'state of mind' by considering a more physical and definite counterpart of it. It is always possible for the computer to break off from his work, to go away and forget all about it, and later to come back and go on with it. If he does this he must leave a note of instructions (written in some standard form) explaining how the work is to be continued. This note is the counterpart of the 'state of mind'. We will suppose that the computer works in such a desultory manner that he never does more than one step at a sitting. The note of instructions must enable him to carry out one step and write the next note. Thus the state of progress of the computation at any stage is completely determined by the note of instructions and the symbols on the tape. That is, the state of the system may be described by a single expression (sequence of symbols), consisting of the symbols on the tape followed by Δ (which we suppose not to appear elsewhere) and then by the note of instructions. This expression may be called the 'state formula'. We know that the state formula at any given stage is determined by the state formula before the last step was made, and we assume that the relation of these two formulae is expressible in the functional calculus. In other words, we assume there is an axiom A which expresses the rules governing the behaviour of the computer, in terms of the relation of the state formula at any stage to the state formula at the preceding stage. If this is so, we can construct a machine to write down the successive state formula, and hence to compute the required number.

6.1.2 Turing machines

We take a finite set A, called the **alphabet**, with elements $a_0, \ldots, a_n$ called **letters**; the element a_0, which plays a special role, is called **blank**.

Consider a tape, infinite in both directions, divided into squares. We place an element of A on each square, requiring all but finitely many squares to be blank. More formally, we could look at a function f from $\mathbf{Z}$ to A, with $fn=a_0$ for $|n|$ large enough. However, the informal approach is easier to follow, and it will normally be used. Instead of an infinite tape, we could consider a finite tape with arrangements for adding on extra squares of tape when required. In describing what symbols are on the tape, there is no need to mention any of the infinite run of consecutive blanks to the left and right of the non-blank portion, but there is no harm in doing so. For instance, if the alphabet consists of a, b, c, and 0 (blank) then the sequences $a0bb00ca$, $00a0bb00ca$, and $0a0bb00ca0$ all refer to the same sequence of symbols on the tape.

In order to transfer information to and from the tape, we need a read–write head. At any moment this is positioned over one square of the tape, which we refer to as the square being **scanned**. We shall consider the head as capable of moving (though in real life it is more likely that it is the tape which moves, as in a cassette recorder). A **tape description** consists of a sequence of symbols telling us what is on the tape (so this sequence must contain all the non-blank symbols) together with a note of which square is being scanned. This is usually done by underlining the scanned symbol. For instance, in the example above, $a\underline{0}bb00ca$ shows the first 0 being scanned, $a0b\underline{b}00ca$ shows the second b being scanned, and $a0bb0\underline{0}ca$ shows the third 0 being scanned. We insist that the sequence is long enough to show the scanned square; thus, in our example, if we wish to scan the 0 which is two squares to the left of the initial a, we must write the tape description as $\underline{0}0a0bb00ca$.

Our machine has to have a finite set Q of **states**. (In a physical machine, these could be the positions of cog-wheels, the positions of switches, and so on.) At any moment the machine is in some state. A **configuration** consists of a tape description and a state. (Note that the words 'state' and 'configuration' are not used in the way Turing used them.) We write $uqav$ to mean the configuration which is in state q with tape description $u\underline{a}v$, where a is a letter and u and v are strings of letters. What the machine does next depends only on its state and the symbol being scanned. We now make a formal definition.

Definition A **Turing machine (by quadruples)** consists of a finite alphabet A as above, a finite set Q of states, and a set of quadruples which can be either of form $qaa'q'$, where q and q' are states and a and a' are letters, or of one of the forms $qaRq'$ or $qaLq'$, where q and q' are states, a is a letter, and R and L are two new symbols. We require also that at most one quadruple begins with any given pair q,a.

More formally, we could define the machine to consists of A, Q, and a subset of $Q \times A \times (A \cup \{R,L\}) \times Q$. However, the informal approach is clearer.

The quadruples are meant to tell us how the machine acts. Let C be the configuration $uqav$. If no quadruple begins with q,a we say C is **terminal** (the machine is given no instructions what to do in this case, so it stops). If there is a quadruple $qaa'q'$ then C **yields** $uq'a'v$ (the machine goes into the new state q' and writes a' on the scanned square instead of a). If there is a quadruple $qaRq'$ then C yields $uaq'bv'$, where the string v is written as bv' for some letter b and some (possibly empty) string v' (the machine changes state to q' and the head moves one square right). If there is a quadruple $qaLq'$ then C yields $u'q'bav$, where u is $u'b$ with b a letter and u' a possibly empty string (the machine changes state to q' and the head moves one square left). The letters R and L are meant to signify 'right' and 'left'. Note that the scanning head is a read–write head, since it must be able to read the letter on the scanned square (in order to determine what action to take), and must also be able to write to the scanned square. Note that if more than one quadruple began with a given pair we would not be able to tell what configuration followed C. Nonetheless, non-deterministic machines, which do not satisfy this condition, are still of interest, though we shall not consider them (see section 8.1.4 for a brief discussion of them). We could also consider machines with several tapes and several heads on each machine. Such machines, as indicated in section 8.1, do not compute any additional functions; however, they may compute faster than the machines defined here.

The **computation** of the Turing machine T starting with the configuration C_1 is defined to be a finite or infinite sequence of configurations $C_1, C_2, \ldots$ such that either the sequence is infinite and C_r yields C_{r+1} for all r or the sequence is finite with its last term C_n being terminal and C_r yielding C_{r+1} for all $r<n$.

We usually require one state to be called the **starting state**. We can then talk of the computation starting with a given tape description, meaning that the starting configuration is this tape description in the starting state.

It is also often convenient to consider machines which have a **halting state** h. This means that a configuration is terminal iff it is in state h.

Seven special examples of Turing machines on the alphabet $\{0,1\}$ will be of special importance. The first of these is the machine R which has quadruples $q0Rq'$ and $q1Rq'$. This has starting state q and halting state q'. When started, it moves one square right and halts. Similarly the machine L with quadruples $q0Lq'$ and $q1Lq'$ has starting state q, halting state q', and moves one square left and then halts.

The machine P_0 has quadruples $q00q'$ and $q10q'$, while P_1 has quadruples $q01q'$ and $q11q'$. Both of these have starting state q and halting state q'. P_0 prints 0 on the scanned square and then halts, while P_1 prints 1 and then halts.

The machine $R*$ has quadruples $q0Rq'$, $q1Rq'$, $q'1Rq'$, and $q'00q''$. This has starting state q and halting state q''. When started, this machine moves one square right (whatever the symbol initially scanned). It then keeps moving right as long as it is scanning 1. As soon as it scans 0 it changes state to q'' and halts. Thus $R*$ moves to the first blank to the right of the initially scanned square. Similarly the machine $L*$ with quadruples $q0Lq'$, $q1Lq'$, $q'1Lq'$, and $q'00q''$ has starting state q, halting state q'', and it moves to the first blank to the left of the initially scanned square and then halts.

The final special machine is Test. This has quadruples $q00q_0$ and $q11q_1$. When started in state q it does not change the tape description, but moves into state q_0 or q_1 according to the scanned symbol being 0 or 1.

Let T be a Turing machine. The names given to the states are not of great importance, and if we rename the states we shall regard the machine as unchanged. For instance, in a machine with two states it does not matter whether the states are called q_0 and q_1, or 'up' and 'down', or 'in' and 'out', or 'on' and 'off'; in a machine with seven states, we could call the states $q_0, \ldots, q_6$ or 'red', 'orange', 'yellow', 'green', 'blue', 'indigo', and 'violet'. Of course, in making such a renaming, we must rename each state in each quadruple in which it occurs. By renaming states, we can assume that two (or more) given Turing machines T and T' have no states in common (or only certain specified states in common).

In particular, let T be a Turing machine with a halting state, and let T' be any Turing machine. Name the states so that the halting state of T is the starting state of T' and the machines have no other state in common. Let TT' be the machine whose states are those of T and T' and which has as quadruples all the quadruples of T and all the quadruples of T'. (Note that, because the only state of T which is a state of T' is the halting state, no pair begins two quadruples.) Let the starting state of TT' be the starting state of T. When this machine is started on a tape description it will follow the computation of T until, if ever, this computation halts. When this happens, because the halting state of T is the starting state of T', it will follow the computation of T'.

More generally, if $T_1, \ldots, T_r$ are Turing machines of which $T_1, \ldots, T_{r-1}$ all have halting states, we can construct the machine $T_1 \ldots T_r$. In particular, if T is a machine with a halting state, we can take each T_i to be T; in this case the product is just written T^r.

Let T_0 and T_1 be Turing machines on the alphabet $\{0,1\}$. Let Test be the machine constructed above. Suppose the starting state of Test is not a state of T_0 or T_1, that the states q_0 and q_1 of Test are the starting states of T_0 and T_1, and that T_0 and T_1 have no states in common. Denote by Test$\{T_0, T_1\}$ the Turing machines whose quadruples are those of Test, of T_0 and of T_1. This machine, started on a given tape description, will follow the computation of T_0 or of T_1 according to the scanned symbols being 0 or 1. These machines will be useful later.

6.1.3 Turing machines by quintuples

The Turing machines we have been considering either print a symbol or move at each step. Turing originally considered machines which could both print and move in the same step. We look at these machines as well as the previous ones, as they are more convenient for use in the next chapter.

Definition A **Turing machines by quintuples** consists of a finite alphabet A, a finite set of states Q, and a set of quintuples of form $qaa'q'R$ or $qaa'q'L$, where q and q' are states and a and a' are letters. A given pair q,a begins at most one quintuple.

As before, the configuration $uqav$ is terminal if no quintuple begins with q,a. If there is a quintuple $qaa'q'R$ then $uqav$ yields $ua'q'bv'$, where v is bv', while if there is a quintuple $qaa'q'L$ then $uqav$ yields $u'q'ba'v$, where u is $u'b$.

Intuitively, we can go from one of these kinds of Turing machines to the other without difficulty. Given a Turing machine by quintuples, we need to split the single write and move step into two steps, one writing and the next moving. Given a Turing machine by quadruples, if we want to write, which requires staying where we are, we can do so using steps which simultaneously move and write if we first write and move right and next write (the symbol that is being scanned) and move left.

More formally, let T be a Turing machine by quadruples. Let T' be the Turing machines by quintuples whose states are the states of T together with an auxiliary state q_L for each state q of T. To each quadruple $qaRq'$ (or $qaLq'$) of T, the machine T' has quintuples $qaaq'R$ (or $qaaq'L$). To each quadruple $qaa'q'$ of T, T' has a quintuple $qaa'q'_LR$. Finally, for every state q of T and for every letter x, T' has a quintuple q_LxxqL. It is easy to check that if C is a terminal configuration of T then C is also a terminal configuration of T'. If the configuration C of T yields C_1, then for T' either C yields C_1 or C yields a configuration C' such that C' yields C_1 and C' is in one of the states q_L. It follows that if the computation of T starting at C is infinite so is the computation of T' starting at C. Also, if the computation of T starting at C is finite and ends with $C*$ then the computation of T' starting at C is also finite and ends at $C*$.

Conversely, suppose T is a Turing machine by quintuples. Let the Turing machine T' by quadruples have the states of T and two additional states q_R and q_L for each state q of T. To each quintuple $qaa'q'R$ (or $qaa'q'L$) of T, let T' have a quadruple $qaa'q'_R$ (or $qaa'q'_L$). Also T' has, for every state q and letter x, quadruples q_RxRq and q_LxLq. It is now easy to check that if C is a terminal configuration of T then C is also terminal for T', while if C yields C_1 for T then for T' C yields C' and C' yields C_1, where C' is in one of the auxiliary states. As above, it follows that the computation of T starting at C is finite and ends at $C*$ iff the computation of T' starting at C is finite and ends at $C*$.

Finally, observe that we can allow Turing machines which have both quadruples and quintuples. If we need to, we can convert all the quadruples to quintuples as above; alternatively, we can convert all the quintuples to quadruples.

It is sometimes convenient to require that, in a Turing machine by quintuples, each state can be approached from only one direction. More precisely, we require the set of all states to be divided into two disjoint sets Q_R and Q_L such that if there is a quintuple ending in qR then q is in Q_R, while if there is a quintuple ending in qL then q is in Q_L. If we have an arbitrary Turing machine by quintuples we can replace each state by two states to get a new Turing machine T' which possesses this property. It will again be true that the computation of T starting with C is finite and ends with $C*$ iff the computation of T' starting with C is finite and ends with one of the two configurations obtained from $C*$ by replacing its state by one of the two corresponding states of T'.

6.2 COMPUTATION BY TURING MACHINES

6.2.1 Simulation of abacus machines

In this section all our Turing machines will have alphabet $\{0,1\}$. We shall consider machines by quadruples; as we have seen, we can go from these to machines by quintuples and back again.

The notation 1^x will denote a block of x 1s. Thus $\underline{0}1^2 01^3$ is short for 0110111. An expression 1^0 is omitted, so that $\underline{0}1000110\overline{1}$ can be written as $\overline{0}1^1 01^0 01^0 01^2 01^1$. The tape description **corresponding to** $\overline{\xi} = (x(1), x(2), \ldots)$ is defined to be $\underline{0}1^{x(1)} 01^{x(2)} \ldots$.

Theorem 6.1 *To each abacus machine M there is a Turing machine T on the alphabet $\{0,1\}$ and with a halting state which simulates M. That is, $\xi M = \eta$ iff*

T, started on the tape description corresponding to ξ, ends on the tape description corresponding to η.

Proof This will be proved by induction. The induction starts with Turing machines Add_k and Sub_k which simulate a_k and s_k. These will be constructed later.

Let M be $M_1 \ldots M_r$. Suppose we have a suitable machine T_i which simulates M_i for $i = 1, \ldots, r$. Then the product machine $T_1 \ldots T_r$ plainly simulates $M_1 \ldots M_r$.

Now suppose M is simulated by T. We want to find a Turing machine which simulates $(M)_k$. We will construct later a Turing machine Test_k which, started on a tape description corresponding to $\xi = (x(1), x(2), \ldots)$, ends on the same tape description in one of two states, ending in the state q_0 if $x(k) = 0$ and in the state q_1 if $x(k) \neq 0$, and for which a configuration is terminal iff it is in one of the states q_0 and q_1.

We may name the states of T so that T starts on q_1, has its halting state q as the starting state of Test_k, and has no other state in common with Test_k. Let T' be the Turing machine whose states and quadruples are those of T together with those of Test_k. Because no quadruple of T begins with q and no quadruple of Test_k begins with q_1, we see that at most one quadruple begins with any pair. No quadruple of T or Test_k begins with q_0. Also, for every letter a, and any state q' other than q, q_0, or q_1, there is a quadruple beginning $q'a$. Hence q_0 is the halting state for T'. Let the starting state of T' be q. We shall show that T' simulates $(M)_k$.

Take any ξ and the corresponding tape description. Start T' on this tape description. The computation of T' will first follow Test_k. Hence, if $x(k) = 0$ then T' will return to the same tape description in the state q_0, and will then halt. In this case $\xi(M)_k = \xi$. If $x(k) \neq 0$ then T' will return to the same tape description in state q_1. As this is the starting state of T, the computation of T' will now simulate the computation of M. In particular if ξM is not defined the computation of T' will be infinite. If, however, $\xi M = \eta$ the computation of T' will reach the configuration consisting of the tape description corresponding to η in the state q. Since this is the starting state of Test_k the procedure now repeats, and so on. It is plain that if $\xi(M)_k$ is not defined then the corresponding computation of T' is infinite. If $\xi(M)_k$ is ξM^r, where r is as small as possible, then the corresponding computation of T' passes through the tape descriptions corresponding to $\xi M, \xi M^2, \ldots, \xi M^r$, on which it ultimately stops (after making one test to be sure the kth coordinate is 0). Thus T' simulates $(M)_k$. The theorem is proved by induction, once we have constructed the machines Add_k, Sub_k, and Test_k.■

Before constructing these machines, we will say what is meant by the function computed by a Turing machine T. Let $\mathbf{x} = (x(1), \ldots, x(k))$ be in $\mathbf{N}^k$. We define $\text{In}_{T,k}\mathbf{x}$ to be the configuration in the starting state of T whose tape description corresponds to $\mathbf{x}$. We will define an output function Out_T from the set of configurations into $\mathbf{N}$. Once this has been done, we can define the function f from $\mathbf{N}^k$ to $\mathbf{N}$ **computed by** T by requiring $f\mathbf{x}$ to be undefined if the computation of T starting with $\mathrm{In}_{T,k}\mathbf{x}$ is infinite, and to be y if this computation ends in a configuration with output y. A function is **Turing computable** if it is computable by some Turing machine.

There are several reasonable ways of defining Out_T. We could define $\mathrm{Out}_T C$ to be the number of 1s on the tape in the tape description of C. Alternatively, we could define it as the number of 1s strictly between the scanned square and the next blank to the right; thus, if C is $uqa1^y0v$, $\mathrm{Out}_T C$ would be y.

We shall take a definition which will be especially useful in the next chapter; on those configurations for which it is defined it will coincide with the above two. We shall require Out_T to be the partial function given by $\mathrm{Out}_T C = y$ if the tape description of C is $\underline{0}1^y$.

Theorem 6.2 *Any abacus computable function is Turing computable.*

Proof Let f be abacus computable. Then we know there is an abacus machine M which, on input $\mathbf{x}$, halts if $f\mathbf{x}$ is defined, and, if it halts has $f\mathbf{x}$ in the first register and all other registers empty.

Let T be a Turing machine which simulates M. Then the computation of T started on $\mathrm{In}_{T,k}\mathbf{x}$ is infinite if $f\mathbf{x}$ is not defined. If $f\mathbf{x} = y$ the computation of T started from $\mathrm{In}_{T,k}\mathbf{x}$ will stop in a configuration whose tape description corresponds to $(y, 0, 0, \ldots)$; that is, the tape description will be $\underline{0}1^y$. Hence f is computed by T. ∎

We still have to construct three particular machines.

Test_k can just be defined as $R{*}^{k-1}R\mathrm{Test}\{L{*}^k, L{*}^k\}$. For $R{*}^{k-1}$ moves to the 0 immediately before the block $1^{x(k)}$. This is followed by R, which moves one square further right. The symbol now scanned is 0 if $x(k) = 0$ and is 1 if $x(k) \neq 0$. In either case we have to move to the kth 0 to the left of the scanned square, which is the 0 originally scanned. The two copies of $L{*}^k$ provide two different states on which we finally halt, depending on whether $x(k)$ is 0 or not.

To define Add_k we first need to find extra room to insert the additional 1. So we need to find a machine Shiftleft which, started on $u0\underline{0}1^x0v$, ends on $u01^x\underline{0}0v$. We can then define Add_k to be $\mathrm{Shiftleft}^{k-1}P_1L{*}^k$. For suppose we start this machine on the tape description corresponding to ξ. When we have performed the portion $\mathrm{Shiftleft}^{k-1}$ the resulting tape description will be $1^{x(1)}0 \ldots 01^{x(k-1)}0\underline{0}1^{x(k)}0 \ldots$. Then P_1 will replace the scanned 0 by 1, and $L{*}^k$ will return to the 0 originally scanned.

Shiftleft is defined as $P_1R{*}LP_0R$. For, started on $u0\underline{0}1^x0v$, this passes through the tape descriptions $u0\underline{1}1^x0v$, $u01^x1\underline{0}v$ (after all, 11^x is the same as 1^x1), $u01^x\underline{1}0v$, $u01^x\underline{0}0v$, and $u01^x0\underline{0}v$.

Similarly, the machine Shiftright, defined as $P_1L{*}RP_0L$, when started on $u01^x\underline{0}0v$ will end on $u0\underline{0}1^x0v$. We use this to define Sub_k, but whether or not it is needed depends on whether or not $x(k)$ is 0, so we have to use Test.

We define Sub_k to be $R{*}^{k-1}R\mathrm{Test}\{L{*}^k, T_k\}$, where T_k is

P_0LShiftright^{k-1}R. Simpler machines would be possible, but we would have to define further auxiliary machines to construct them.

We know, as with Test$_k$, that when we apply $R*^{k-1}R$ the final tape description has 0 scanned if $x(k) = 0$ and has 1 scanned if $x(k) \neq 0$. In the first case, we simply move back to the 0 originally scanned. In the second case, we begin by replacing the 1 scanned by 0. We then have to move the first $k-1$ blocks one square to the right to get the correct form. Because of the way Shiftright works, we first have to move one square left, so that we are on the 0 immediately to the right of $1^{x(k-1)}$, then apply Shiftright $k-1$ times. When this is done, we have overshot by one square, and we find the tape description is $\underline{0}01^{x(1)} \ldots$; so we have to finish by moving one square right.

We have seen that functions computable by register programs are partial recursive; hence they can be computed by Turing machines. We can prove directly that register programs can be simulated by Turing machines. We simply take one machine Add$_k$, Sub$_k$, or Test$_k$ for each line a_k, s_k, or J_k (and a machine with one state and no quadruples for each STOP line). We then have to identify the final states of these machines with the starting states of other such machines, according to the ways the program takes us from one line to another. Details are left to the reader.

6.2.2 Variations on the above definitions

If we look carefully, we see that Add$_k$ and Sub$_k$ do not finish on the square originally scanned. This is sometimes inconvenient. In particular, suppose we could always finish on the square originally scanned and without ever moving to the left of this square. Then we could work with a tape which was only infinite in one direction, since the portion of the tape to the left of the originally scanned square could be deleted. It is often convenient to have this possibility.

In our process, in order to add 1 to (or subtract 1 from) $x(k)$, we have moved the first $k - 1$ blocks. We might try to move the blocks after block k instead, so as to get the above property. But this cannot be done without modifications in our definitions. The problem is that, for instance, a tape description beginning $\underline{0}1^2 01^3 000$ could correspond to (2,3), to (2,3,0,0,15), or to (2,3,0,0,0,0,4). If we look at a finite part of the tape, even if we have found a long sequence of blanks, we can never be sure that no further non-blank symbols occur.

There are two standard ways round this. Instead of finding a Turing machine T which simulates the abacus machine M, we can obtain, for each n, a Turing machine which simulates M on those ξ for which $x(i) = 0$ for all $i > n$. Choose any m so that register i is not used if $i > m$. Suppose first that $n \geqslant m$. Then, on such ξ, as we build the Turing machine inductively, we only need Add$_k$ and Sub$_k$ to act correctly on sequences η with $y(i) = 0$ for $i > n$. Thus we can safely move the $n - k$ blocks following block k, and we will have what we need; the fact that this does not perform Add$_k$ (or Sub$_k$) for all tape descriptions does not matter, as we never come across the ones which create problems. If $n < m$, we may come across sequences η with

$y(i) \neq 0$ for some $i > n$. But in this case every η we meet has $y(i) = 0$ if $i > m$, and this time we move the $m - k$ blocks following block k.

An alternative approach, which is often used, is to change the tape description corresponding to ξ. Given ξ, choose any n such that $x(i) = 0$ for $i > n$. We then make the tape description $0\underline{1}1^{1+x(1)}01^{1+x(2)}0 \ldots 0^{1+x(n)}$ correspond to ξ. Hence there are infinitely many tape descriptions corresponding to any ξ, since there are infinitely many choices of n. We can then recognise the end of a tape description corresponding to ξ by two consecutive blanks. With care, we can use this to shift the blocks following block k. There are some technical problems in the simulation which can occur if the chosen n is less than m, where m is the largest register used by M. We shall not pursue this matter further.

Another possibility is to use a larger alphabet. Observe that any x in $\mathbf{N}$ can be uniquely written as $x_0+x_1n+ \ldots x_rn^r$, where $1 \leqslant x_i \leqslant n$ for all i (provided we take the empty sum to represent 0). We can then express x by the string $x_r, \ldots, x_1, x_0$ on $\{1, \ldots, n\}$. Thus, on the alphabet $\{0,1, \ldots, n\}$ (or, more generally, on any alphabet with $n+1$ elements, by renaming the elements) we could require the tape description corresponding to ξ to be $\underline{0}u(1))0u(2)0\ldots$, where $u(i)$ is the string representing $x(i)$. We could then compute functions using these machines. A brief indication is given in the next chapter that the functions computed by these machines are all partial recursive. If we wish to show that all abacus computable functions (and hence all partial recursive functions) can be computed by such machines, whichever alphabet we choose, we will need to find machines which simulate a_k and s_k. This is somewhat trickier than before. Firstly, if the string u corresponds to x the expression for the strings corresponding to $x+1$ and $x-1$ are more complicated than before. Second, we are no longer certain that adding 1 to (or subtracting 1 from) a number increases (or decreases) the length of the corresponding string, so we must sometimes use a shift and sometimes not. Finally, the machines which shift right or left have to be more complicated than before. All these problems can be overcome; no major theory is involved, just a mass of detail. Further, we might want to ensure that we could use a tape infinite only in one direction, in which case we would have to deal with the previous considerations as well.

Exercise 6.1 Suppose we use the tape configuration $\underline{0}1^{x(1)+1}01^{x(2)+1} \ldots$ to correspond to $(x(1), x(2), \ldots)$. Find Turing machines Add_k and Sub_k which simulate a_k and s_k and which end on the square they started on, and which never move to the left of the original square. (You will need a machine which moves to the right until it reaches two consecutive blanks. You may also find the machine given by $q0Rq_1$, $q1Rq_1$, q_110q_2R, q_21Rq_2, q_201q_1R, q_100q_3 useful.)

In the remaining exercises u and v are strings on $\{a_0, a_1, \ldots, a_n\}$, and w is a string on $\{a_1, \ldots, a_n\}$. Our Turing machines will have both quadruples and quintuples.

Exercise 6.2 What does the Turing machine with quintuples $qa_ia_iq_iR$ for all i, $q_ia_ja_iq_jR$ for all $j \neq 0$ and all i, and $q_ia_0a_iq'R$ do when started in the configuration $uqa_0wa_0a_0v$?

Exercise 6.3 What does the Turing machine with quintuples $qa_ia_iq_1R$, $q_1a_na_nq_1R$, $q_1a_ia_iq_2R$ for $0<i<n$, $q_2a_ia_iq_3R$ for $i \neq 0$, and the quadruples $q_1a_0a_0q_3$ and $q_2a_0a_0q_4$ do when started in uqa_0wa_0v (i) when w is a_0^k for some k and (ii) for other w?

Exercise 6.4 What does the Turing machine with quintuples $qa_ia_iq_1L$ for all i, $q_1a_ia_{i+1}q_2L$ for all $i \neq n$, $q_1a_na_1q_1L$, $q_2a_ia_iq_2L$ for $i \neq 0$, and quadruple $q_2a_0a_0q_3$ do (i) when started on $ua_0a_0a_n^kqa_0v$ for some k and (ii) when started on ua_0wqa_0v, where w is not of form a_n^k?

Exercise 6.5 With the help of the previous machines and other similar machines, construct machines Add_k and Sub_k which simulate a_k and s_k when the alphabet is $\{0,1,\ldots,n\}$.

Exercise 6.6 Define, for a Turing machine T with alphabet $\{a_0,\ldots,a_n\}$, total functions Out_T and $\mathrm{In}_{T,k}$ which enable us to say what is meant by the function computed by T.

6.2.3 Halting problems and related results

Many problems for Turing machines can be shown to be undecidable by clever manipulation of machines. They can also be looked at using already proved results about partial recursive functions (or intuitively computable functions).

For instance, there is a Turing machine such that $\{n;\ T$ eventually halts when started on $0\underline{1}^n\}$ is undecidable. We need only choose T to compute the partial characteristic function of a set which is r.e. but not recursive.

We can number Turing machines (at least, provided we require the states to belong to a fixed countable set $\{q_0, q_1, \ldots\}$, which is no real restriction). We can then obtain a universal function as in Theorem 2.4. The Turing machine which computes this function can be referred to as a **universal Turing machine**. On the input $01^m0\underline{1}^n$, it finishes with the output of T_m on input $0\underline{1}^n$. To formalise this without going into great detail about Turing machines, it is necessary to use the results of the next chapter. It is easy to proceed from a numbering of Turing machines to the numbering of the corresponding modular machines, and then to use Kleene's Normal Form Theorem (which is proved in the next chapter) to obtain a partial recursive function universal for Turing computable functions.

With this numbering $\{n;\ T_n$ eventually halts when started on a blank tape$\}$ is not decidable. For the numbering of Turing machines provides an indexing of computable functions as in section 2.4. Our set is the same as $\{n;\ \phi_n(0)$ is defined$\}$, and this set is undecidable, by Rice's Theorem. Another proof of this result will be given later.

Many other properties can be shown undecidable by similar techniques. As a contrast, we show $\{n;\ T_n$ when started on a blank tape eventually prints $1\}$ is decidable. For, given any Turing machine T, we run T until it either halts or a state occurs twice (one or the other must happen by $k+1$ steps if the machine has k states). We can see if the computation has printed 1 by this stage. If not, and the computation has not halted, the remainder of the computation just keeps repeating a cycle of configurations, remaining throughout on a blank tape.

One of the most interesting undecidable problems is known as the busy beaver problem. For any Turing machine T, let $f(T) = m$ be defined iff T has a halting state and eventually halts on $\underline{0}1^m$ when started on a blank tape. Let $b(n)$ be $\max\{f(T);\ T$ has n states$\}$. Notice that we may assume that the states of T are $q_1, \ldots, q_n$ if T has n states; consequently there are only finitely many machines to look at for each n. If we could decide whether or not a machine halted when started on a blank tape, we could simply look at all those machines with n states which did halt when started on a blank tape, determine for each of these whether they halt on $\underline{0}1^m$ for some m, and find $f(T)$ for each of them. Hence b would be computable if we could decide whether or not a machine halted when started on a blank tape. We shall show that b is not computable.

Suppose that b is computable by a Turing machine B which has k states. Let $P(n)$ be a Turing machine which halts on $\underline{0}1^n$ when started on a blank tape. Then $f(P(n)BB) = b(b(n))$. Since $P(n)BB$ can be chosen to have $n+2k$ states, we see that $b(n+2k) \geqslant b(b(n))$.

Now $b(n+1) > b(n)$. For let T be a machine with n states which halts on $\underline{0}1^m$ when started on a blank tape. Because T has a halting state, by adding one state and two quintuples we get a machine T' with $n+1$ states, one of which is a halting state, such that T' halts on $\underline{0}1^{m+1}$ when started on a blank tape. We can choose T so that m is $b(n)$. The result follows, as $b(n+1) \geqslant f(T') = m+1$. Then $b(i) > b(j)$ if $i > j$. Since $b(n+2k) \geqslant b(b(n))$, we see that $n+2k \geqslant b(n)$ for all n.

Now let D be a machine which ends on $\underline{0}1^{2n}$ when started on $\underline{0}1^n$. The machine $P(n)D$ has $n+d$ states for some fixed d, and $f(P(n)D) = 2n$. Hence $b(n+d) \geqslant 2n$ for all n.

From the previous two paragraphs we find that $n+d+2k \geqslant b(n+d) \geqslant 2n$ for all n, which gives the absurdity $d+2k \geqslant n$ for all n. Hence b is not computable.

7

Modular machines

The concept of Turing machines comes from a profound intuition into the nature of computation (and was developed before electronic computers existed). Unlimited register machines also are a natural notion if one's intuition is deep enough. By contrast an understanding of unlimited register machines was needed before abacus machines were created, although, once developed, abacus machines are fairly natural. The modular machines which are discussed in this chapter, however, are not at all intuitive. They come from looking at a numerical coding of Turing machines, and observing how the Turing machine computation affects the coding. They require an understanding of Turing machines before they could be invented. As such they might be described as 'second generation' machines. Like physical computing machines, this second generation nature makes them easier and faster than first generation machines such as Turing machines. We will find that the fact that Turing computable functions are computable by modular machines is almost immediate, while the proof that functions computable by modular machines are partial recursive, and even Kleene's Normal Form Theorem, is also straightforward.

7.1 TURING MACHINES

We consider a Turing machine defined by quintuples. Renaming the states and the tape symbols if necessary, we may assume that the tape symbols are the natural numbers 0 (blank), $1, \ldots, n$, and that the states are also natural numbers, at most one of which is also a tape symbol. Let m be a number greater than n and also greater than any state.

We can represent any configuration $C = \ldots b_1 b_0 q a c_0 c_1 \ldots$ by the four numbers q, a, $u = \Sigma b_i m^i$, $v = \Sigma a_i m^i$. These numbers are usefully combined into two in one of the following two ways. The pair $(um+q, vm+a)$ is called the **right-associate** of C, while $(um+a, vm+q)$ is the **left-associate** of C. Not all members of $\mathbf{N}^2$ are associates of configurations, but we can tell whether or not a pair is an associate (right or left) of some configuration, and, if so, can find the configuration. This would not hold if we allowed two numbers to be both tape symbols and states; if one number is both a tape symbol and a state then some pairs will be both right and left associates of the same configuration (for most of the applications we can assume the tape symbols and states have no number in common, but nothing is simplified by assuming this).

Suppose C yields C' using the quintuple $qaa'q'R$ so that C' is $\ldots b_1b_0a'q'c_0c_1\ldots$. Then the four numbers corresponding to C' are q', c_0, $u' = a' + b_0m + b_1m^2 + \ldots$, $v' = c_1 + c_2m + \ldots$. Hence the right-associate of C' is $(um^2 + a'm + q', v)$; we do not look at the left-associate of C' in this case. Similarly if C had yielded C' using the quintuple $qaa'q'L$ the left-associate of C' would have been $(u, vm^2 + a'm + q')$ and we would not have looked at the right associate of C'. This behaviour motivates the strange definition of modular machines in the next section.

7.2 MODULAR MACHINES

A **modular machine** consists of positive integers m,n with $m>n$ together with a (possibly empty) set of quadruples (a,b,c,R) or (a,b,c,L) where a,b,c are numbers with $0 \leqslant a<m$, $0 \leqslant b<m$, and $0 \leqslant c<m^2$, and R and L are direction symbols; we require that at most one quadruple begins with a given pair a,b. When $n=1$ we call the modular machine **special**.

The set of configurations of the modular machine M is just $\mathbf{N}^2$. For any $(\alpha,\beta)\in\mathbf{N}^2$ we define u,v,a,b by $\alpha = um+a$, $\beta = vm+b$ where $0 \leqslant a<m$ and $0 \leqslant b<m$. We call (α,β) **terminal** if no quadruple begins with a,b. If there is a quadruple (a,b,c,R) we say that (α,β) **yields** (um^2+c,v) while if there is a quadruple (a,b,c,L) we say (α,β) yields (u,vm^2+c). The integer n will not be used yet.

The reason these machines are called modular machines is that the behaviour of M on (α,β) depends on the remainders of α and β modulo m.

Now let T be a Turing machine of the kind considered in section 7.1. The modular machine M **corresponds** to T if its defining integers are those m and n which are used for T and if, in addition, to each quintuple $qaa'q'D$ of T (where D is either R or L) the modular machine M has two quadruples $(q,a,a'm+q',D)$ and $(a,q,a'm+q',D)$. Since m can be any large enough integer, there are infinitely many modular machines corresponding to a given T. The definition of modular machines and the discussion in section 7.1 make the following lemma obvious.

Lemma 7.1 *Let T be a Turing machine and M a corresponding modular machine. Let C be a configuration of T, and (α,β) an associate (left or right) of C. If C is terminal for T, then (α,β) is terminal for M. If C yields C' then (α,β) yields an associate of C'.*

Let M be any modular machine. Let (α_1,β_1), (α_2,β_2), ... be either an infinite sequence such that (α_i,β_i) yields $(\alpha_{i+1},\beta_{i+1})$ for all i or a finite sequence whose last pair is terminal and such that (α_i,β_i) yields $(\alpha_{i+1},\beta_{i+1})$ if the latter is defined. This sequence (which is obviously unique once (α_1,β_1) is given) is called the **computation of M starting with** (α_1,β_1). The following lemma is obvious, iterating lemma 7.1.

Lemma 7.2 *Let M be a modular machine corresponding to the Turing machine T, and let (α,β) be an associate of the configuration C. If the computation of T starting from C is infinite then so is the computation of M starting from (α,β). If the computation of T starting from C terminates with C' then the computation of M starting with (α,β) terminates with an associate of C'.*

We define a function g_M:$\mathbf{N}^2\to\mathbf{N}^2$ by $g_M(\alpha,\beta)=(\alpha',\beta')$ if the computation of M starting with (α,β) is finite and has (α',β') as last pair. This function is rather strange. If it is defined anywhere then there is a pair a,b beginning no quadruple, and then g_M is the identity on any pair $(um+a, vm+b)$. We shall use the integers m and n (but not the quadruples) to define the **input function** $\text{In}_M:\mathbf{N}\to\mathbf{N}^2$ and the **output function** $\text{Out}_M:\mathbf{N}^2\to\mathbf{N}$. We then call the composite $\text{Out}_M g_M \text{In}_M$ the function **computed by** M. More generally we could define a function $\text{In}_{M,r}:\mathbf{N}^r\to\mathbf{N}^2$, and call all the functions $\text{Out}_M g_M \text{In}_{M,r}$ functions computed by M. If these functions are given a suitable definition the following lemma will be obvious.

Lemma 7.3 *Let M be a modular machine corresponding to the Turing machine T.*

(*i*) *For any $\mathbf{x}\in\mathbf{N}^r$ $In_{M,r}\mathbf{x}$ is an associate of the configuration of T corresponding to $\mathbf{x}$ as input, the starting state being $n+1$.*
(*ii*) *If (α,β) is an associate of the configuration C of T, then the output (if any) of C is $Out_M(\alpha,\beta)$.*

Combining Lemmas 7.2 and 7.3 we obtain the following theorem.

Theorem 7.4 *Let M be a modular machine corresponding to the Turing machine T. Then, for any r, M and T compute the same function from $\mathbf{N}^r$ to $\mathbf{N}$.*

Theorem 7.4 depends on defining the input and output functions so as to make Lemma 7.3 correct. The formulas are not deep, but they are somewhat messy. The general case will be left to the reader, and only the formulas for special modular machines and one-variable input will be given; the theory of this case extends without difficulty to the general case.

Let M be a special modular machine. Since $n=1$, the starting state is the number 2. Define $\text{In}_M x$ to be $(2,\Sigma^{i\leq x}m^i) = (2,(m^{x+1}-1)/(m-1))$ and $\text{Out}_M(\alpha,\beta)$ to be $(\mu y\leq\beta)(m^{y+1}>\beta)$. If M corresponds to the Turing machine T then $\text{In}_M x$ is the right associate of 01^x in state 2, while if (α,β) is an associate of $\underline{0}1^y$ in any state then $\text{Out}_M(\alpha,\overline{\beta})$ is y. Hence Lemma 7.3, and consequently also Theorem 7.4, holds in this case. If we had defined the output for all configurations in Chapter 6 (for instance, as in Exercise 6.6) a more complicated definition of Out_M would be needed for Lemma 7.3(ii) to hold.

Exercise 7.1 Define $\mathrm{In}_{M,r}$ for the special modular machine M and any r so that Lemma 7.3(i) holds.

Exercise 7.2 Define In_M and $\mathrm{In}_{M,r}$ for any modular machine so that Lemma 7.3(i) holds.

Exercise 7.3 Define Out_M for any modular machine M so that Lemma 7.3(ii) holds.

Exercise 7.4 If you have given a definition of output for all configurations of Turing machines T (either with tape symbols 0,1 or more generally) define Out_M for (special) modular machines so that Lemma 7.3(ii) holds.

7.3 PARTIAL RECURSIVE FUNCTIONS AND MODULAR MACHINES

Let M be a special modular machine with quadruples (a_i, b_i, c_i, R) for $1 \leqslant i \leqslant r$ and quadruples (a_i, b_i, c_i, L) for $r < i \leqslant r+k$ (where r and k may be 0). Define $\mathrm{Next}_M : \mathbf{N}^2 \to \mathbf{N}^2$ by $\mathrm{Next}_M(\alpha,\beta) = (\alpha',\beta')$ if (α,β) yields (α',β') and $\mathrm{Next}_M(\alpha,\beta) = (\alpha,\beta)$ if (α,β) is terminal. Let $\mathrm{Comp}_M : \mathbf{N}^3 \to \mathbf{N}^2$ be the iterate of Next_M. Let Term_M be the set of terminal pairs.

Lemma 7.5 *The set Term_M and the functions In_M, Out_M, Next_M, and Comp_M are primitive recursive.*

Proof The formulas already given for In_M and Out_M show they are primitive recursive.

The numbers u,v,a,b defined by $\alpha = um + a$, $\beta = vm + b$ with $a,b < m$ are primitive recursive functions of (α,β). Now (α,β) is terminal iff no pair (a_i, b_i) equals (a,b). Hence the set Term_M is given by the primitive recursive condition $(a \neq a_1 \vee b \neq b_1) \wedge \ \ldots \ \wedge (a \neq a_{r+k} \vee b \neq b_{r+k})$. The function Comp_M, being the iterate of Next_M, will be primitive recursive if Next_M is. Also Next_M is given by the following definition by cases, and so is primitive recursive.

If $a = a_1$ and $b = b_1$ then $\mathrm{Next}_M(\alpha,\beta) = (um^2 + c_1, v)$
......
If $a = a_r$ and $b = b_r$ then $\mathrm{Next}_M(\alpha,\beta) = (um^2 + c_r, v)$
If $a = a_{r+1}$ and $b = b_{r+1}$ then $\mathrm{Next}_M(\alpha,\beta) = (u, vm^2 + c_{r+1})$
......
If $a = a_{r+k}$ and $b = b_{r+k}$ then $\mathrm{Next}_M(\alpha,\beta) = (u, vm^2 + c_{r+k})$.
Otherwise $\mathrm{Next}_M(\alpha,\beta) = (\alpha,\beta)$. ■

Theorem 7.6 *The function computed by a special modular machine is partial recursive.*

Proof Let f be computed by M. Let $\phi,\psi:\mathbf{N}^2\to\mathbf{N}$ be the functions $\text{Out}_M\text{Comp}_M(\text{In}_M x,t)$ and $\chi_M\text{Comp}_M(\text{In}_M x,t)$, where χ_M is the characteristic function of Term_M. By Lemma 7.5 both ϕ and ψ are primitive recursive. By definition the function f computed by M is given by $fx = \phi(x,t)$ for any t such that $\psi(x,t) = 0$ (bearing in mind that $\text{Next}_M(\alpha,\beta) = (\alpha,\beta)$ if (α,β) is terminal). In particular f is partial recursive, since
$fx = \phi(x,\mu t(\psi(x,t) = 0))$.■

Combining Theorems 5.10, 6.2, 7.4, and 7.6 we have the following theorem.

Theorem 7.7 *The following conditions on a function $f:\mathbf{N}\to\mathbf{N}$ are equivalent.*

(*i*) *f is partial recursive.*
(*ii*) *f is abacus computable.*
(*iii*) *f is computable by a Turing machine with tape symbols* 0 *and* 1.
(*iv*) *f is computable by a special modular machine.*

More generally, as indicated in the exercises, these conditions are also equivalent to f being computable by an arbitrary Turing machine or by an arbitrary modular machine, and similar results hold for functions from $\mathbf{N}^r$ to $\mathbf{N}$ for any r.

The following result has already been stated as Theorem 4.2.

Theorem 7.8 *Let $f:\mathbf{N}\to\mathbf{N}$ be partial recursive. Then there are primitive recursive $u:\mathbf{N}\to\mathbf{N}$ and $v:\mathbf{N}^2\to\mathbf{N}$ such that $fx=u(\mu y\{v(x,y)=0\})$.*

Proof Take the function ϕ and ψ used in the proof of Theorem 7.7. Define u to be ϕJ^{-1} and $v(x,y)$ to be $\psi J^{-1}y + |x-Ky|$. If fx is defined there will be some t with $fx=\phi(x,t)$ and $\psi(x,t)=0$. Let y be $J(x,t)$ and we will have $fx=uy$ and $v(x,y)=0$ since $\psi J^{-1}y=\psi(x,t)=0$ and $Ky=x$.

Conversely, suppose there is some y with $v(x,y)=0$. Then $Ky=x$ and $\psi J^{-1}y=0$. Define t to be Ly. Then we have $\psi(x,t)=0$ and so fx is defined and $fx=\phi(x,t)=uy$.

Consequently fx is defined iff there is some y with $v(x,y)=0$, and then $fx=uy$ for any such y, and, in particular, for the least such y.■

7.4 KLEENE'S NORMAL FORM THEOREM

We now assign a number to each special modular machine M, called its **Gödel number**, and look at the material of the previous section using the Gödel number of the machine as a parameter.

Let M be a special modular machine with quadruples (a_i,b_i,c_i,D_i) for

$i \leqslant k$, where each D_i is either R or L. The Gödel number of M (which will depend on the order in which the quadruples are given) is

$$2^m \Pi p_{4i+1}^{a_i} p_{4i+2}^{b_i} p_{4i+3}^{c_i} p_{4i+4}^{d_i},$$

where $d_i = 1$ if $D_i = R$ and $d_i = 2$ if $D_i = L$.

For any $g \in \mathbf{N}$, whether or not it is the Gödel number of a machine, we make the following definitions. We define m to be $\exp_0 g$, k to be quo-$(4, \text{length} g) \dot{-} 1$, and $a_i = \exp_{4i+1} g$, $b_i = \exp_{4i+2} g$, $c_i = \exp_{4i+3} g$, and $d_i = \exp_{4i+4} g$. Then m and k are primitive recursive functions of g, while a_i, b_i, c_i, and d_i are primitive recursive functions of g and i.

Now take any g, α, and β. If we let $u = \text{quo}(m,\alpha)$, $v = \text{quo}(m,\beta)$, $a = \text{rem}(m,\alpha)$ and $b = \text{rem}(m,\beta)$, then u, v, a, and b will be primitive recursive functions of g, α, and β. If $j = (\mu i \leqslant k)\ (a = a_i \wedge b = b_i)$, then j is also a primitive recursive function of g, α, and β. Also $j = k+1$ iff there is no $i \leqslant k$ such that $a = a_i$ and $b = b_i$.

We define a primitive recursive set TERM, and primitive recursive functions IN, OUT, NEXT, and COMP. The function IN:$\mathbf{N}^2 \to \mathbf{N}^3$ is given by $\text{IN}(g,x) = (g, 2, (m^{x+1} - 1)/(m-1))$ and OUT:$\mathbf{N}^3 \to \mathbf{N}$ is $(\mu y \leqslant \beta)\ (m^{y+1} > \beta)$.

The set TERM is $\{(g,\alpha,\beta);\ j = k+1\}$. The function NEXT:$\mathbf{N}^3 \to \mathbf{N}^3$ is given by the following definition by cases:

$$\begin{aligned} &\text{if } d_j = 1 \text{ then NEXT}(g,\alpha,\beta) &= (g, um^2 + c_j, v), \\ &\text{if } d_j = 2 \text{ then NEXT}(g,\alpha,\beta) &= (g, u, vm^2 + c_j), \\ &\text{otherwise NEXT}(g,\alpha,\beta) &= (g,\alpha,\beta). \end{aligned}$$

The function COMP is the iterate of NEXT.

Now let g be the Gödel number of the machine M. By the definition of j and of TERM, $(g,\alpha,\beta) \in \text{TERM}$ iff $(\alpha,\beta) \in \text{Term}_M$. Also $\text{In}_M x$ is the last two entries of $\text{IN}(g,x)$ and $\text{OUT}(g,\alpha,\beta) = \text{Out}_M(\alpha,\beta)$. Further, bearing in mind that $d_j = 1$ if $D_j = R$ and $d_j = 2$ if $D_j = L$ while $d_j = 0$ if $j = k+1$, we see that $\text{NEXT}(g,\alpha,\beta) = (g,\alpha',\beta')$ iff $\text{Next}_M(\alpha,\beta) = (\alpha',\beta')$. It follows that $\text{COMP}(g,\alpha,\beta,t) = (g,\alpha',\beta')$ iff $\text{Comp}_M(\alpha,\beta,t) = (\alpha',\beta')$.

Now define Φ to be OUTCOMP(IN$(g,x),t)$ and Ψ to be XCOMP-(IN$(g,x),t)$, where X is the characteristic function of TERM. Let g be the Gödel number of M. Let ϕ,ψ be as defined in Theorem 7.6. Then $\Phi(g,x,t) = \phi(x,t)$ and $\Psi(g,x,t) = \psi(x,t)$. Thus the function f computed by M has fx defined iff there is some t with $\Psi(g,x,t) = 0$ and then $fx = \Phi(g,x,t)$ for any such t. This property could be used instead of Kleene's Normal Form Theorem for most applications. The following theorem is equivalent to Kleene's Normal Form Theorem, as partial recursive functions are exactly those computed by special modular machines.

Theorem 7.9 *There are primitive recursive functions* U:$\mathbf{N} \to \mathbf{N}$ *and* V:$\mathbf{N}^3 \to \mathbf{N}$ *with the following property. If f is the function computed by the* *special modular machine M, and g is the Gödel number of M, then* $fx = U(\mu t\{(V(g,x,t)=0)\})$.

Proof As in Theorem 7.8, we define U to be ΦJ_3^{-1} and $V(g,x,y)$ to be $\Psi J_3^{-1}y + |Ky-g| + |KLy-x|$. Since $J_3^{-1}y = (Ky, KLy, LLy)$ we see that if $V(g,x,y)=0$ then we have $Ky=g$, $KLy=x$, and $\Psi(g,x,t)=0$, where t is defined to be LLy. It then follows that $fx = \Phi(g,x,t) = Uy$. Conversely, if fx is defined there must be some t with $\Psi(g,x,t)=0$ and then $\Phi(g,x,t)=fx$. Hence, defining y to be $J_3(g,x,t)$ we have $fx = Uy$ and $V(g,x,y)=0$.

So we have shown that fx is defined iff there is y with $V(g,x,y)=0$ and that $fx=Uy$ for any such y, and, in particular, for the least such y, as required.■

The following result provides a formal proof of Lemma 2.6, since the parameter g in Kleene's Normal Form Theorem can be taken as the Gödel number of a special modular machine computing the function.

Proposition 7.10 *Let* $F:\mathbf{N}^2 \to \mathbf{N}$ *be partial recursive. Then there is a primitive recursive* $g:\mathbf{N}\to\mathbf{N}$ *such that, for all x, gx is the Gödel number of a special modular machine computing the function* f_x, *where* f_x *is given by* $f_x y = F(x,y)$.

Proof Let T be a Turing machine computing F. Let T have states 3(starting state), ..., r. Let $P(x)$ be a Turing machine that starting with the tape description $\underline{0}u$ ends with $\underline{0}1^x0u$ for any u. We may assume $P(x)$ starts with state 2, ends with state 3 and has $r+1, \ldots, r+x$ as its remaining states. Then f_x is computed by the Turing machine $P(x)T$, and we define gx to be the Gödel number of the modular machine corresponding to $P(x)T$. It is easy to check that g is primitive recursive.■

The **halting set** $H(M)$ of a modular machine M is defined to be the set of those (α,β) for which the computation of M starting with (α,β) is finite. If $(0,0)$ is terminal we let $H_0(M)$ denote the set of those (α,β) for which the computation starting with (α,β) terminates with $(0,0)$.

Theorem 7.11 *(i) For any M, H(M) is r.e. (ii) There is a special modular machine* M_0 *such that* $H(M_0)$ *is not recursive. (iii)* M_0 *can be chosen so that* $H_0(M_0)$ *is not recursive.*

Proof. (i) Let χ be the characteristic function of the terminal set of M. Then $(\alpha,\beta)\in H(M)$ iff there is some t with $\chi\mathrm{Comp}(\alpha,\beta,t)=0$. Since the function $\chi\mathrm{Comp}(\alpha,\beta,t)$ is (primitive) recursive, $H(M)$ is r.e. by Exercise 4.3.

(ii, iii) There is a set K which is r.e. but not recursive (Theorem 4.9, which used Kleene's Normal Form Theorem). Let T_0 be a Turing machine computing the partial characteristic function of K, and let M_0 be a special modular machine corresponding to T_0. Then $x\in K$ iff $\mathrm{In}x\in H(M_0)$, where In is the input function of M_0. Further we can assume that T_0 has a halting state and that this state is the number 0. With this extra assumption $x\in K$ iff

In$x \in H_0(M_0)$. Since In is (primitive) recursive and K is not recursive, both $H(M_0)$ and $H_0(M_0)$ are not recursive.■

Exercise 7.5 Show that the set of Gödel numbers of special modular machines is a primitive recursive set.

Exercise 7.6 Define the Gödel number of an arbitrary modular machine and prove that Theorem 7.9 (with new functions U,V) holds for functions defined by any modular machine.

Exercise 7.7 Suppose T is a Turing machine whose set of states can be divided into two as in the last paragraph of section 6.1. Show that we can define a modular machine M for which Lemmas 7.2 and 7.3 hold but such that M has only one quadruple corresponding to each quintuple of T.

Exercise 7.8 Obtain a formula for the function g of Theorem 7.10 and show that g is primitive recursive. (This is most easily done using the previous exercise.)

8

Church's Thesis and Gödel numberings

In the first section of this chapter we review the evidence that the intuitively computable functions are exactly the partial recursive functions. In the second section we see how to define recursive and r.e. subsets of an arbitrary countable set.

8.1 CHURCH'S THESIS

8.1.1 Church's Thesis

Church's Thesis states that

> the intuitively computable functions are exactly the partial recursive functions.

This is called a thesis, rather than a conjecture, because it is essentially unprovable. That is, anyone who wanted to prove it would have to begin by specifying precisely what was meant by an intuitively computable function. It would then be open to other people to deny that this precise specification defined what they considered to be intuitively computable. Nevertheless, there is a great deal of evidence for Church's Thesis.

We have already seen that partial recursive functions are all intuitively computable. Now, many suggested notions of computable functions have turned out to define the partial recursive functions, even though they are very different in appearance. Other suggested notions of computability have led to subsets of the set of partial recursive functions; in such cases, it is usually easy to accept that certain processes that intuitively lead from computable functions to computable functions do not preserve this class (for instance, minimisation on the set of primitive recursive functions), so that this class must be a proper subset of the set of all computable functions. Also, no definition of computable functions has been suggested which does not make a computable function partial recursive. One might wonder if this was due to our lack of ingenuity in constructing functions. However, it seems that there is a common pattern to most proofs that a suggested definition gives only partial recursive functions. This pattern, which we shall discuss in the following subsections, is sufficiently general that it seems likely that any new suggestions for a definition of computability will fit into the same pattern.

Church's Thesis is more than a philosophical statement about the nature of computability. It is a useful tool in proofs. We shall often find, in the following chapters, that it is easy to give an informal description of a function from which it appears that the function is computable. Again, we may give an informal description of how to decide whether or not an element is in a given set. To show that the function is partial recursive, or that the set is recursive, might involve some long and messy calculations. Such calculations, which are not likely to give any insight as to what is happening, are usually replaced by an appeal to Church's Thesis. That is, since we have given an intuitive argument that the function is computable (or that the set is decidable), we then claim that Church's Thesis tells us that the function is partial recursive. This saves tedious calculations; readers should convince themselves, however, that any time Church's Thesis is used, a formal proof can be made by anyone who is sufficiently industrious. We shall give, in Chapter 12, one detailed example of how an appeal to Church's Thesis can be replaced by a detailed formal argument.

8.1.2 Register programs

We have seen that partial recursive functions, abacus computable functions, Turing computable functions, and functions computed by a modular machine are all the same. We know that very many specific functions are partial recursive (indeed many are primitive recursive). This does suggest that the intuitively computable functions may be partial recursive. On the other hand, abacus machines are so simple that the reader may wish to argue that abacus computable functions could not include all partial recursive functions. In this sub-section, we shall show that functions computed by register programs are partial recursive, and we shall show that certain, apparently more complicated, extensions of register programs also compute only partial recursive functions.

Let R be a register program. We suppose its lines are numbered $0, 1, \ldots, r$. Suppose that, for all n, register n contains the number $x(n)$, and that we are about to perform line i of the program. To this situation corresponds the number $g = 2^i \Pi p_n^{x(n)}$, which we call the code, or Gödel number, of the given configuration. Evidently, g is a code iff $\exp_0 g \leqslant r$. Also, g is a code of a terminal configuration iff $\exp_0 g$ is one of a given number of values (corresponding to the STOP lines). If we wish to use R to compute functions from $\mathbf{N}^k$ to $\mathbf{N}$ we define $\mathrm{In}(x(1), \ldots, x(k))$ to be $2^0 \Pi p_n^{x(n)}$, which is the code of $x(n)$ in register n for all n on line 0. The output must be the contents of register m for some m. Hence we define $\mathrm{Out}g$ to be $\exp_m g$, which is the output corresponding to a configuration with code g.

We must now see how the next step function behaves when we consider the codes. We have a function $\mathrm{Next}: \mathbf{N} \to \mathbf{N}$ such that $\mathrm{Next}g = g$ if g is not a code or if g is the code of a terminal configuration. If g is the code of the non-terminal configuration C then $\mathrm{Next}g$ is the code of the next configuration C'

yielded by C. This will depend on the instructions on line i, where $i = \exp_0 g$. If we look at the definitions, we find that Next is primitive recursive, given by a definition by cases. (Namely, we first need to know the value of $\exp_0 g$; if this is j, we need to look at the instruction on line j. If this instruction is a_n, we need only multiply the code by p_n, but if it is s_n, we need also to consider two cases depending on whether or not $\exp_n g$ is 0. If the instruction is of form J_n, we again have two cases depending on whether or not $\exp_n g$ is 0.)

Thus Comp, which is defined to be the iterate of Next, is also primitive recursive. The function computed by R is OutComp(In**x**,t) for any t for which Comp(In**x**,t) is terminal. As in the previous chapter, this shows the function computed by R is partial recursive.

Now consider a more complicated type of register program. We might permit lines whose instructions are

replace x_n by $f(x_{m(1)}, \ldots, x_{m(q)})$ for some function f, or
Test whether or not property P holds for $(x_{m(1)}, \ldots, x_{m(q)})$; if so, go to line j, and if not go to line j'.

Here the function f must be taken from some set of initially given functions, and the property P from some set of initially given properties. Since the function f can be computed, and the property P can be tested, in one step of the computation, they must be fairly simple (for instance, we might allow f to be addition, multiplication, or exponentiation, and we might allow P to be the property 'divides' or 'between'). Bearing in mind that the set of primitive recursive functions is quite complex, it seems reasonable to assume that all these functions and properties are primitive recursive.

Making this assumption, the function Next will be given by a definition by cases which will be more complex than before. But, because each f and P is primitive recursive, Next will still be primitive recursive. The function computed by the program will be given by the same expression as before, and so it will still be partial recursive.

We can go further and add lines such as

check the contents of register n; if this is m, perform a_m, or
check the contents of register n; if this is m, go to line m.

The function Next will still be given by a definition by cases, of a yet more complicated form (if the first of these, for instance, occurs on line i, the corresponding case is 'if $\exp_0 g = i$ and $\exp_n g = m$ then multiply by p_m'). However, Next is still primitive recursive, and so the function computed by this program is still partial recursive.

8.1.3 Other models of computation

Both modular machines and register programs fit into the following model of computation. Turing machines also fit into this model, as do the generalisations of Turing machines which allow several tapes and several heads on each tape.

We are given a set Γ of configurations. Each C in Γ has a code, which is in

$\mathbf{N}$ (or possibly in $\mathbf{N}^m$ for some m). The set of codes is recursive. Certain configurations are called terminal. These are reasonably easy to recognise, and so the set Term of codes of terminal configurations is (primitive) recursive. If C is not terminal, it yields a next configuration C'. This function is reasonably simple, and so the function Next is partial recursive, where Next is defined by Next$g = g'$ iff g is the code of C and C yields C' whose code is g' (alternatively, we may extend Next in some convenient way on terminal codes and non-codes so as to make Next total and recursive, or even primitive recursive).

To compute functions from $\mathbf{N}^k$ to $\mathbf{N}$ using this computing system we need an input function from $\mathbf{N}^k$ to Γ and an output function from a subset of Γ to $\mathbf{N}$. These should be fairly simple, so we would expect In$:\mathbf{N}^k \rightarrow \mathbf{N}$, given by In$\mathbf{x}$ is the code of the input, to be recursive, and Out$:\mathbf{N} \rightarrow \mathbf{N}$, given by Out$g$ is the output of C if g is the code of C, to be partial recursive. (Frequently, the output is defined on all configurations; if so, we may wish to define Out in some convenient way for non-codes, so that the resulting total function is recursive, or even primitive recursive.)

The assumption that the relevant functions are all partial recursive is very reasonable. It just amounts to requiring that each individual step in the computation is fairly simple. Hence many suggested models of computation fit into this pattern, and it seems likely that any model proceeding by single steps will fit this pattern.

With the above assumptions, the function computed by this system is OutComp(In$\mathbf{x}$,s), where $s = \mu t$(Comp(In$\mathbf{x}$,t) is in Term). Here Comp is the iterate of Next, so Comp is partial recursive. As before, this makes the function computed by the system a partial recursive function.

Our original abacus machines do not fit into this approach, since they do not work by single steps. But we saw directly that they compute partial recursive functions only. Again, one can reasonably hope that any such model will have a fairly simple way of generating new computer programs from old ones. If so, as in the case of abacus machines, it seems likely that the function computed by such a system is obtained from functions computed by simpler systems using at most composition, iteration, and minimisation. The starting computers are likely to be simple enough that they compute only (primitive) recursive functions. Thus, inductively, the function computed by such a system is likely to be partial recursive.

We cannot claim any general argument that all partial recursive functions are computed by any of the systems we have just looked at. In each individual case it is necessary to see whether or not all partial recursive functions can be computed. Usually this is done by showing that an approach that is known to compute all partial recursive functions can be mimicked in the system that we want to look at.

8.1.4 Non-determinism

When we considered Turing machines and modular machines, we insisted that no pair began more than one quadruple. If we remove this condition we would not have a unique next configuration to a given one. However, we could still refer to a computation as being a sequence of configurations in which each could be reached in one step from the previous one.

For Turing and modular machines, this non-determinism is inessential (although it becomes very important in the theory of complexity of computation); that is, deterministic machines are as natural to look at as non-deterministic ones. Some other models of computation are essentially non-deterministic, in the sense that any attempt to make them deterministic is unnatural. We shall show that non-deterministic models of computation still compute only partial recursive functions.

The coding, and Term and In are just as before. Next now has to be replaced by a subset NEXT of $\mathbf{N}^2$ such that (g,g') is in NEXT iff g and g' are the codes of configurations C and C' such that C' is one of the possibilities for the next configuration of C. As before, if the possibilities are reasonably simple, NEXT is recursive.

It is not convenient to use such a model directly to compute a function. For then we would probably need to look at all computation starting with In$\mathbf{x}$ and check that those which terminated all gave the same output. This would raise problems, as we would have to look at an unknown number of computations.

Instead, we use the system to *accept* inputs. We take a subset of the terminal configurations, which we call the accepting subset. This should be simple enough that its set of codes is recursive. We can then say that a subset S of $\mathbf{N}^k$ is accepted by the system if $\mathbf{x} \in S$ iff there is at least one computation starting from In$\mathbf{x}$ which terminates with an accepting configuration. We can then say that a function is computed by the system if its graph is accepted. To show that computable functions are partial recursive, it is enough to show that accepted sets are r.e.

We cannot conveniently iterate NEXT, so we use a different technique. We have a coding of the configurations. This leads to a coding of strings of configurations. Namely, if $g(i)$ is the code of C_i for $i = 1, \ldots, n$, the string $C_1, \ldots, C_n$ is coded by $2^n \Pi p_i^{g(i)}$. Not all integers code strings, but the set of codes of strings is recursive (either by Church's Thesis, or by a direct proof in the next section). Then G is the code of a string which is a computation which terminates in an accepting configuration iff G codes a string and, with n being $\exp_0 G$, $\exp_n G$ is the code of an accepting string, and for all $i < n$ $(\exp_i G, \exp_{i+1} G)$ is in NEXT. Hence the set A of these codes is recursive. Finally, the set accepted by this system is $\{\mathbf{x} \in \mathbf{N}^k; \exists G \text{ with } \exp_1 G = \text{In}\mathbf{x} \text{ and } G \text{ in } A\}$. This set is r.e., since it is obtained by existential quantification of a recursive set.

8.2 GÖDEL NUMBERINGS

Let X be a countable set. We wish to say what is meant by recursive and r.e. subsets of X. This is done by taking a bijection $\phi: X \to \mathbf{N}$, and defining the subset A to be recursive (or r.e.) if ϕA is recursive or r.e. In particular, it is clear that A is recursive iff both A and $X - A$ are r.e., because the similar property holds for $\mathbf{N}$.

Plainly which sets are recursive or r.e. depends on the choice of ϕ. If ψ is another bijection they define the same recursive and r.e. subsets iff $\psi\phi^{-1}:\mathbf{N}\to\mathbf{N}$ is recursive (in which case its inverse $\phi\psi^{-1}$ is also recursive). If we have two countable set X and Y, and corresponding bijections ϕ_X and ϕ_Y, we say $f:X\to Y$ is partial recursive if $\phi_Y f\phi_X^{-1}$ is partial recursive. In the cases we look at, there will be several natural possibilities for the bijections (or, rather, for related injections); these will all give the same notions of recursive and r.e. sets and partial recursive functions.

Definition A **Gödel numbering** of the countable set X is a one–one map from X into $\mathbf{N}$ such that ϕX is recursive.

Let ϕ be a Gödel numbering of X. Then we know, by Proposition 4.7, that there is a strictly increasing recursive function f from $\mathbf{N}$ onto ϕX. Then $f^{-1}\phi$ will be a bijection from X onto $\mathbf{N}$. We use this bijection to define recursive and r.e. subsets of X. This would not be particularly useful if we always had to refer to the function f. However, we shall see that we need only look at ϕ.

Proposition 8.1 *Let ϕ be a Gödel numbering of X. Then the subset A of X is r.e. iff ϕA is r.e., and is recursive iff ϕA is recursive.*

Proof By definition A is r.e. iff $f^{-1}\phi A$ is r.e., where f is as above. Since f is recursive, $f^{-1}\phi A$ will be r.e. if ϕA is r.e., and $f(f^{-1}\phi A)$ will be r.e. if $f^{-1}\phi A$ is r.e. Since f maps onto ϕX, $\phi A = f(f^{-1}\phi A)$. Hence A is r.e. iff ϕA is r.e.

Now A is recursive iff A and $X-A$ are r.e. By the above this holds iff ϕA and $\phi X-\phi A$ are r.e. Similarly ϕA is recursive iff both ϕA and $\mathbf{N}-\phi A$ are r.e. At this point we use the fact that ϕX is recursive, and so both ϕX and $\mathbf{N}-\phi X$ are r.e.

As $\phi X-\phi A=\phi X\cap(\mathbf{N}-\phi A)$, and ϕX is r.e., we see that $\phi X-\phi A$ is r.e. if $\mathbf{N}-\phi A$ is r.e. Conversely, as $\mathbf{N}-\phi A=(\mathbf{N}-\phi X)\cup(\phi X-\phi A)$, and $\mathbf{N}-\phi X$ is r.e., we see that $\mathbf{N}-\phi A$ is r.e. if $\phi X-\phi A$ is r.e.■

Readers might well think that a further condition would be useful in the definition of Gödel numberings. Namely, it seems reasonable at first sight to require that ϕ is intuitively computable. But this is an illusion. The only way we could make this precise is if we knew what was meant by a recursive function from X to $\mathbf{N}$. But the whole purpose of ϕ is to talk about recursiveness for X, so we would not need ϕ if we already knew about this.

We now give some examples of Gödel numberings.

Let n be any positive integer. Then any positive integer a can be written uniquely as $\Sigma\alpha_i n^i$, where $1\leq\alpha_i\leq n$ for all i. This is a Gödel numbering of the set of all strings on $\{1,\ldots,n\}$. Let S be any set of cardinality n. Then any bijection from S to $\{1,\ldots,n\}$ gives a bijection from the strings on S to the strings on $\{1,\ldots,n\}$, and hence gives a Gödel numbering of the strings on S. A different bijection will give a different Gödel numbering, but will give the same recursive and r.e. subsets. The easiest way of showing that the relevant map is recursive is to use Church's Thesis; a direct proof is indicated in the exercises.

We have frequently looked at a Gödel numbering of the set of strings on **N**, namely the map sending $(x(1),\ldots,x(n))$ to $2^n\Pi p_i^{x(i)}$. An alternative Gödel numbering is to send the string to $\Sigma 2^{y(i)}$, where $y(i)$ is $x(1)+\ \ldots\ +x(i)+i-1$. Again, these Gödel numberings can be shown to give the same recursive and r.e. subsets, either by using Church's Thesis or by an explicit proof.

Given any bijection from a countable set S to **N**, we get a bijection from the strings on S to the strings on **N**, and hence a Gödel numbering of the strings on S. This will depend on the bijection used. Two bijections ϕ and ψ will give the same recursive subsets iff $\psi\phi^{-1}:\mathbf{N}\rightarrow\mathbf{N}$ is recursive. Another Gödel numbering of the strings on S is obtained by writing the members of S as $a_0, a_1, \ldots$ (using a bijection from S to **N**), and then mapping the set of strings on S into the set of strings on the two-element set $\{a,'\}$ by sending a_0 to a, a_1 to a', a_2 to a'', and so on. This gives the same recursive and r.e. sets as before.

Proposition 8.2 *Let X be a countable recursive subset of a countable set Y. Then a subset A of X is recursive (or r.e.) as a subset of X iff it is recursive (or r.e.) as a subset of Y.*

Proof Since we are talking about recursive subsets of Y, we are implicitly referring to a bijection ϕ from Y to **N**. As X is a recursive subset of Y, by definition ϕX is a recursive subset of **N**. Hence ϕ can be regarded as a Gödel numbering of X. Now, by definition, A is recursive (or r.e.) as a subset of Y iff ϕA is recursive (or r.e.). The result follows from the previous proposition.■

The above result will be useful later. It enables us to use such phrases as 'A is recursive' without worrying too much about which set A is regarded as a subset of.

Proposition 8.3 *Let A be a recursive subset of X. Then the strings on A are a recursive subset of the strings on X.*

Proof We have a bijection from X to **N**. Let B be the image of A under this bijection. Then it is enough to show that the strings on B form a recursive subset of the strings on **N**.

Take the standard Gödel numbering of the strings on **N**, using the prime decomposition. Then g is a Gödel number of a string on B iff it is a Gödel number of some string and $\forall i\leqslant \exp_0 g(i{=}0\vee\exp_i g\in B)$, which is clearly a recursive property.■

If we want a Gödel numbering of the finite subsets of **N** there are three obvious candidates, which give the same recursive and r.e. subsets of this set. One is to regard a finite set $\{x(1),\ldots,x(n)\}$ as having its elements given in increasing order, and then identifying the subset with the corresponding string. Another is to take the Gödel number of the subset as the smallest Gödel number of any string obtained by taking the elements of the set in some order. The third is to let the Gödel number of this subset be $\Sigma 2^{x(i)}$. With any of these, we can extend this to a Gödel numbering of the finite subsets of a countable set S. We can also prove that if A is a recursive subset of S then the finite subsets of A form a recursive subset of the set of finite subsets of S.

Exercise 8.1 For any x, write $x+1$ as $\alpha_0+\alpha_1 n+ \ldots +\alpha_r n^r$, where n is a fixed positive integer and $1 \leqslant \alpha_i \leqslant n$. Show that r is a primitive recursive function of x, and that α_i is a primitive recursive function of x and i. Use this to show that any two bijections from a set A to $\{1,\ldots,n\}$ give the same recursive and r.e. subsets of the strings on A.

Exercise 8.2 Let A be a r.e. subset of the countable set X. Show that the strings on A are a r.e. subset of the strings on X.

9

Hilbert's Tenth Problem

Hilbert, in an address to the International Congress of Mathematicians in 1900, gave a list of problems he considered important. The interest aroused by this list led to several developments in mathematics, and many of the problems have now been solved. The one that concerns us is the tenth on his list, in which he asked if it was possible to decide for an arbitrary polynomial with integer coefficients whether or not it had an integer solution. We shall see that there is no such decision procedure.

The material in this chapter provides an alternative proof of Kleene's Normal Form Theorem, with a stronger conclusion. In this approach we do not need to look at Turing machines and modular machines, though these are of interest in their own right. We will also be able to give two different proofs of the Undecidability and Incompleteness Theorems, one using modular machines and the other using diophantine sets.

9.1 DIOPHANTINE SETS AND FUNCTIONS

Definition An **exponential polynomial** in the variables $x(1), \ldots, x(n)$ is a polynomial (in the ordinary sense; polynomials are required to have integer coefficients, but these may be negative) in the variables $x(1), \ldots, x(n)$, $y(1), \ldots, y(m)$, where each $y(i)$ is either of the form $x(k)^{x(j)}$ for some k and j (depending on i) or of the form $r^{x(j)}$ for some j where r is a natural number (and j and r depend on i). If each $y(i)$ is $r^{x(j)}$ for some r the exponential polynomial is called a **unary exponential polynomial**, while if each $y(i)$ is $2^{x(j)}$ for some j it is called a **base-2 exponential polynomial**.

Definition The subset A of $\mathbf{N}^k$ is called **diophantine** if there is a polynomial P such that $\mathbf{x} \in A$ iff $\exists \mathbf{y}(P(\mathbf{x}, \mathbf{y}) = 0)$. Also A is called **exponential diophantine**, **unary exponential diophantine**, or **base-2 exponential diophantine** if there is P which is, respectively, an exponential polynomial or a unary exponential polynomial or a base-2 exponential polynomial such that $\mathbf{x}\in A$ iff $\exists\mathbf{y}(P(\mathbf{x},\mathbf{y}) = 0)$.

A property is called diophantine if the set it defines is diophantine, while a function is diophantine if its graph is diophantine. Similar conventions apply to the other concepts.

Before we can give many examples, we need to show some simple properties. Plainly the condition $P{=}0 \vee Q{=}0$ can be written as $PQ{=}0$, while the condition $P{=}0 \wedge Q{=}0$ can be written as $P^2{+}Q^2{=}0$. If P and Q are polynomials, exponential polynomials, unary exponential polynomials, or base-2 exponential polynomials so are PQ and $P^2{+}Q^2$. In future only the polynomial case will be mentioned explicitly; the other cases will be equally obvious.

The conditions $\exists\mathbf{y}(P(\mathbf{x},\mathbf{y}){=}0) \vee \exists\mathbf{y}(Q(\mathbf{x},\mathbf{y}){=}0)$ can be written as $\exists\mathbf{y}(P{=}0 \vee Q{=}0)$, while $\exists\mathbf{y}(P(\mathbf{x},\mathbf{y}){=}0) \vee \exists\mathbf{z}(R(\mathbf{x},\mathbf{z}){=}0)$ can be written as $\exists\mathbf{y},\mathbf{z}(P{=}0 \vee R{=}0)$; consequently the union of diophantine sets is diophantine. The condition $\exists\mathbf{y}(P(\mathbf{x},\mathbf{y}){=}0) \wedge \exists\mathbf{y}(Q(\mathbf{x},\mathbf{y}){=}0)$ is the same as $\exists\mathbf{y}(P(\mathbf{x},\mathbf{y}){=}0) \wedge \exists\mathbf{z}(Q(\mathbf{x},\mathbf{z}){=}0)$ and so can be written as $\exists\mathbf{y}, \mathbf{z}(P^2(\mathbf{x},\mathbf{y}){+}Q^2(\mathbf{x},\mathbf{z}){=}0)$; hence the intersection of diophantine sets is diophantine. Further, the condition $\exists\mathbf{y}(P(\mathbf{x},\mathbf{y}){=}0 \wedge \mathbf{y}\in B)$, where B is a diophantine set, can be written as $\exists\mathbf{y}(P{=}0 \wedge \exists\mathbf{z}(Q(\mathbf{y},\mathbf{z}){=}0))$, which is the same as $\exists\mathbf{y},\mathbf{z}(P(\mathbf{x},\mathbf{y}){=}0 \wedge Q(\mathbf{y},\mathbf{z}){=}0)$.

Let A be a diophantine subset of $\mathbf{N}^k$, B a diophantine subset of $\mathbf{N}^r$, and f a diophantine function from $\mathbf{N}^k$ to $\mathbf{N}^r$. Then fA and $f^{-1}B$ are diophantine, since they are, respectively, $\{\mathbf{z};\exists\mathbf{y}(\mathbf{y}\in A \wedge \mathbf{z}{=}f\mathbf{y})\}$ and $\{\mathbf{x};\exists\mathbf{u}(\mathbf{u}\in B \wedge \mathbf{u}{=}f\mathbf{x})\}$.

One might expect that there are more complicated sets, satisfying conditions in which there are expressions of form P^Q, where P and Q are polynomials, or even more complicated systems of exponents upon exponents. However, a condition of form $\ldots P^Q \ldots$ is the same as the condition $\exists u,v(\ldots u^v \ldots \wedge u{=}P \wedge v{=}Q)$, where u and v are two new variables. Proceeding like this, any such complicated condition can be reduced to an exponential diophantine condition.

Examples The set of positive integers is diophantine, being $\{x;\exists y(x{=}y{+}1)\}$.

The predicates $\leqslant$ and $<$ are diophantine, since $x{\leqslant}y$ iff $\exists z(y{=}x{+}z)$, while $x{<}y$ iff $\exists z(y{=}x{+}z{+}1)$.

The predicate $a{\equiv}b \bmod c$ is diophantine, since it can be written as $\exists x((a{-}b)^2{=}c^2x^2)$.

The function rem is diophantine, since $\text{rem}(m,n){=}r$ iff $\exists q(n{=}mq{+}r) \wedge r{<}m$. Similarly quo is diophantine.

The functions J and J_k are diophantine. More generally, it is easy to check that if P is any polynomial with rational coefficients such that the values of P are always natural numbers (when the variables are natural numbers) then P is diophantine.

Theorem 9.1 (Main Theorem on diophantine sets) *Let A be a subset of* $\mathbf{N}^k$. *Then the following are equivalent*:

(1) *A is r.e.*,
(2) *the partial characteristic function of A is partial recursive*,
(3) *A is exponential diophantine*,
(4) *A is base-2 exponential diophantine*,
(5) *A is diophantine*.

Start of proof Observe first that it is enough to prove the theorem when $k=1$. For A will satisfy any of (1) to (5) iff $J_k A$ satisfies the same property, since J_k is diophantine.

Plainly, (3), and so also (4) and (5), implies (1). That (1) implies (2) was shown in Proposition 4.4. The same result also showed that (2) implied (1). However, this latter fact used Theorem 4.2. The material in this chapter will provide an alternative approach to Theorem 4.2 and Kleene's Normal Form Theorem. It is therefore important to notice that we actually prove that (2) implies (4). We shall soon see that this result is enough to prove Theorem 4.2.

The proof that (2) implies (4) will occupy the next two sections. In section 9.4 we shall prove that the exponential function is diophantine, which will complete the proof of the Main Theorem. For the moment, we will abandon the proof of the Main Theorem, and we will establish some important consequences of it.

Theorem 9.2 *Let* $f:\mathbf{N}\rightarrow\mathbf{N}$ *be partial recursive. Then the partial characteristic function of the graph of f is partial recursive.*

Proof The function g given by $g(m,n)=\mu r(r+|n-fm|=0)$ is partial recursive. If fm is not defined, then $g(m,n)$ is not defined. If fm is defined then $g(m,n)$ is defined iff $n=fm$, since otherwise there is no suitable r. Hence g is the required partial characteristic function.■

We can now prove a stronger form of Theorem 4.2.

Theorem 9.3 *Let* $f:\mathbf{N}\rightarrow\mathbf{N}$ *be partial recursive. Then there is a primitive recursive function* $v:\mathbf{N}^2\rightarrow\mathbf{N}$ *such that* $fm=K(\mu r(v(x,r)=0))$.

Proof By the previous lemma and the Main Theorem, the graph of f is diophantine. (In fact it is enough to know this graph is exponential diophantine; hence this theorem can be proved without using section 9.4.) Thus there is a polynomial P in $k+2$ variables for some k such that $n=fm$ iff there is $\mathbf{x}$ with $P(m,n,\mathbf{x})=0$. As P does not map into $\mathbf{N}$, we do not want to refer to P itself as primitive recursive. However, there are polynomials P_1 and P_2 with non-negative coefficients such that $P=P_1-P_2$, and then $|P|=|P_1-P_2|$ is plainly primitive recursive.

Let $v(m,r)$ be $|P(m,J_{k+1}^{-1}r)|$. Then v is primitive recursive. Since the first component of J_{k+1}^{-1} is K, we see that if $v(m,r)=0$ then $Kr=fm$. Conversely, if $fm=n$, take a suitable $\mathbf{x}$ and define r to be $J_{k+1}(n,\mathbf{x})$, and we will have $v(m,r)=0$. Thus $fm=Kr$ for any r such that $v(m,r)=0$, and, in particular, this holds for the least such r, as required.■

Theorem 9.4 *The set of polynomials P with integer coefficients for which $P=0$ has a solution in natural numbers is undecidable.*

Proof Let K be a set which is r.e. but not recursive. By the Main Theorem K is diophantine, so there is a polynomial P such that $k \in K$ iff there are $x_1, \ldots, x_n$ with $P(k,x_1,\ldots,x_n)=0$. If we could decide for all polynomials whether or not they had a solution in natural numbers, then, in particular, we could decide for any k whether or not $P(k,x_1,\ldots,x_n)=0$ had a solution, and so we could decide whether or not k was in K. Since K is not recursive, this is impossible.■

Notice that, although this result is nearly a negative solution to Hilbert's Tenth Problem, it is not quite what is needed. For Hilbert's Problem asks about integer solutions, and we have only answered about non-negative solutions.

However, we can easily derive the result we want. It is a theorem of Lagrange (a proof is given in the last section of this chapter) that every natural number is the sum of four squares. Take any polynomial P. Obtain from it a polynomial Q by replacing each variable x_i by $u_i^2+v_i^2+y_i^2+z_i^2$, where u_i, v_i, y_i, and z_i are new variables. Then $P=0$ has a solution in natural numbers iff $Q=0$ has an integer solution. Since we cannot decide whether or not an arbitrary polynomial has a solution in natural numbers, we cannot decide whether or not an arbitrary polynomial has an integer solution. This provides the negative answer to Hilbert's Tenth Problem.

We can now provide a striking characterisation of r.e. sets.

Theorem 9.5 *Let A be any r.e. subset of* **N**. *Then there is a polynomial P such that A is the set of non-negative values of P.*

Proof There is a polynomial Q such that $n \in A$ iff there is $\mathbf{x}$ with $Q(n,\mathbf{x})=0$. Let $P(n,\mathbf{x})$ be $(n+1)(1-Q^2)-1$. If $Q(n,\mathbf{x}) \neq 0$, plainly $1-Q^2 \leqslant 0$, and so $P(n,\mathbf{x})$ is negative. Thus the non-negative values of P come from those $(n,\mathbf{x})$ with $Q(n,\mathbf{x})=0$. By the definition of Q, given n there is such an $\mathbf{x}$ iff $n \in A$, and for such an n and $\mathbf{x}$ we have $P(n,\mathbf{x})=n$, giving the result.■

It is an interesting problem to find suitable polynomials when A is given. For instance, we could ask for a polynomial whose non-negative values were exactly the prime numbers. Explicit formulae for such polynomials are known.

9.2 CODING COMPUTATIONS

Let A be a subset of **N** whose partial characteristic function is partial recursive. Then this function is abacus computable. Since the value of the function is zero on its domain, by Lemma 5.4 the abacus machine can be chosen so that the output (if any) of the computation has all registers empty. By Proposition 5.13, this abacus machine can be replaced by a register program performing the same computation. Further, the register program can be chosen so that its only STOP instruction is on the last line, any line $j:J_i(b,c)$ has b and c different from j, and so that an instruction s_k is never applied to an empty register. Let the instructions be numbered $1,\ldots,r$. Since only finitely many registers are used, we may assume the registers used are numbered $1,\ldots,m$.

Consider the computation starting with n in the first register, and all other registers empty. Let x_{it} be the contents of register i at time t. Let p_{jt} be 1 if the computation is on line j at time t and 0 otherwise.

Suppose $n \in A$. Then the computation halts at some time s, and the numbers x_{it} and p_{jt} are defined for $t \leqslant s$. They satisfy the following conditions.

$$\Sigma_j p_{jt} = 1 \text{ for all } t. \tag{1}$$

This equation is plainly equivalent to saying that for all t there is exactly one j with $p_{jt}=1$, the rest being 0.

$$p_{10} = 1. \tag{2}$$

$$p_{rs} = 1, \text{ and } p_{rt} = 0 \text{ for } t \neq s. \tag{3}$$

$$x_{10} = n, \text{ and } x_{i0} = 0 \text{ for } i \neq 1. \tag{4}$$

The condition relating x_{it} and $x_{i,t+1}$ can be written as

$$x_{i,t+1} = x_{it} + \Sigma' p_{jt} - \Sigma'' p_{jt}, \tag{5}$$

where Σ' denotes the sum over those j for which the instruction on line j is a_i, and Σ'' is the sum over those j for which the instruction on line j is s_i. This uses the assumption that an instruction s_i is performed only if register i is not empty.

The remaining conditions are, for any j for which the instruction on line j is either a_i or s_i for some i,

$$p_{j+1,t+1} = 1 \text{ provided } p_{jt} = 1, \tag{6}$$

while if line j is $J_i(b,c)$, we have the condition

$$\text{if } p_{jt} = 1 \text{ then } p_{b,t+1} = 1 \text{ if } x_i = 0 \text{ and } p_{c,t+1} = 1 \text{ if } x_i \neq 0. \tag{7}$$

Conversely, suppose conditions (1) to (7) hold. Then, by induction on t, it is easy to see that, for $t \leqslant s$, the contents of register i at time t is x_{it}, and that $p_{jt} = 1$ iff the computation is on line j at time t. In particular, the computation starts with n in register 1, the remaining registers being empty, and reaches line r (and so halts) at time s. Hence n is in A.

At first sight, conditions (1) to (7) might appear to be exactly what we need. However, the number of variables x_{it} and p_{jt} depends on s, which is itself a variable. We need a fixed number of variables. This is achieved by regarding the x_{it} and p_{jt} as the digits in the Q-ary expansion of numbers X_i and P_j, where Q is a large power of 2.

Thus we put

$$q = n+r+s+1, \text{ and } Q = 2^q. \tag{8}$$

We also define I by

$$1+(Q-1)I = Q^{s+1}. \tag{9}$$

It follows that

$$I = 1+Q+ \ \ldots \ +Q^s.$$

We define X_i to be $\Sigma x_{it}Q^t$ and P_j to be $\Sigma p_{jt}Q^t$. Conversely, given numbers X_i and P_j, we define x_{it} and p_{jt} by these formulas. Notice, though, that if X_i and P_j are given, there is no guarantee that x_{it} and p_{jt} are zero for $t>s$. This will, however, follow from the conditions we use.

We will need an auxiliary relation $\preccurlyeq$. Let the numbers a and b be written in binary notation as $a = \Sigma\alpha_i 2^i$ and $b = \Sigma\beta_i 2^i$. Then we write $a \preccurlyeq b$ iff $\alpha_i \leqslant \beta_i$ for all i. Notice that, for any c, $a \preccurlyeq 2^c - 1$ iff $a \leqslant 2^c - 1$, because all the binary digits of $2^c - 1$ are 1. Because Q is a power of 2, it is easy to change from the Q-ary representation of a number to its binary representation, and vice versa.

The property that each p_{jt} is 0 or 1 is equivalent to

$$P_j \preccurlyeq I. \tag{10}$$

We can then express (1) as

$$\Sigma P_j = I. \tag{11}$$

This uses the fact that, in forming ΣP_j, there is no carry from one power of Q to a higher power, since in forming the sum we add at most r terms equal to 1 in each power of Q, and $r<Q$.

Condition (2) can be written as

$$1 \preccurlyeq P_1\,, \tag{12}$$

while (3) can be written as

$$P_r = 2^s. \tag{13}$$

Condition (6) is evidently equivalent to

$$QP_j \leqslant P_{j+1}. \tag{14}$$

Conditions (5) and (7) are harder to express than the others. Suppose first that the x_{it} come from a computation. Then, for all t, $x_{i,t+1} \leqslant x_{it} + 1$. Hence $x_{it} \leqslant n + s$, and so, by the definition of Q,

$$X_i \leqslant (Q/2 - 1)I. \tag{15}$$

Conversely, if (15) holds, we have $x_{it} = 0$ for $t > s$, and $x_{it} \leqslant Q/2 - 1$ for $t \leqslant s$. This fact will be useful later.

Now consider the equations

$$X_1 = QX_1 + n + \Sigma' QP_j - \Sigma'' QP_j, \tag{16}$$

$$X_i = QX_i + \Sigma' QP_j - \Sigma'' QP_j \text{ for } i \neq 1, \tag{17}$$

where Σ' and Σ'' have the previously defined meanings (which depend on i). The right-hand side of (17) can be written as $\Sigma Q^{t+1}(x_{it} + \Sigma' p_{jt} - \Sigma'' p_{jt})$, and similarly for (16). Consequently, if (4) and (5) hold, (16) and (17) will also hold. The converse is not so obvious, since we want the above expression to be the Q-ary expansion of X_i, and we have to worry about whether or not there is any carry, either from a higher power of Q or to a higher power. Since $\Sigma' p_{jt}$ is 0 or 1, and $x_{it} \leqslant (Q/2 - 1)$, there can be no carry to a higher power. There will be no carry from a higher power provided that $\Sigma'' p_{jt}$, which can only be 0 or 1, is 1 only if $x_{it} \neq 0$. This certainly holds if x_{it} and p_{jt}, defined from X_i and P_j, are actually the data of a computation.

We show, by induction on t, that the conditions (8) to (20), where (18) to (20) are defined shortly, ensure that this property holds. Suppose the property holds for all $t \leqslant u$. Then (16) and (17) will give, for all i, that $x_{i,t+1} = x_{it} + \Sigma' p_{jt} - \Sigma'' p_{jt}$, for all $t \leqslant u$. In particular, $x_{i,u+1}$ is the contents of register i at time $u+1$, as needed.

Finally we need to consider condition (7). If $b = c$ this is replaced by

$$QP_j \leqslant P_b. \tag{18}$$

Otherwise it is replaced by

$$QP_j \leqslant P_b + P_c, \tag{19}$$

$$QP_j \leqslant P_c + QI - 2X_i. \tag{20}$$

Now (19) amounts to saying that if $p_{jt}=1$, then either $p_{b,t+1}$ or $p_{c,t+1}$ is 1 (bearing in mind that at most one of them is 1, by (11)).

We now have to consider (20). We know that (1) to (7), in the presence of (8) and (9), imply (10) to (19). Thus we have to show that (20) is equivalent to (7), given (8) to (19). Suppose we have u such that, for $t \leqslant u$ and all i and j, x_{it} and p_{jt} come from a computation. Consider a line $j:J_i(b,c)$, with b and c different from j, and look at the relevant formulas (7) and (20). If $p_{ju}=0$, (20) imposes no condition on the coefficient of Q^{u+1} on the right-hand side, and (7) also imposes no condition. Hence we assume $p_{ju}=1$. As $c \neq j$, we have $p_{cu}=0$. We also know, from (16) and (17), that $x_{i,u+1}=x_{iu}$.

The right-hand side of (20) is $\Sigma Q^t(p_{ct}+Q-2x_{it})$. We know, from (15), that $x_{it}<Q/2$, so that all the coefficients are non-negative. It is then easy to check that, because $p_{cu}=0$, the only terms which can contribute to the coefficient of Q^{u+1} in the Q-ary expansion of the right-hand side of (20) are the terms $Q^u(Q-2x_{iu})$ and $Q^{u+1}(p_{c,u+1}+Q-2x_{i,u+1})$. Hence if $x_{iu}=0$, this coefficient is $1+p_{c,u+1}$. Now the relation $\ll$ is a condition on the binary expansions, not on the Q-ary expansion. As $p_{ju}=1$, the coefficient of Q^{u+1} in the binary expansion of the right-hand side of (20) must be 1. This is equivalent to $p_{c,u+1}=0$, which, by (19), amounts to $p_{b,u+1}=1$.

If $x_{iu}\neq 0$, then the coefficient of Q^{u+1} in the Q-ary expansion is $p_{c,u+1}+Q-2x_{i,u+1}$, and hence the coefficient of Q^{u+1} in the binary expansion is $p_{c,u+1}$. Hence (20) amounts to saying that $p_{c,u+1}=1$ if $x_{iu}\neq 0$. So we have shown that (20) and (7) are equivalent, as needed.

We have now shown that $n\in A$ iff the conditions (8) to (20) hold. These are conditions on a fixed number of variables. However they involve the relation $\ll$. In the next section we express this relation by a base-2 exponential diophantine condition, so showing that A is base-2 exponential diophantine.

9.3 REMOVAL OF THE RELATION $\ll$

Lemma 9.6 *$a \ll b$ iff rem$(2, \binom{b}{a})=1$, where $\binom{b}{a}$ is the binomial coefficient.*

Proof We write $P\equiv Q$, where P and Q are polynomials, to mean that every coefficient of $P-Q$ is divisible by 2. By an easy induction, starting from $(1+x)^2=1+x^2+2x$, we see that, for all i,

$$(1+x)^{2^i} \equiv 1+x^{2^i}, \text{ and hence } (1+x)^b \equiv \Pi(1+x^{2^i})^{\beta_i},$$

where $b=\Sigma\beta_i 2^i$. It is easy to see that the term x^a occurs in the product on the right iff $a\ll b$. If it occurs its coefficient is plainly 1. Since the coefficient of x^a on the left side is the binomial coefficient $\binom{b}{a}$, the result follows.■

Lemma 9.7 *Let* $u = 2^{b+1}$. *Then* $\binom{b}{a} = rem(u, quo(u^a, (u+1)^b))$.

Proof We know that $(u+1)^b = \Sigma\binom{b}{i}u^i$. The result follows if we can show that $\Sigma^{i<a}\binom{b}{i}u^i < u^a$. It is plainly enough to show that $\Sigma^{i \leqslant b}\binom{b}{i} < u$. But this is immediate, since the sum is just $(1+1)^b$.■

We have now shown that the relation $\leqslant$ is exponential diophantine, and hence we have shown that, in the Main Theorem, (2) implies (3). We shall now show that the function x^y is base-2 exponential diophantine, thus showing that (2) implies (4).

Lemma 9.8 *The function* x^y *is base-2 exponential diophantine*

Proof If $y > 1$, we see that $x^y < 2^{xy} - x$. Also $2^{xy} \equiv x \bmod (2^{xy} - x)$. It follows, raising this congruence to the yth power, that $x^y = z$ iff

$$(y = 0 \wedge z = 1) \vee (y = 1 \wedge z = x) \vee \exists u, v(y > 1 \wedge u + x = 2^{xy} \wedge v = xy^2 \wedge rem(u, 2^v) = z),$$

which gives the result.■

In the next section we show that x^y is diophantine (it turns out that it is no easier to show just that 2^y is diophantine). We use methods of elementary number theory. The techniques used are not at any great depth; however they are rather intricate (in the expository paper by Davis which I follow for most of the next section he calls this material 'Twenty-four easy lemmas'). Readers who are primarily interested in the use of the Main Theorem to prove results about recursive functions could omit this section, and use only the fact that r.e. sets are base-2 exponential diophantine, which we have now proved.

9.4 EXPONENTIATION

In this section we show that the function m^n is diophantine, thus completing the proof of the main theorem. We have to begin by considering the behaviour of the solutions to some difference equations.

Let a be an integer greater than 1. Define x_n and y_n by

$$x_{n+1} = 2ax_n - x_{n-1},\ x_0 = 1,\ x_1 = a,$$
$$y_{n+1} = 2ay_n - y_{n-1},\ y_0 = 1,\ y_1 = 1.$$

When the parameter a needs consideration we refer to these as $x_n(a)$ and $y_n(a)$.

Let $d = a^2 - 1$; then $\sqrt{d}$ is irrational. It follows that if $m + n\sqrt{d} = u + v\sqrt{d}$ then $m = u$ and $n = v$, and so we also have $m - n\sqrt{d} = u - v\sqrt{d}$.

(I) It is easy to see, by induction, that $x_n \pm y_n\sqrt{d} = (a \pm \sqrt{d})^n$. Since $(a + \sqrt{d})(a - \sqrt{d}) = 1$, we also have that $x_n - y_n\sqrt{d} = (a + \sqrt{d})^{-n}$. Hence $x_n^2 - dy_n^2 = 1$.

(II) Conversely, we show that any solution of $x^2 - dy^2 = 1$ with x and y in **N** must satisfy $x = x_n$ and $y = y_n$ for some n. Observe first that there must be some n with $(a + \sqrt{d})^n \leqslant x + y\sqrt{d} < (a + \sqrt{d})^{n+1}$. Let $u + v\sqrt{d}$ be $(x + y\sqrt{d})(a - \sqrt{d})^n$, so both u and v are integers. We also have $u - v\sqrt{d} = (x - y\sqrt{d})(a + \sqrt{d})^n$, and so $u^2 - dv^2 = x^2 - dy^2 = 1$.

Since $(a - \sqrt{d})^n = (a + \sqrt{d})^{-n}$, we have $1 \leqslant u + v\sqrt{d} < a + \sqrt{d}$. Taking the inverse of these inequalities and changing sign, we find that $-1 \leqslant -u + v\sqrt{d} < -a + \sqrt{d}$. Adding these to the original inequalities then gives $0 \leqslant 2v\sqrt{d} < 2\sqrt{d}$. Hence $v = 0$, and so $u = 1$, since $u^2 - dv^2 = 1$. It follows that $x + y\sqrt{d} = (a + \sqrt{d})^n$, so $x = x_n$ and $y = y_n$, as required.

(III) It follows easily from the equalities $x_n \pm y_n\sqrt{d} = (a \pm \sqrt{d})^n = (a + \sqrt{d})^{\pm n}$ that $x_{m \pm n} + y_{m \pm n}\sqrt{d} = (x_m + y_m\sqrt{d})(x_n \pm y_n\sqrt{d})$, and hence that we have the addition formulas $x_{m \pm n} = x_m x_n \pm d y_m y_n$, and $y_{m \pm n} = x_n y_m \pm x_m y_n$. In particular $x_{n \pm 1} = a x_m \pm d y_m$, $y_{m \pm 1} = a y_m \pm x_m$. It follows that $y_{m+1} > y_m$, and hence that $y_m \geqslant m$ for all m.

(IV) We denote the greatest common divisor of two integers r and s by (r,s). Since $x_n^2 - dy_n^2 = 1$, we see that $(x_n, y_n) = 1$. By induction on k, using the addition formulas, we see that $y_n | y_{nk}$ for all n and k, where $r|s$ means that r divides s.

(V) We can now deduce that $y_n | y_t$ iff $n|t$. The previous remark gives the result if $n|t$. Assume that $y_n | y_t$ and that n does not divide t. Write t as $nq + r$, where $0 < r < n$. Since $y_t = x_r y_{nq} + x_{nq} y_r$, and we already know that y_n divides y_{nq}, we see that $y_n | x_{nq} y_r$. As $(y_{nq}, x_{nq}) = 1$, and $y_n | y_{nq}$, it follows that $y_n | y_r$. But we know that $y_r < y_n$, a contradiction.

(VI) We next show that $y_{nk} \equiv k x_n^{k-1} y_n \bmod y_n^3$. To see this, we write $x_{nk} + y_{nk}\sqrt{d}$ as $(x_n + y_n\sqrt{d})^k$. Expanding this by the binomial theorem, and equating the terms involving $\sqrt{d}$ gives the result.

(VII) It follows immediately that if $k = y_n$ then $y_n^2 | y_{nk}$. As a partial converse to this we show that $y_n | t$ if $y_n^2 | y_t$. We already know that $n|t$, since $y_n | y_t$. Write t as nk. Then we know that y_n^2 must divide $k x_n^{k-1} y_n$, so that y_n divides $k x_n^{k-1}$. Since $(y_n, x_n) = 1$, this tells us that y_n divides k.

(VIII) From the difference equation, easy inductions (the initial cases when n is 0 or 1 being obvious) give the following results. $y_n \equiv n \bmod 2$, and $y_n \equiv n \bmod (a-1)$. Also if $a \equiv b \bmod c$ we find that, mod c, $x_n(a) \equiv x_n(b)$ and $y_n(a) \equiv y_n(b)$.

(IX) We now show that $x_{2n\pm j} \equiv -x_j \bmod x_n$, and hence $x_{4n\pm j} \equiv x_j \bmod x_n$. For the addition formula tells us that $x_{2n\pm j} = x_n x_{n\pm j} + dy_n y_{n\pm j} \equiv dy_n(y_n x_j \pm x_n y_j) \equiv dy_n^2 x_j = (x_n^2 - 1)x_j \equiv -x_j$.

(X) Thg most difficult of our results, is the following. Let $x_i \equiv x_j \bmod x_n$, with $i \leqslant j \leqslant 2n$ and $n > 0$. Then $i = j$, unless $a = 2$, $n = 1$, $i = 0$, and $j = 2$.

We first suppose that x_n is odd, and let $q = (x_n - 1)/2$. Then no two of the nubers $-q, \ldots, q$ are congruent mod x_n, and every integer is congruent to one of them. Now we know that $1 = x_0 < x_1 < \ldots, < x_{n-1}$, and that $x_{n-1} \leqslant x_n/a \leqslant x_n/2$. In particular $x_{n-1} \leqslant q$. Also we know that the number $x_{n+1}, \ldots, x_{2n}$ are congruent mod x_n to $-x_{n-1}, \ldots, -x_0$, by the previous result. Hence the numbers $x_0, \ldots, x_{2n}$ are mutually incongruent mod x_n.

When x_n is even, we let q be $x_n/2$. In this case $-q \equiv q \bmod x_n$. The result will follow as before unless $x_{n-1} = q$; this possibility would give $x_{n+1} \equiv -q$. However this case requires $x_n = 2x_{n-1}$, and, as we know that $x_n = ax_{n-1} + dy_{n-1}$, we must have $a = 2$ and $y_{n-1} = 0$, so that $n = 1$, as needed.

The previous result can be extended to show that if $x_j \equiv x_i \bmod x_n$, with $0 < i \leqslant n$ and $0 \leqslant j < 4n$, then either $j = i$ or $j = 4n - i$. For if $j \leqslant 2n$, we get $i = j$ unless we are in the exceptional case. Since i is not 0, we would then have $i = 2, j = 0$, and $n = 1$, contradicting $i \leqslant n$, so this case cannot happen. If $j > 2n$ define k as $4n - j$. Then $k > 0$, and $x_k \equiv x_j \equiv x_i \bmod x_n$. As $0 < k < 2n$, we cannot be in the exceptional case, and so must have $k = i$.

It follows at once that if $0 < i \leqslant n$ and $x_i \equiv x_j \bmod x_n$ then $j \equiv \pm i \bmod 4n$. For we need only write j as $4nq + k$, where $0 \leqslant k < 4n$ and use the previous result and the fact that $x_{4n+r} \equiv x_r$ for all r.

We are now in a position to prove the main results.

Proposition 9.9 *The function $y_k(a)$ is diophantine.*

Proof We show that, provided k is non-zero (the general case can be obtained from this, but is not needed) $y = y_k(a)$ iff $a > 1$ (else $y_k(a)$ is not defined) and there are positive integers x, u, v, s, t, b, r, p, q, c, and d satisfying the following eight equations.

$$x^2 - (a^2 - 1)y^2 = 1 \tag{1}$$

$$u^2 - (a^2 - 1)v^2 = 1 \tag{2}$$

$$s^2 - (b^2 - 1)t^2 = 1 \tag{3}$$

$$v = ry^2 \tag{4}$$

$$b = 1 + 4py = a + qu \tag{5}$$

$$s = x + cu \tag{6}$$

$$t = k + 4(d - 1)y \tag{7}$$

$$y \geqslant k. \tag{8}$$

First suppose the equations have a solution. As the variables are not zero, there will, by (II), be positive i, j, and n such that $x=x_i(a)$, $y=y_i(a)$, $u=x_n(a)$, $v=y_n(a)$, $s=x_j(b)$, and $t=y_j(b)$. By (4), $y\leqslant v$, and so $i\leqslant n$. From (5), $b\equiv a \bmod x_n(a)$, from which we know, by (VII), that $x_j(b)\equiv x_j(a) \bmod x_n(a)$. From (6) we have $x_j(b)\equiv x_i(a) \bmod x_n(a)$. Hence we have $x_j(a)\equiv x_i(a) \bmod x_n(a)$, so (X) tells us that $j\equiv \pm i \bmod 4n$.

Equation (4) tells us that $y_i(a)^2|y_n(a)$, from which we know that $y_i(a)|n$. Hence $j\equiv \pm i \bmod 4y_i(a)$.

We know that $y_j(b)\equiv j \bmod (b-1)$. By (5), $b\equiv 1 \bmod 4y_i(a)$, so that $y_j(b)\equiv j \bmod 4y_i(a)$. By (7), we have $y_j(b)\equiv k \bmod 4y_i(a)$. Consequently $k\equiv \pm i \bmod 4y_i(a)$.

However, we know that $i\leqslant y_i(a)$, and condition (8) tells us that $k\leqslant y_i(a)$. The congruence then requires that $i=k$. As i is defined to satisfy $y=y_i(a)$, we have, as required, $y=y_k(a)$.

Conversely, suppose that $y=y_k(a)$. Set x to be $x_k(a)$, so that (1) holds. Let $m=2ky_k(a)$, and define $u=x_m(a)$, $v=y_m(a)$. Then (2) holds. We find that $y^2|v$, since we have that $y^2|y_{m/2}(a)$, and the latter divides $y_m(a)$. Hence we can find r satisfying (4). Since m is even, we know from (VIII) that v is even, and hence u is odd.

As u is odd and $(u,v)=1$, and y divides v, we find that $(u,4y)=1$. By the Chinese Remainder Theorem (Theorem 9.11 below) there is a non-negative b_0 satisfying $b_0\equiv 1 \bmod 4y$ and $b_0\equiv a \bmod u$. Adding a suitable multiple of $4uy$ to b_0, we can find positive b, p, and q satisfying (5). Then (3) is satisfied by putting s and t to be $x_k(b)$ and $y_k(b)$.

Since $b>a$, we have $s>x$, as $s=x_k(b)$ and $x=x_k(a)$. It also follows, since $b\equiv a \bmod u$, that $s\equiv x \bmod u$. Hence we can choose c to satisfy (6).

Since t is $y_k(b)$, we know that $t\geqslant k$ and $t\equiv k \bmod (b-1)$. By (5) we have that $t\equiv k \bmod 4y$, so that (7) can be satisfied. Finally (8) holds, since $y=y_k(a)$.■

Theorem 9.10 *The function a^n is diophantine.*

Proof It is most convenient to assume that $n>0$ and $a>1$. We do not need the general case; it can be obtained by a simple modification.

We define Near(u,v) to be the nearest integer to v/u. The function Near is diophantine, since Near$(u,v)=w$ iff $(2\text{rem}(u,v)\leqslant u \wedge w=\text{quo}(u,v)) \vee (2\text{rem}(u,v)>u \wedge w=\text{quo}(u,v)+1)$.

We shall show that $a^n=\text{Near}(y_{n+1}(k),y_{n+1}(ka))$ provided k is large enough. By the previous proposition, a^n will be diophantine if the requirement that k is large enough can be expressed as a diophantine condition.

It is easy to check, by induction, that, for any $b>1$ and any n, $(2b-1)^n \leqslant y_{n+1}(b) \leqslant (2b)^n$. Apply this for $b=ka$ and for $b=k$, and divide the inequalities. We find that $y_{n+1}(ka)/y_{n+1}(k)$ lies between $(2ka-1)^n/(2k)^n = a^n(1-1/2ka)^n$ and $(2ka)^n/(2k-1)^n = a^n(1-1/2k)^{-n}$. Hence Near$(y_{n+1}(k),y_{n+1}(ka)) = a^n$ if $(1-1/2ka)^n > 1-1/2a^n$ and $(1-1/2k)^{-n} < 1+1/2a^n$.

Since $y_{n+1}(a) \geqslant (2a-1)^n > a^n$, it is enough to require that $(1-1/2ka)^n > 1-1/2y_{n+1}(a)$ and $(1-1/2k)^{-n} < 1+1/2y_{n+1}(a)$, and the latter condition can be written as $(1-1/2k)^n > (1+1/2y_{n+1}(a))^{-1}$.

Now $(1-x)^n > 1-nx$ for any real x between 0 and 1 (because $(1-x)^n + nx$ has a positive derivative). Hence the required conditions hold provided that $1-n/2ka > 1-1/2y_{n+1}(a)$ and $1-n/2k > (1+1/2y_{n+1}(a))^{-1}$. These two conditions, when cleared of fractions, are diophantine conditions, since $y_{n+1}(a)$ is diophantine.■

9.5 GÖDEL'S SEQUENCING FUNCTION AND MIN-COMPUTABLE FUNCTIONS

9.5.1 Gödel's sequencing function

The **Gödel sequencing function** $\gamma:\mathbf{N}^3 \to \mathbf{N}$ is defined by $\gamma(i,t,u) = \text{rem}(1+(i+1)t,u)$. We shall see that this enables us to replace any property of a variable number of variables by a property with a fixed number of variables. (This could also be done using the known coding of strings. However, γ is a much simpler function than the latter; in particular, it uses only addition and multiplication, not exponentiation.) The key example of this is in Lemma 9.13 and its applications in Theorems 9.15 and 9.16 and Lemma 13.13. We need a classical result from number theory before obtaining this property of γ.

Theorem 9.11 (Chinese Remainder Theorem) *Let $n_1, \ldots, n_k$ be integers greater than* 1, *and suppose that, for $i \neq j$, the greatest common divisor of n_i and n_j is* 1. *Let a_i be an integer less than n_i, for $i = 1, \ldots, k$. Then there is an integer a such that* $\text{rem}(n_i,a) = a_i$ *for $i = 1, \ldots, k$.*

Proof Map $\{x; x < n_i \ldots n_k\}$ to $\{x; x < n_1\} \times \ldots \times \{x; x < n_k\}$ by sending x to $(x_1, \ldots, x_k)$, where x_i is $\text{rem}(n_i,x)$. We want to show that this function is onto. Since both sets have $n_1 \ldots n_k$ elements, it is enough to show the function is one–one. So suppose $\text{rem}(n_i,x) = \text{rem}(n_i,y)$ for all $i \leqslant k$. Then $x-y$ is divisible by all n_i. Hence $x-y$ is divisible by the product $n_1 \ldots n_k$, by the assumption about greatest common divisors. Since both x and y are less than $n_1 \ldots n_k$, we must have $x = y$.■

Proposition 9.12 *Let $a_0, \ldots, a_k$ be in* **N**. *Then there exist t and u in* **N** *such that $\gamma(i,t,u) = a_i$ for all $i \leqslant k$.*

Proof Take any s greater than all of $k, a_0, \ldots, s_k$. Let t be $s!$. Let n_i be $1+(i+1)t$, for all $i \leqslant k$. By the previous theorem it is enough to show that the greatest common divisor of n_i and n_j is 1 when $i \neq j$.

So let d divide n_i and n_j. Then d divides $(j+1)n_i - (i+1)n_j = j-i$. So d is at most k, and hence d divides t. Since d divides both t and $1+(i+1)t$, d must be 1, as needed.■

Lemma 9.13 *Let $F:\mathbf{N}^2 \to \mathbf{N}$ be the iterate of $f:\mathbf{N} \to \mathbf{N}$. Then $F(m,n) = r$ iff* *there are t and u such that* $\gamma(0,t,u)=m$, $\gamma(n,t,u)=r$, *and, for all* $i<n$, $\gamma(i+1,t,u)=f\gamma(i,t,u)$.

Proof Plainly, $F(m,n)=r$ iff there are a_i for all $i\leqslant n$ such that $a_0=m$, $a_n=r$, and $a_{i+1}=fa_i$ for all $i<n$. The result follows by applying the proposition to this sequence.■

An interesting use of the sequencing function occurs in the next result.

Proposition 9.14 *Let A be a r.e. subset of* **N**, *and let* $X=\{n;\forall x\leqslant n(x\in A)\}$. *Then X is r.e.*

Proof By Exercise 4.3 there is a recursive subset B of $\mathbf{N}^2$ such that $x\in A$ iff $\exists y\{(x,y)\in B\}$. Hence $n\in X$ iff for all $i\leqslant n$ there is some y_i with $(i,y_i)\in B$. This holds iff there are t and u such that $(i,\gamma(i,t,u))\in B$ for all $i\leqslant n$. This condition on n, t, and u is recursive, being obtained by bounded quantification from a recursive set. Since we are applying existential quantifiers to this to obtain X, we see that X is r.e.■

9.5.2 min-computable functions

Definition The set of **min-computable** functions is the smallest set closed under composition and minimisation and containing addition, multiplication, all projections, and the function $c:\mathbf{N}^2\to\mathbf{N}$ given by $c(m,n)=1$ if $m=n$ and $c(m,n)=0$ otherwise. The set of **regular min-computable** functions is the smallest set containing these functions and closed under composition and regular minimisation.

As usual (look at Lemma 3.1 and its proof) a function is in one of these sets iff there is a sequence of functions satisfying relevant conditions.

We begin by showing that various simple functions are regular min-computable. First, the functions from **N** to **N** which are constantly 0 and 1 are both regular min-computable. For $0=c(x,\mu y(c(x,y)=0))$, and $1=\mu y(c(0,y)=0)$. By addition, it follows that any constant function from **N** to **N** is regular min-computable, and, by composition with a projection, so is any constant function from any $\mathbf{N}^k$ to **N**.

The cosign and sign functions are regular min-computable. For $\operatorname{co}x=c(x,0)$ and $\operatorname{sg}x=\operatorname{co}(\operatorname{co}x)$.

We define a property to be regular min-computable if its characteristic function is. It then follows that if P and Q are regular min-computable so are $\neg P$, $P\vee Q$, and $P\wedge Q$. This true because $\chi_{\neg P}=\operatorname{co}\chi_P$, $\chi_{P\vee Q}=\chi_P\chi_Q$, and $P\wedge Q$ is $\neg(\neg P\vee\neg Q)$. Also $f^{-1}P$ is regular min-computable if P and the components of f are regular min-computable.

The function $|x-y|$ is regular min-computable, being $\mu z(x+z=y\vee y+z=x)$. The predicate $x\leqslant y$ is regular min-computable, since $x\leqslant y$ iff $x+|x-y|=y$. It follows that if f and g are regular min-computable so is the predicate $f=g$, since its characteristic function is $\operatorname{sg}|f-g|$, and similarly so is the predicate $f\leqslant g$.

Bounded minimisation and quantification preserves regular min-computability. For let P be a regular min-computable predicate of $k+1$ variables. Then $(\mu z \leqslant y)P$ is $\mu z(z=y+1 \vee P)$, while $(\exists z \leqslant y)P$ iff $\mu z(z=y+1 \vee P) \leqslant y$. Finally, $(\forall z \leqslant y)P$ is $\neg(\exists z \leqslant y)\neg P$.

The functions J, K, and L are regular min-computable. For $J(m,n) = m+\mu y(y+y=(m+n)(m+n+1))$, and $Kr=(\mu m \leqslant r)(\exists n \leqslant r)(J(m,n)=r)$, and similarly for L.

The functions rem and γ are regular min-computable. Plainly γ is regular min-computable if rem is. Now $\text{rem}(m,n)=\mu r(n=mq+r)$, where $q=\mu x(m(x+1) \geqslant n+1)$.

We can now prove the main theorem on min-computable functions, using the diophantine properties. We also give a proof not using this concept, but involving the sequencing function.

Theorem 9.15 *A function is min-computable iff it is partial recursive, and is regular min-computable iff it is recursive.*

Proof Plainly min-computable functions are partial recursive, and regular min-computable functions are recursive.

Suppose $f:\mathbf{N}^k \to \mathbf{N}$ is partial recursive. We know there are polynomials P and Q with non-negative coefficients such that $fn = K(\mu y(P(n,J_r^{-1}y) = Q(n,J_r^{-1}y))$. Since J_r^{-1} is obtained from K and L by compositions, and P and Q are obviously regular min-computable, the result follows.

In case the reader has not covered the material on diophantine functions, we give another proof. By Theorem 4.2, it is enough to show that every primitive recursive function is regular min-computable. The initial functions are all regular min-computable. Since J, K, and L are regular min-computable, it is enough to show that if $f:\mathbf{N} \to \mathbf{N}$ is regular min-computable so is its iterate F.

Let $\beta(x,y)$ be $\gamma(x,J^{-1}y)$, so β is regular min-computable. By Lemma 9.13, we can write $F(m,n)$ as $\beta(n,x)$ where $x = \mu y(\beta(0,y) = m \wedge (\forall i < n)(\beta(i+1,y) = f\beta(i,y)))$. The result follows. ■

9.6 UNIVERSAL DIOPHANTINE PREDICATES AND KLEENE'S NORMAL FORM THEOREM

Theorem 9.16 *For any k, the r.e. subsets of $\mathbf{N}^k$ can be numbered as $D_0, D_1, \ldots$ in such a way that $\{(\mathbf{x},y); \mathbf{x} \in D_y\}$ is diophantine.*

Remark Readers who have omitted section 9.4 will have to be content with a numbering such that $\{(\mathbf{x},y); \mathbf{x} \in D_y\}$ is base-2 exponential diophantine. This is still enough to get all the results we need about Kleene's Normal Form Theorem.

Proof By the Main Theorem, it is enough to give a numbering such that $\{(\mathbf{x},y); \mathbf{x} \in D_y\}$ is r.e. Also, using the bijection J_k, it is enough to prove the result when k is 1.

We begin by numbering the polynomials with positive coefficients. Let $P_0 = 1$, $P_{3i+1} = x_i$, $P_{3i+2} = P_{Ki} + P_{Li}$, $P_{3i+3} = P_{Ki} \cdot P_{Li}$. Plainly any polynomial with positive coefficients is P_n for some n. We shall regard P_n as a polynomial in $x_0, \ldots, x_n$; these variables need not all occur in P_n, but plainly no other variables can occur. We now define D_n to be the set $\{x_0; \exists x_1 \ldots \exists x_n(P_{Kn} = P_{Ln})\}$. By the Main Theorem every r.e. subset of **N** is D_n for some n. (Note that 0 is not of the form P_r; however a condition $P = 0$ can be replaced by $P + 1 = 1$.)

If we are content with the base-2 exponential diophantine case, we number the base-2 exponential polynomials with positive coefficients as $P_0 = 1$, $P_{4i+1} = x_i$, $P_{4i+2} = 2^{x_i}$, $P_{4i+3} = P_{Ki} + P_{li}$, $P_{4i+3} = P_{Ki} \cdot P_{Li}$, and then proceed in an almost identical manner.

Now $x \in D_n$ iff there are $a_0, \ldots, a_{3n}$ such that $a_0 = 1$, $a_1 = x$, $a_{Kn} = a_{Ln}$, and, for all $i \leqslant n$ we have $a_{3i+2} = a_{Ki} + a_{Li}$, $a_{3i+3} = a_{Ki} \cdot a_{Li}$. (In fact we only need to have a_j defined for all $j \leqslant n$, but there is no harm in requiring their existence for other j, and this makes the formulas easier.)

Using Gödel's sequencing function, this holds iff there are t and u such that $\gamma(0,t,u) = 1$, $\gamma(1,t,u) = x$, $\gamma(Kn,t,u) = \gamma(Ln,t,u)$, and, for all $i \leqslant n$, we have $\gamma(3i+2,t,u) = \gamma(Ki,t,u) + \gamma(Li,t,u)$, $\gamma(3i+3,t,u) = \gamma(Ki,t,u) \cdot \gamma(Li,t,u)$. Now γ, K, and L are primitive recursive, and we are using bounded minimisation. Hence we have $x \in D_n$ iff there are t and u such that (x,n,t,u) is in a known primitive recursive set. Thus $\{(x,n); x \in D_n\}$ is r.e., and hence diophantine, being the projection of a primitive recursive set.■

It is now easy to obtain a strengthening of Kleene's Normal Form Theorem.

Theorem 9.17 *There is a primitive recursive function* $V: \mathbf{N}^3 \to \mathbf{N}$ *such that for any partial recursive* $f: \mathbf{N} \to \mathbf{N}$ *there is* k *with* $fx = K(\mu y(V(k,x,y) = 0))$.

Proof By the previous theorem there is a numbering of the r.e. subsets of $\mathbf{N}^2$ as $D_0, D_1, \ldots$ and a (base-2 exponential) polynomial P such that $(x,y) \in D_k$ iff there are $z_1, \ldots, z_n$ with $P(k,x,y,z_1,\ldots,z_n) = 0$. Define V by $V(k,x,w) = |P(k,x,J_{n+1}^{-1}w)|$. Then V is primitive recursive, and $(x,y) \in D_k$ iff there is w such that $V(k,x,w) = 0$ and $y = Kw$ (recall that K is the first component of J_{n+1}^{-1}). For if there is such a w we simply let $z_1, \ldots$ be the last n components of $J_{n+1}^{-1}w$, while if we are given $z_1, \ldots$ we need only define w to be $J_{n+1}(y,\mathbf{z})$.

Now let f be partial recursive. Then $\{(x,y); y = fx\}$ is r.e. Hence there is k such that $y = fx$ iff $(x,y) \in D_k$. So we have $y = fx$ iff there is w with $y = Kw$ and $V(k,x,w) = 0$. The result follows at once.■

Consider what properties of K were used in the above proof. All that we used is that K is the first component of J_{n+1}^{-1}. And the only fact we needed about J_{n+1}^{-1} is that it is a surjection from **N** to $\mathbf{N}^{n+1}$. Since this map is obtained from the functions K and L by compositions, all we need to ensure it is a surjection is that the map J_2^{-1}, whose components are K and L, is a surjection. Thus we immediately obtain the following version of Kleene's Theorem.

Theorem 9.18 *Let $U:\mathbf{N}\rightarrow\mathbf{N}$ be primitive recursive, and suppose there is another primitive recursive function $U':\mathbf{N}\rightarrow\mathbf{N}$ such that the function sending n to $(Un, U'n)$ is a surjection from $\mathbf{N}$ to $\mathbf{N}^2$. Then there is a primitive recursive $V:\mathbf{N}^3\rightarrow\mathbf{N}$ such that for any partial recursive $f:\mathbf{N}\rightarrow\mathbf{N}$ there is k with $fx = U(\mu y(V(k,x,y)=0))$.*

9.7 THE FOUR SQUARES THEOREM

We have seen that in order to obtain the negative solution to Hilbert's Tenth Problem for **Z** is is necessary to show that any non-negative integer is the sum of four squares. There are subtle proofs of this property, which give a deep insight into why it holds. We will follow the simple direct proof due to Lagrange.

We begin with an identity due to Euler, which is proved by multiplying out and checking;
$(a^2+b^2+c^2+d^2)(x^2+y^2+z^2+t^2) = (ax+by+cz+dt)^2 +$
$(ay-bx+ct-dz)^2+(az-cx-bt+dy)^2+(at-dx+bz-cy)^2$.

It follows that the set of numbers which are the sum of four squares is closed under multiplication. Hence it is enough to show that all primes are the sum of four squares. Since $2=1^2+1^2+0^2+0^2$, the result is true for 2. So we will look at odd primes p.

We first show that there is some m with $0<m<p$ such that mp is the sum of four squares. Consider the $p+1$ numbers a^2 and $-1-b^2$ where $0\leqslant a,b\leqslant(p-1)/2$. Two of these must have the same remainder when divided by p. However a^2 and x^2 have the same remainder when divided by p iff p divides either $a+x$ or $a-x$. Thus the numbers a^2 must all have different remainders. Similarly the numbers $-1-b^2$ also have different remainders. Hence there must be a and b such that $a^2+b^2+1^2+0^2$ is divisible by p. Write this as mp, and we see that $m<p$, because a and b are at most $(p-1)/2$.

Finally we show that if mp is the sum of four squares with $1<m<p$ then there is n with $1\leqslant n<m$ such that np is also the sum of four squares. This will prove the result by induction.

Let $mp=\Sigma x_i^2$, and suppose first that m is even. Then either every x_i is even, or they are all odd, or exactly two of them are even; in the last case we may assume that x_1 and x_2 are even. In all three cases each of $x_1\pm x_2$ and $x_3\pm x_4$ are even, So we can write $(m/2)p$ as $((x_1+x_2)/2)^2 + ((x_1-x_2)/2)^2 + ((x_3+x_4)/2)^2 + ((x_3-x_4)/2)^2$, as required.

Now let m be odd. Define y_i by $x_i\equiv y_i$ mod m and $|y_i|<m/2$. Then $\Sigma y_i^2\equiv\Sigma x_i^2\equiv 0$ mod m, so we can define n by $\Sigma y_i^2=nm$. We cannot have $n=0$. For this would make every y_i zero, and so would make every x_i divisible by m. But then mp would be divisible by m^2, which is impossible since p is prime and $1<m<p$.

Plainly $n<m$, since each y_i is less than $m/2$. Also $m^2np=(\Sigma x_i^2)(\Sigma y_i^2)$. Use Euler's identity to write m^2np as the sum of four squares. We shall show that each integer involved is divisible by m, whence np itself is the sum of four squares. One of the relevant integers is $\Sigma x_i y_i$. This is congruent mod m to Σy_i^2, since $x_i \equiv y_i$ mod m. But, as needed, $\Sigma y_i^2 \equiv 0$ mod m. Another of the integers is $x_1y_2-x_2y_1+x_3y_4-x_4y_3$. This is congruent to $y_1y_2-y_2y_1+y_3y_4-y_4y_3$,; that is, it is congruent to 0 mod m, as needed. Similarly the other two integers involved are divisible by m, completing the proof of the theorem.

10

Indexings and the recursion theorem

In this chapter we shall look at ways of numbering the partial recursive functions, showing that all reasonable ways are essentially the same. We shall also prove the recursion theorem, which is a very useful tool in obtaining functions with special properties.

10.1 PAIRINGS

We have previously defined the bijections $J_n:\mathbf{N}^n \to \mathbf{N}$ (with J_1 being the identity). In this chapter $J_n(x_1, \ldots, x_n)$ will be denoted by $[x_1, \ldots, x_n]$ or by $[\mathbf{x}]$. We know that the ith component of $J_n^{-1}x$ is $KL^{i-1}x$ for $i < n$ and is $L^{n-1}x$ for $i = n$.

By definition, $[x_1, \ldots, x_{n+1}] = [x_1, [x_2, \ldots, x_{n+1}]]$. An easy induction shows that $[x_1, \ldots, x_{n+1}]$ also equals $[x_1, \ldots, x_{n-1}, [x_n, x_{n+1}]]$.

Inductively, we can easily see that $[x_1, \ldots, x_n] \leqslant [y_1, \ldots, y_n]$ if $x_i \leqslant y_i$ for all i, and that $x_i \leqslant [x_1, \ldots, x_n]$ for all i. In particular, if $x_i \leqslant y$ for all i then $[x_1, \ldots, x_n] \leqslant \mathrm{Large}(n,y)$, where $\mathrm{Large}(n,y)$ is $[y, \ldots, y]$, with y occurring n times. From the previous paragraph, $\mathrm{Large}(n+1,y) = J(y, \mathrm{Large}(n,y))$. As $\mathrm{Large}(1,y) = y$, Large is primitive recursive. (Strictly speaking, we cannot call Large primitive recursive, as we have not defined Large $(0,y)$. However, we never need to use $\mathrm{Large}(0,y)$, and so we give it any convenient definition. The same applies to other functions which we define later.)

Since L^ix is a primitive recursive function of i and x (being the value at (x,i) of the iterate of L), it follows that we can define a primitive recursive function $\mathrm{Seq}:\mathbf{N}^3 \to \mathbf{N}$ such that $\mathrm{Seq}(i,n,x) = KL^{i-1}x$ for $0 < i < n$ and $\mathrm{Seq}(n,n,x) = L^{n-1}x$. It then follows that if $x = [x_1, \ldots, x_n]$ then $\mathrm{Seq}(i,n,x) = x_i$ for $1 \leqslant i \leqslant n$.

We now show how to define a primitive recursive function $\mathrm{Cat}:\mathbf{N}^3 \to \mathbf{N}$ such that $\mathrm{Cat}(m,x,y) = [x_1, \ldots, x_m, y_1, \ldots, y_n]$ when $x = [x_1, \ldots, x_m]$ and $y = [y_1, \ldots, y_n]$. Note that we do not require n as an additional variable. From the definition of [ ], an easy induction shows that if Cat is defined so that this formula holds when $n = 1$ then it holds for all n.

So let x be $[x_1, \ldots, x_m]$, whence $x_i = \text{Seq}(i,m,x)$ for $1 \leqslant i \leqslant m$. We must define $\text{Cat}(m,x,y)$ to be $[x_1, \ldots, x_m, y]$. Thus $\text{Cat}(m,x,y)$ is the unique z such that $\text{Seq}(i,m+1,z) = \text{Seq}(i,m,x)$ for $1 \leqslant i \leqslant m$ and $\text{Seq}(m+1, m+1, z) = y$. This condition is primitive recursive, since Seq is primitive recursive. As z is the only number satisfying this condition, it is the least such. Further z is at most $\text{Large}(m+1, x+y)$, and this primitive recursive bound is enough to show Cat is primitive recursive.

10.2 INDEXINGS

Any way of numbering the partial recursive functions from **N** to **N** as $\phi_0, \phi_1, \ldots$ is called an **indexing** (or **Gödel numbering**), and if $f = \phi_i$ we call i an **index** of f. The indexing is called **universal** if there is a partial recursive function Φ such that $\Phi(i,x) = \phi_i x$ for all i and x. A universal indexing is called **acceptable** if there is a recursive function $c{:}\mathbf{N}^2 \to \mathbf{N}$ such that the composite $\phi_i\phi_j$ has $c(i,j)$ as an index.

Any model of computation will provide an indexing, as each program (of whatever kind) is a string on some countable alphabet, and we know how to number strings. By Church's Thesis any such indexing will be acceptable. To obtain an acceptable indexing without using Church's Thesis we shall have to be specific about our model of computation. We will only need one explicit indexing to work with.

Suppose we define ϕ_i by $\phi_i x = U(\mu t(V(i,x,t) = 0))$, where U and V are primitive recursive. Kleene's Normal Form Theorem tells us that U and V can be chosen so that $\phi_0, \ldots$ is a universal indexing. Each proof of the theorem will lead to an indexing. In particular, the proof in section 7.4 gives a universal indexing which we call the **modular indexing**, while the proof in section 9.6 gives a universal indexing which we call the **diophantine indexing**.

By Lemma 7.10 and the lemma below, the modular indexing is acceptable. The lemma below also shows that the diophantine indexing is acceptable; however, it will take some work to show that the diophantine indexing satisfies the conditions of the lemma, and we will leave this verification till later.

Lemma 10.1 *A universal indexing is acceptable if for every partial recursive* $f{:}\mathbf{N}^2 \to \mathbf{N}$ *there is a recursive* $s{:}\mathbf{N} \to \mathbf{N}$ *such that* $\phi_{si}x = f(i,x)$ *for all* i *and* x.

Proof Since the indexing is universal, we can write $\phi_i\phi_j x$ as $\Phi(i, \Phi(j,x))$. The function sending (n,x) to $\Phi(Kn, \Phi(Ln,x))$ is partial recursive. So, by hypothesis, there is a recursive function s such that this equals $\phi_{sn}x$. The required function c is plainly sJ.■

This lemma is a converse of a special case of the next result.

We can use the bijections J_n to obtain an indexing of the partial recursive functions from $\mathbf{N}^n$ to **N**. This indexing will also be denoted by $\phi_0, \phi_1, \ldots$. More precisely, we define $\phi_i{:}\mathbf{N}^n \to \mathbf{N}$ by requiring $\phi_i\mathbf{x}$ to be $\phi_i[\mathbf{x}]$. Thus the same notation ϕ_i can be used to refer to a function of any number of variables, the number depending on the context.

Theorem 10.2 (Kleene's *s-m-n* Theorem) *Let $\phi_0, \phi_1, \ldots$ be an acceptable indexing. Then there is a recursive function $s{:}\mathbf{N}^3 \to \mathbf{N}$ such that, for any i, m, n, and $x_1, \ldots, x_m$, the function sending $(y_1, \ldots, y_n)$ to $\phi_i(x_1, \ldots, x_m, y_1, \ldots, y_n)$ has $s(i, m, [\mathbf{x}])$ as an index. In particular, if $f{:}\mathbf{N}^{m+n} \to \mathbf{N}$ is any recursive function, then the function sending $\mathbf{y}$ in $\mathbf{N}^n$ to* $f(\mathbf{x}, \mathbf{y})$ can be given an index recursive in $\mathbf{x}$.

Proof The second sentence is immediate from the first, since we can choose an index i_0 for f, and the function we want is then $s(i_0, m, [\mathbf{x}])$.

Using the function Cat of the previous section and the conventions about indexing functions of several variables, we are then asked to show that $\phi_{s(i,m,x)}y = \phi_i \mathrm{Cat}(m, x, y)$. Now Cat is partial recursive, so we may take an index k for it.

Suppose we can find a recursive function $R{:}\mathbf{N} \to \mathbf{N}$ such that $\phi_{Rx}y = [x, y]$ for all x and y. Then $\phi_i \mathrm{Cat}(m, x, y) = \phi_i \phi_k \phi_{Rm} \phi_{Rx} y$. Using the recursive function c occurring in the definition of an acceptable indexing, this function has as an index $c(i, u)$ where $u = c(k, v)$ and $v = c(Rm, Rx)$, and so this index is the required $s(i, m, x)$.

Define P and Q from $\mathbf{N}$ to $\mathbf{N}$ by $Py = [0, y]$ and $Qz = J(1 + Kz, Lz)$, so that $Q[x, y] = [x + 1, y]$. These functions are partial recursive. Let p and q be indexes for P and Q. Define R by $R0 = p$ and $R(x+1) = c(q, Rx)$. Then R will be recursive, since it is obtained by primitive recursion from the recursive function sending z to $c(q, z)$. It is easy to check by induction that R has the required property.■

This theorem is often stated as ensuring the existence, for given m and n, of a recursive function s of $m + 1$ variables such that $\phi_i(\mathbf{x}, \mathbf{y})$ has $s(i, \mathbf{x})$ as an index, where $\mathbf{x}$ is in $\mathbf{N}^m$ and $\mathbf{y}$ is in $\mathbf{N}^n$. This explains the name of the theorem. Because of the particular pairings we have used, there is no need to mention the integer n, and all the values of m can be treated at once.

Let $\phi_0, \phi_1, \ldots$ be any acceptable indexing, and let $\psi_0, \psi_1, \ldots$ be any indexing. Suppose we can translate from either to the other; that is, suppose there are recursive f and g such that $\phi_i = \psi_{fi}$ and $\psi_i = \phi_{gi}$ for all i. Then, for all i and x, $\psi_i x = \phi_{gi} x = \Phi(gi, x)$, so that $\psi_0, \ldots$ is a universal indexing. It is acceptable, since $\psi_i \psi_j x = \phi_{gi} \phi_{gj} x$ has $c(gi, gj)$ as ϕ-index, and so has $f(c(gi, gj))$ as ψ-index.

Conversely, from the next lemma we see that any two acceptable indexings can be translated into each other. We shall see later that the translations can be chosen to be bijections.

Lemma 10.3 *Let $\phi_0, \ldots$ be any universal indexing, and let $\psi_0, \ldots$ be any indexing such that for any partial recursive $f{:}\mathbf{N}^2 \to \mathbf{N}$ there is a recursive function $s{:}\mathbf{N} \to \mathbf{N}$ with $f(x, y) = \psi_{sx} y$ for all x and y. Then there is a recursive function t such that $\phi_i = \psi_{ti}$ for all i.*

Proof We have $\phi_i x = \Phi(i, x)$ for all i and x. Since Φ is partial recursive, we need only let t be the function s which corresponds to Φ.■

10.3 THE RECURSION THEOREM AND ITS APPLICATIONS

Throughout this section we shall work with an acceptable indexing $\phi_0, \ldots$ whose universal function is Φ, and with the function c ensuring that the indexing is acceptable.

Theorem 10.4 (Recursion Theorem) *Let f:$\mathbf{N} \to \mathbf{N}$ be recursive. Then there is some number n such that $\phi_{fn} = \phi_n$.*

Proof By the *s-m-n* theorem, there is a recursive function g such that $\phi_{gx}y = \Phi(\Phi(x, x), y)$ for all x and y. Let m be an index for fg, and let n be gm.

Then $\Phi(m, m) = \phi_m m = fgm = fn$. Now $\phi_n y = \Phi(\Phi(m, m), y)$, by definition, and this equals $\Phi(fn, y)$ which in turn equals $\phi_{fn}y$, as needed.■

We can extend this theorem by including a parameter.

Theorem 10.5 *There is a recursive function θ such that, for any recursive f and any x with $f = \phi_x$, we have $\phi_{f\theta x} = \phi_{\theta x}$.*

Proof Let i be an index for the function g of the previous theorem. We defined n to be gm, where m was an index of fg. Here fg is $\phi_x\phi_i$, so we may take m to be $c(x, i)$. Hence we define θx to be $gc(x, i)$.■

The recursion theorem is sometimes referred to as a 'pseudo fixed-point' theorem. We do not have $fn = n$, which would make n a fixed point of f. Nevertheless, the functions with indexes fn and n are the same, so we have something like a fixed point. There is another recursion theorem, which provides a genuine fixed point for a certain operator.

As an example, there is an n such that the function with index n is constant and equal to n (regarding an index as defining a program, we could refer to this as a program which simply prints its own number on any input). To see this, let $f(i, x) = i$ for all i and x. Use the *s-m-n* theorem to find a recursive function s such that $\phi_{si}x = f(i, x)$, and obtain n using the recursion theorem on the function s. Hence, for all x, $\phi_n x = \phi_{sn}x$ by definition of n, and the latter is just $f(n, x)$, which is n.

As another example, we will show that there is a recursive function f such that $f0 = 1$ and $f(x+1) = (x+1)fx$ for all x. Of course, we already know that this is just the factorial function, which is primitive recursive. The point is that the method can be used for harder cases. Most situations where a function is defined in terms of itself (or by means of an expression involving its index) can be treated as in the two examples. (Also, if the reader is used to a programming language using recursive calls, a similar technique can be used to ensure that only partial recursive functions are obtained by such models.)

First note that there is exactly one function satisfying the given conditions, and it is total. Next, for any i and x, define $g(i, x)$ by $g(i, 0) = 1$ and

$g(i,x+1) = (x+1)\Phi(i,x)$. Then g is a partial recursive function, and we can find a recursive function s such that $\phi_{si}x = g(i,x)$ for all i and x. Let n satisfy the conclusions of the recursion theorem for s. Then $\phi_n x = \phi_{sn}x = g(n,x)$, and so $\phi_n 0 = 1$ and $\phi_n(x+1) = g(n,x+1) = (x+1)\Phi(n,x) = (x+1)\phi_n x$. Hence ϕ_n is the function we want, and so the function is both total and partial recursive, as needed.

Lemma 10.6 *Let u:$\mathbf{N} \to \mathbf{N}$ be any recursive function. Then there is a one–one recursive function v:$\mathbf{N} \to \mathbf{N}$ such that $\phi_{vx} = \phi_{ux}$ for all x.*

Proof Let s:$\mathbf{N}^3 \to \mathbf{N}$ be the function of the *s-m-n* theorem. Define a function g:$\mathbf{N}^3 \to \mathbf{N}$ by

$$
\begin{aligned}
g(z,j,y) &= 0 \text{ if } \exists k<j(s(z,1,k) = s(z,1,j)),\\
g(z,j,y) &= 1 \text{ if } \forall k<j(s(z,1,k) \neq s(z,1,j))\\
&\qquad \text{and } \exists k \leqslant y(k>j \wedge s(z,1,k) = s(z,1,j)),\\
g(z,j,y) &= \Phi(uj,y) \text{ otherwise.}
\end{aligned}
$$

Then g is partial recursive, being given by a definition by cases, the condition for each case being recursive (see Proposition 4.8). Hence there is a recursive function f such that $\phi_{fz}(j,y) = g(z,j,y)$ for all z,j, and y. Let n come from f by the recursion theorem. Then $\phi_n(j,y) = g(n,j,y)$ for all j and y.

Define v by $vj = s(n,1,j)$. If v is one–one, then, since $s(n,1,j)$ is vj plainly only the third condition in the definition of $g(n,j,y)$ applies. We will then have $\phi_n(j,y) = g(n,j,y) = \Phi(uj,y) = \phi_{uj}y$ for all j and y. By the definition of s, we have $\phi_{vj}y = \phi_n(j,y) = \phi_{uj}y$, as needed.

We still have to show that v is one–one. Suppose not. Take the smallest j such that $vj = vk$ for some $k<j$, and then take the smallest such k. By definition, since $s(n,1,j) = s(n,1,k)$, we have $g(n,j,y) = 0$ for any y, and we also have $g(n,k,y) = 1$ for all $y \geqslant j$. Since $\phi_{vj}y = g(n,j,y)$ and $\phi_{vk}y = g(n,k,y)$, this contradicts the assumption that $vj = vk$.■

We have seen that there exist recursive translation functions between any two acceptable indexings. With the aid of the previous theorem, we can make such translations one–one.

Theorem 10.7 *Let $\phi_0, \ldots$ and $\psi_0, \ldots$ be acceptable indexings. Then there is a one–one recursive function f such that $\phi_i = \psi_{fi}$ for all i.*

Proof By Lemma 10.3 there is a recursive function g such that $\phi_i = \psi_{gi}$ for all i. Applying the previous lemma to $\psi_0, \ldots$, there is a one–one recursive function f such that $\psi_{fi} = \psi_{gi}$ for all i, as needed.■

A recursive function p:$\mathbf{N}^2 \to \mathbf{N}$ such that $\phi_i = \phi_{p(i,r)}$ for all i and r and such that $\{p(i,r)$; all $r\}$ is infinite for each i is called a **padding function** for the indexing. Many models of computation have obvious paddings obtained by adding unnecessary steps to the program. For instance, the abacus machines M and $(a_1s_1)^rM$ compute the same functions whatever r and M are.

We shall look at the diophantine indexing later, and now consider the modular indexing.

Proposition 10.8 *The modular indexing has a padding function.*

Proof Let g be a non-zero number, and define m, k, a_i and so on as in section 7.4. For any $h \neq 0$ let $m(h), k(h), a_i(h)$ and so on be the corresponding numbers for h. Recall that NEXT(h, α, β) depends only on $m(h)$ and the smallest i such that rem$(m(h), \alpha) = a_i(h)$ and rem$(m(h), \beta) = b_i(h)$, and that whether or not $(h, \alpha, \beta) \in$TERM also depends only on these two.

Let h be such that $m(h) = m, a_i(h) = a_i, b_i(h) = b_i, c_i(h) = c_i$, and $d_i(h) = d_i$ for all $i \leqslant k$, and such that for all i with $k < i \leqslant k(h)$ there is some $j \leqslant k$ with $a_i(h) = a_j$ and $b_i(h) = b_j$. From the previous sentence and the definition of ϕ_h, it follows that $\phi_h = \phi_g$. Write a and b for a_1 and b_1.

The required function p can therefore be defined for $g \neq 0$ by $p(g, r) = gp^a_{4k+5}p^b_{4k+6}p^r_{4k+8}$. We define $p(0, r)$ to be $p(g_0, r)$, where g_0 is the Gödel number of a special modular machine computing ϕ_0, so that $g_0 \neq 0$.■

The next result shows that the above padding function may be taken to be one–one.

Lemma 10.9 *Let f:$\mathbf{N}^2 \to \mathbf{N}$ be a recursive function such that for every i $\{f(i, n); all\ n\}$ is infinite. Then there is a one–one recursive function g:$\mathbf{N}^2 \to \mathbf{N}$ such that for every i and j there is some n with $g(i, j) = f(i, n)$.*

Proof Define h by $h(r, u) = f(K(r+1), v)$, where $v = \mu t(\forall s \leqslant r(f(K(r+1), t) > \exp_s u))$. Then h is partial recursive by construction, and h is total by the condition on f. Then we may define a function G by $G0 = f(0, 0)$ and $G(r+1) = h(r, Hr)$, where H is the history of G. Thus G is recursive, being defined by a course-of-values recursion from h. By construction, $G(r+1) > Gs$ for all $s \leqslant r$, and so G is one–one. For any r there is some n with $G(r+1) = f(K(r+1), n)$, by construction. Then GJ is the required function g.■

Theorem 10.10 *If one acceptable indexing has a padding function (a one-one padding function) then every acceptable indexing has a padding function (a one–one padding function).*

Proof Let $\phi_0, \ldots$ be an acceptable indexing with a padding function p and let $\psi_0, \ldots$ be any acceptable indexing. By Theorem 10.7 there are one–one recursive functions f and g such that $\psi_i = \phi_{fi}$ and $\phi_i = \psi_{gi}$ for all i.

Then $\psi_i = \phi_{fi} = \phi_{p(fi, r)} = \psi_{gp(fi, r)}$ for all r. So we define the recursive function q by $q(i, r) = gp(fi, r)$. Because f and g are one–one, we find that $\{q(i, r); \text{all } r\}$ is infinite for any i, and q will be one–one if p is. Hence q is the required padding function for $\psi_0, \ldots$■

Lemma 10.11 *Let $\phi_0, \ldots$ and $\psi_0, \ldots$ be acceptable indexings. Then there is a recursive function g such that $\phi_i = \psi_{gi}$ for all i and $0 < gi < g(i+1)$ for all i.*

Proof Take any recursive function t such that $\phi_i = \psi_{ti}$ for all i; changing the value of $t0$ if necessary, we may assume $t0 \neq 0$. We define gi by padding ti suitably.

Precisely, let p be a padding for the ψ-indexing. Define a function h which is both partial recursive and total by $h(i,u) = p(t(i+1),x)$, where $x = \mu y(p(t(i+1),y) > u)$. Hence $h(i,u) > u$, and the functions with ψ-indexes $t(i+1)$ and $h(i,u)$ are the same, since p is a padding. We now define g by $g0 = t0$ and $g(i+1) = h(i,gi)$. Then g will be recursive, being defined by primitive recursion from h, and g satisfies the required conditions.■

We can now prove that any two acceptable indexings are essentially the same. Readers who know the Cantor–Schroeder–Bernstein Theorem (which says that there is a bijection between two sets if there is an injection from each to the other) will see that the proof below is just a version of the proof of that theorem, with due attention being paid to recursiveness.

Theorem 10.12 (Rogers' Isomorphism Theorem) *Let* $\phi_0, \ldots$ *and* $\psi_0, \ldots$ *be acceptable indexings. Then there is a recursive bijection* $t:\mathbf{N} \rightarrow \mathbf{N}$ *such that* $\phi_i = \psi_{ti}$ *for all* i.

Remark Note that the inverse t^{-1} of t is also recursive, since it is total and $t^{-1}x = \mu y(ty = x)$ for all x.

Proof Let f and g be recursive functions such that $\phi_i = \psi_{fi}$ and $\psi_i = \phi_{gi}$ and with $0 < fi < f(i+1)$ and $0 < gi < g(i+1)$ for all i. Then $fi > i$ and $gi > i$ for all i.

Given x, we can define n and y as recursive functions of x by $n = \mu' m \leqslant x(\exists z \leqslant x((gf)^m z = x))$ and $y = \mu z((gf)^n z = x)$. Now $(gf)^r u \geqslant r+u$ for all r and u. It follows that $(gf)^n y = x$, by definition, but that y is not in $gf\mathbf{N}$. We call y the *founder* of x.

Define t by $tx = fx$ if the founder of x is not in $g\mathbf{N}$ and $tx = g^{-1}x$ if the founder of x is in $g\mathbf{N}$. Plainly x is its own founder unless $x \in gf\mathbf{N}$, and so $g^{-1}x$ is defined if the founder of x is in $g\mathbf{N}$. Hence t is total. Also $g\mathbf{N}$ is recursive, since g is strictly increasing. Since g^{-1} is partial recursive, because $g^{-1}x = \mu z(gz = x)$, it follows that t is partial recursive. Hence t is recursive.

Suppose $tx_1 = tx_2$. If we have $tx_i = fx_i$ for $i = 1$ and 2, or $tx_i = g^{-1}x_i$ for $i = 1$ and 2, plainly $x_1 = x_2$. Otherwise we may assume that $tx_1 = fx_1$ and $tx_2 = g^{-1}x_2$. As $tx_1 = tx_2$, we then find that $gfx_1 = x_2$. But then x_1 and x_2 have the same founder, which contradicts the fact that tx_1 is fx_1 while tx_2 is $g^{-1}x_2$. Hence t is one–one.

Now take any x, and let u be the founder of gx, so that $gx = (gf)^n u$ for some n. If $u \in g\mathbf{N}$ then $t(gx) = g^{-1}gx = x$. This happens in particular if n is 0, since then u is gx. If $u \notin g\mathbf{N}$ we can write n as $m+1$. Let v be $(gf)^m u$. Then $gfv = gx$, so $x = fv$. In this case v and gx have the same founder u. As $u \notin g\mathbf{N}$, we then have $tv = fv = x$. Hence t is onto.■

When we use Church's Thesis, we can only conclude that certain functions are recursive. Detailed computations often show that relevant functions are primitive recursive, and we can then show that various constructed functions are primitive recursive. Thus, if the function s of 10.1 is primitive recursive then so is the function c. Conversely, if c is primitive recursive, so is the function s of the *s-m-n* theorem. The translation function t of 10.3 will also be primitive recursive. The function θ of Theorem 10.5 will then also be primitive recursive, as will the function v of Lemma 10.6, and the function f of 10.7.

We plainly cannot require the function t of Rogers' Isomorphism Theorem to be primitive recursive, since any recursive bijection t will enable us to define a new acceptable indexing function from a given one. This fact is associated with the use of unbounded minimisation in the padding function. The padding function for the modular indexing is primitive recursive and plainly satifies $p(g,r) > r$ for all r. Then the one–one padding function corresponding to this by 10.9 is primitive recursive, since the minimisation involved can be bounded. Then 10.10 shows that any acceptable indexing which has a primitive recursive function c will also have a primitive recursive one–one padding function. In 10.11 for the function g to be primitive recursive we need p to be primitive recursive but we also require a primitive recursive function θ such that $p(i,j) > y$ if $j > \theta(i,y)$, in order that the minimisation should be bounded. It then follows that the function t of Rogers' Isomorphism Theorem is primitive recursive provided both indexings have a primitive recursive function for composition and both padding functions have corresponding functions θ. Bearing in mind how each padding function is built, it is enough to show, for both indexings, that for each i there is a primitive recursive function h such that $s(i,1,x) > y$ if $x > hy$; this, in turn, can be replaced by a condition on the functions giving an index for composition.

We know, by Rice's Theorem, that $\{n;\phi_n$ is a finite function$\}$ is not r.e. We can define an indexing $\theta_0, \theta_1, \ldots$ of the finite functions as follows. It is easy to see that for any finite function f there are n and x such that fi is defined iff $i \leqslant n$ and $\mathrm{Seq}(i+1,n+1,x) \neq 0$ and then $fi = \mathrm{Seq}(i+1,n+1,x) \dot{-} 1$. We give this function the index $J(n,x)$. We can prove the following extension of the Rice–Shapiro Theorem.

Theorem 10.13 *Let A be a set of partial recursive functions. Then $\{n;\phi_n \in A\}$ is r.e. iff there is a r.e. set B such that A consists of all partial recursive functions which extend θ_n for some $n \in B$.*

Proof By the *s-m-n* theorem, there is a recursive function g such that $\theta_n = \phi_{gn}$. Let $\{n;\phi_n \in A\}$ be r.e. Then $\{n;\theta_n \in A\}$ will also be r.e. By the Rice–Shapiro Theorem, a function is in A iff it is the extension of some finite function in A, so that the condition holds.

Conversely, suppose the condition holds. It is enough to show that $\{(m,n);\phi_m$ extends $\theta_n\}$ is r.e. This can be done by the usual dovetailing argument. We suppose that the universal function Φ has the form $U(\mu t(V(n,x,t) = 0))$ with U and V primitive recursive, using Kleene's

Normal Form Theorem. For each m, n, and r, we can first find the domain of θ_n, and for each x in this domain we can check whether or not there is some $t \leqslant r$ with $V(m,x,t) = 0$, and, if so, we can check whether or not $\theta_n x = U(\mu t(V(m,x,t) = 0))$ for all such x.■

10.4 INDEXINGS OF R.E. SETS

Any acceptable indexing of the partial recursive functions leads to two indexings of the r.e. subsets of **N**. Namely, we define W_i to be the domain of ϕ_i and R_i to be the range of ϕ_i. By Rogers' Isomorphism Theorem, any change of acceptable indexings can be achieved by a recursive bijection. Consequently, if W'_i and R'_i are the indexings of r.e. sets coming from another acceptable indexing, there is a recursive bijection t such that $W'_i = W_{ti}$ and $R'_i = R_{ti}$. As a result, we may as well use the modular or diophantine indexings, for which Kleene's Normal Form Theorem can be used.

Proposition 10.14 *There is a recursive function f such that* $W_i = R_{fi}$.

Proof Let Φ be the universal function, and let $g(i,x) = x + \Phi(i,x)$. Then g is partial recursive. Let f be a recursive function such that $\phi_{fi}x = g(i,x) = x + \phi_i x$.

As in Proposition 2.1, we see that $\phi_{fi}\mathbf{N} = \mathrm{dom}\phi_i$, as needed.■

The converse is harder, and requires the use of the Normal Form Theorem.

Lemma 10.15 *There is a primitive recursive function* Step:$\mathbf{N}^3 \to \mathbf{N}$ *such that* $\mathrm{Step}(i,m,x) = \phi_i x + 1$ *if* $\mathrm{Step}(i,m,x) \neq 0$, *and there is some m with* $\mathrm{Step}(i,m,x) \neq 0$ *iff* $\phi_i x$ *is defined.*

Proof We know that there are primitive recursive functions U and V such that $\phi_i x = U(\mu t(V(i,x,t) = 0))$. Define Step$(i,m,x)$ to be 0 if $\forall t \leqslant m(Vi,x,t) \neq 0)$ and to be $1 + U(\mu t \leqslant m(V(i,x,t) = 0))$ otherwise. Then Step is primitive recursive, being given by a definition by cases. If $\mathrm{Step}(i,m,x) \neq 0$, evidently there is some $t \leqslant m$ with $V(i,x,t) = 0$. In this case $\mu t \leqslant m(V(i,x,t) = 0)$ is the same as $\mu t(V(i,x,t) = 0)$, and so $\mathrm{Step}(i,m,x) = 1 + \phi_i x$. Conversely, if $\phi_i x$ is defined, there is some m with $V(i,x,m) = 0$ and then $\mathrm{Step}(i,m,x) \neq 0$.■

Lemma 10.16 *There is a partial recursive function* a:$\mathbf{N} \to \mathbf{N}$ *such that* $a(i)$ *is in* R_i *if* R_i *is not empty.*

Proof R_i is not empty iff there are m and x with $\mathrm{Step}(i,m,x) \neq 0$, and then $\mathrm{Step}(i,m,x) \dot{-} 1$ is in R_i. Hence we may define $a(i)$ to be $\mathrm{Step}(i,r) \dot{-} 1$, where r is defined to be $\mu t(\mathrm{Step}(i, Kt, Lt) \neq 0)$.■

Lemma 10.17 *There is a recursive function f such that for all i we have* $R_i = R_{fi}$ *and* ϕ_{fi} *is either total or nowhere defined.*

Proof We formalise the arguments of Proposition 2.1. Thus we define a partial recursive function $\theta:\mathbf{N}^3 \to \mathbf{N}$ by $\theta(i,m,x) = U(\mu t \leqslant m(V(i,x,t) = 0))$ if $\exists t \leqslant m(V(i,x,t) = 0)$ and $\theta(i,m,x) = a(i)$ otherwise. By the *s-m-n* theorem, there is a recursive function f with $\phi_{fi}(m,x) = \theta(i,m,x)$. By the usual convention, ϕ_{fi} can be regarded either as a function of one variable or as a function of two variables.

If R_i is empty, then $a(i)$ is not defined. In this case, for any x and m, we have $\forall t \leqslant m(V(i,x,t) \neq 0)$, and so ϕ_{fi} is nowhere defined. If R_i is not empty, then $a(i)$ is defined and ϕ_{fi} is total. The arguments of Proposition 2.1 now show that ϕ_{fi} and ϕ_i have the same range.■

Proposition 10.18 *There is a recursive function g such that* $R_i = W_{gi}$ *for all i.*

Proof We continue to follow the arguments of Proposition 2.1. Let f be as in the previous lemma, and define $h(i,x)$ to be $\mu y(\Phi(fi,y) = x)$. This is a partial recursive function. If R_i is empty then $h(i,x)$ is not defined for any x. If R_i is not empty, then ϕ_{fi} is total. In this case $h(i,x) = \mu y(\phi_{fi}y = x)$ is defined iff $x \in R_{fi}$, which equals R_i. It follows that $R_i = W_{gi}$, where g is a recursive function such that $\phi_{gi}x = h(i,x)$.■

By Propositions 10.14 and 10.18 we can translate from either of the two indexings to the other. In particular, in the proposition below, we may use either indexing in each case. The statement is made in the form which is easiest to prove.

Proposition 10.19 *There are recursive functions m and u from* $\mathbf{N}^2$ *to* $\mathbf{N}$ *such that* $W_i \cap W_j = W_{m(i,j)}$ *and* $R_i \cup R_j = R_{u(i,j)}$.

Proof Let $\theta(i,j,x)$ be $\Phi(i,x) + \Phi(j,x)$, and let m be a recursive function such that $\phi_{m(i,j)}x = \theta(i,j,x)$. Then $\phi_{m(i,j)}x$ is defined iff both $\phi_i x$ and $\phi_j x$ are defined. Hence $W_{m(i,j)} = W_i \cap W_j$, as needed.

Now define $h(i,j,x)$ to be $\Phi(i,\text{quo}(2,x))$ if $\text{rem}(2,x) = 0$ and to be $\Phi(j,\text{quo}(2,x))$ if $\text{rem}(2,x) = 1$. Then h is partial recursive, and we take a recursive function u such that $\phi_{u(i,j)}x = h(i,j,x)$. It is easy to check that $R_{u(i,j)} = R_i \cup R_j$.■

Since the complement of an r.e. set need not be r.e., we cannot find a recursive function c such that $W_{ci} = \mathbf{N} - W_i$ for all i. More strongly, it is not possible to find a partial recursive function c such that ci is defined and $W_{ci} = \mathbf{N} - W_i$ whenever W_i is recursive.

For we may define a partial recursive function $\theta:\mathbf{N}^2 \to \mathbf{N}$ by $\theta(i,x) = 0$ if $i \in K$ and $\theta(i,x)$ undefined if $i \notin K$, where K is a r.e. set which is not recursive. We can then define a recursive function h such that $\phi_{hi}x = \theta(i,x)$. It then follows that $W_{hi} = \mathbf{N}$ if $i \in K$ and is empty otherwise. In particular, W_{hi} is recursive for all i. If there were such a function c, we would then have W_{chi} non-empty iff $i \notin K$. Hence $\mathbf{N} - K$ would be the counter-image under the partial recursive function ch of the set $\{j; W_j \text{ non-empty}\}$. This set is r.e., being $\pi_{31}\{(j,x,t); V(j,x,t) \neq 0\}$, giving a contradiction.

10.5 THE DIOPHANTINE INDEXING

We need to consider one indexing explicitly. Those readers who know about modular machines can use the modular indexing and ignore this section. Readers who have not looked at the material on modular machines will know about diophantine properties, and they will need the material on the diophantine indexing in this section. (A similar analysis can be given for the exponential diophantine indexing, if any readers of this section have looked at exponential diophantine properties but not at diophantine ones. The details will be left to the reader.)

Throughout this section, the word 'polynomial' will mean a polynomial with positive coefficients. An indexing for polynomials was given in section 9.6 together with a related indexing for r.e. sets. The explicit details of these indexings will be needed.

We know that $\{(i,r); r \in D_i\}$ is r.e. and hence so is $\{(i,x,y); [x,y] \in D_i\}$. Hence there are polynomials P and Q such that $[x,y] \in D_i$ iff $\exists x_1 \ldots \exists x_n(P(i,x,y,x_1, \ldots, x_n) = Q(i,x,y,x_1, \ldots, x_n))$. The notation is slightly different from that in Theorem 9.16, as we allowed negative coefficients there; the current formula is just another way of expressing the same property. The diophantine indexing is obtained by defining $\phi_i x$ as $K(\mu w(P(i,x,J_{n+1}^{-1}w) = Q(i,x,J_{n+1}^{-1}w)))$.

It is then immediate that $[x,y] \in D_i$ if $y = \phi_i x$. Conversely, if $[x,y] \in D_i$ then $\phi_i x$ is defined. In this case $\phi_i x$ need not equal y; but if $[x,z] \in D_i$ only for $z = y$ then plainly $y = \phi_i x$.

Lemma 10.20 *Let $P(i,x_0, \ldots, x_n)$ be a polynomial not divisible by i. Then there is a strictly increasing primitive recursive function p such that $p(i)$ is an index of $P(i,x_0, \ldots, x_n)$ for all i.*

Proof Note that we require P not to be divisible by i to cover the case $i = 0$, because the zero polynomial is not given an index.

The constant $i + 1$ has an index $k(i)$ given by $k(0) = 0$, $k(i + 1) = 3[k(i), 0] + 2$, looking at the way polynomials are indexed. Then k is clearly a strictly increasing primitive recursive function.

By induction on r and the formula for the index of a product, it follows that $(i + 1)^r R(x_0, \ldots, x_n)$ has a strictly increasing primitive recursive index for any polynomial R and positive r. It then follows, by induction on the degree, that $Q(i+1,x_0,\ldots,x_n)$ has a strictly increasing primitive recursive index $q(i)$ for any Q.

Now let $P(i,x_0, \ldots, x_n) = Q(i,x_0, \ldots, x_n) + R(x_0, \ldots, x_n)$, where Q consists of those monomials involving i and R consists of those not involving i. By hypothesis R is non-zero. Let r be an index of R. The function p defined by $p(0) = r, p(i+1) = 3[q(i),r]+2$ is strictly increasing and primitive recursive. By construction, $p(i)$ is an index of $P(i,x_0, \ldots, x_n)$ for all i. ■

Lemma 10.21 *Let D be a r.e. subset of $\mathbf{N}^2$. Then there is a strictly increasing primitive recursive function d such that $D_{di} = \{x; (i,x) \in D\}$ for all i.*

Proof There are polynomials P and Q such that $(i,x_0) \in D$ iff $\exists x_1 \ldots \exists x_n(P(i,x_0,x_1,\ldots,x_n) = Q(i,x_0,x_1,\ldots,x_n))$. We may assume that P and Q both have non-zero constant term, adding 1 to both sides if necessary. Let p and q be the functions corresponding to P and Q by the previous lemma.

Then $(i,x_0) \in D$ iff $\exists x_1 \ldots \exists x_n(P_{pi}(x_0,\ldots,x_n) = P_{qi}(x_0,\ldots,x_n))$. Define $d(i)$ to be $[p(i),q(i)]$. Then d is a strictly increasing primitive recursive function. Also, by definition, $\{x_0; (i,x_0) \in D\}$ has index $d(i)$.■

Lemma 10.22 *Let $f{:}\mathbf{N}^2 \to \mathbf{N}$ be partial recursive. Then there is a strictly increasing primitive recursive function d such that $\phi_{di}x = f(i,x)$ for all i and x.*

Proof The set $\{(i,x,y); y = f(i,x)\}$ is r.e. Hence so is the set $D = \{(i,u); Lu = f(i,Ku)\}$. Let d be the function corresponding to D by the previous lemma.

If $\phi_{di}x = y$ then $[x,y] \in D_{di}$, and so $(i,[x,y]) \in D$. It follows that $f(i,x) = y$. Conversely, if $f(i,x) = y$ then $(i,[x,z]) \in D$ iff $z = y$. Hence $[x,z] \in D_{di}$ iff $z = y$. We know that this ensures that $y = \phi_{di}x$.■

By Lemma 10.1 we now know that the diophantine indexing is acceptable, and hence that the *s-m-n* theorem holds. The function c which occurs in the definition of acceptability is given by $c(i,j) = d[i,j]$. Consequently, c is primitive recursive and strictly increasing in each variable when the other is kept constant; also $c(i,j) \geqslant [i,j]$. It is then easy to check that the function R occurring in the proof of the *s-m-n* theorem is primitive recursive and strictly increasing. The definition of the function s then shows that it is primitive recursive and strictly increasing in each variable when the others are held constant.

Lemma 10.23 *Let D be a r.e. subset of* **N**. *Then there is a strictly increasing primitive recursive function g such that $D_{gr} = D$ for all r.*

Proof There are polynomials P and Q such that $x_0 \in D$ iff $\exists x_1 \ldots \exists x_n(P(x_0,\ldots,x_n) = Q(x_0,\ldots,x_n))$. We may assume that P and Q have non-zero constant terms, adding 1 to both if necessary. Let p and q be indexes of P and Q.

Then $x_0 \in D$ iff $\exists x_1 \ldots \exists x_n(P+r+1 = Q+r+1)$. Let $k(r)$ be the index for $r+1$ which was constructed in Lemma 10.20. Now $P + r + 1$ has index $3[p,k(r)] + 2$, and similarly for $Q + r + 1$. It follows that the set defined by the condition has index gr, where $gr = [3[p,k(r)] + 2, 3[q,k(r)] + 2]$.■

Lemma 10.24 *Let $f{:}\mathbf{N} \to \mathbf{N}$ be partial recursive. Then there is a primitive recursive strictly increasing function g such that $\phi_{gr}x = fx$ for all r and x.*

Proof The set $D = \{u;Lu = fKu\}$ is r.e. Let g be the function which corresponds to D by the previous lemma.

Then, for any r, $y = fx$ iff $[x,y] \in D$ iff $[x,y] \in D_{gr}$. In particular, given r

and x there is at most one y with $[x, y] \in D_{gr}$. It follows that $[x, y] \in D_{gr}$ iff $y = \phi_{gr}x$, as needed.■

Proposition 10.25 *The diophantine indexing has a primitive recursive padding function which is strictly increasing in each variable.*

Proof We can apply the previous lemma to functions of two variables, using the standard translation between functions of one variable and functions of two variables.

Hence there is a strictly increasing primitive recursive function g such that $\Phi(i, x) = \Phi_{gr}(i, x)$ for all r, i, and x. Let s be the function of the s-m-n theorem. Define $p(i, r)$ to be $s(gr, 1, i)$. Then p is primitive recursive and strictly increasing in each variable. By definition, $\phi_i x = \Phi(i, x) = \Phi_{gr}(i, x) = \phi_{p(i, r)}x$, so p is the required padding function.■

Part II
LOGIC

11

Propositional Logic

In this chapter we look at propositional logic, and we look at the deeper and more important predicate logic in the next chapter. Propositional logic is essentially the study of how connectives such as 'not', 'and', 'or', and 'if... then...' are used. We begin with some background, and then go on to a precise definition of the formulae we study. We look at truth and proof and some of their properties. Truth and provability are very different concepts. However, we shall see that true formulae and provable formulae are the same; this result is Gödel's Completeness Theorem, which will take two sections to prove. (I feel this informal statement of Gödel's theorem is helpful, but I must warn readers that my attempt to give an account of the theorem before introducing the precise definitions has distorted the statement noticably.)

11.1 BACKGROUND

Logic is, in its origins, the study of arguments to see which arguments are correct and which are not. Some examples of arguments, both correct and incorrect, are given below.

(1) London is in England and Paris is in France. Therefore Paris is in France.
(2) London is in England and Paris is in Germany. Therefore Paris is in Germany.
(3) London is in England or Paris is in France. Therefore Paris is in France.
(4) London is in England or Paris is in Germany. Therefore Paris is in Germany.
(5) If it is raining and I am not wearing a hat then I am either carrying an umbrella or getting wet. But it is raining and I am not carrying an umbrella. Therefore I am either wearing a hat or getting wet.

(6) Socrates is a man. All men are mortal. Therefore Socrates is mortal.
(7) All cows have four legs. This object has four legs. Therefore this object is a cow.
(8) Some cows are black. This is a cow. Therefore this is black.

The first five arguments are arguments of propositional logic. The last three, which refer to 'some', 'all', and particular objects, are arguments of predicate logic. Propositional logic just looks at how sentences are connected by words like 'or', 'and', and so on, and does not have to look at the sentences connected in any more detail, whereas predicate logic has to look at the internal structure of sentences.

The first and sixth argument are obviously correct, while the fifth is easily seen to be correct.

The seventh and eighth arguments are obviously incorrect. Incorrect arguments such as these are often used in politics to try and discredit opponents. As an example, consider 'Some members of the Campaign for Nuclear Disarmament are communists. He is a member of the Campaign for Nuclear Disarmament. Therefore he is a communist'.

The fourth argument is obviously incorrect, since its assumption is true but its conclusion is false. The third argument has both assumption and conclusion true; nonetheless it is incorrect. This will be made clear when we say in section 11.3 precisely what is meant by a correct argument. For the moment, observe that correctness of an argument depends only on the ways we use words like 'or', but does not depend on the individual sentences which are connected . Since the third argument has exactly the same form as the fourth, they must both be correct or both incorrect. Of course, it should not be particularly surprising that an incorrect argument should have a true conclusion; one can be lucky on some occasions.

The second argument, although both assumption and conclusion are false, is a correct argument. This holds for the same reasons as in the previous paragraph. The second argument has the same form as the first, so they must both be correct or both incorrect.

It is surprising, at first, that the second argument is correct. But, if we look more closely, all that an argument's correctness can tell us is that the conclusion is true provided that the assumptions are true; it cannot tell us that the assumptions are in fact true. This has some consequences in the everyday world. If someone says to you 'Be logical' it is extremely likely that what they really mean is 'Argue from my assumptions' (their hidden assumption, which they never mention, might be, for instance, 'Animals have no feelings' if they were claiming your objections to battery farming were illogical). There is, of course, no reason to accept the other person's assumptions, which often get hidden in a smoke-screen of alleged logicality. Even if you are shown that a conclusion you do not want to accept follows logically from assumptions you have accepted, that does not mean in practice that you have to accept the conclusion; it is always open to you to reconsider the assumptions, and which you do will depend on how strongly you have previously accepted the assumptions and how strongly you don't want to accept the conclusion.

The other, and more formal, aspect of logic is an analysis of what exactly we mean by saying that a statement is true or that it is provable. These turn out to be very different notions (especially in predicate logic, where truth is an essentially infinite concept, whereas a proof is something finite; this distinction does not appear in propositional logic). Nonetheless, one of the major results of formal logic is Gödel's theorem which shows that a statement is true iff it is provable (but see the warning in the introduction to this chapter).

Readers may feel some disquiet at the notion of a formal analysis of logic. After all, aren't we going to have to use logic in talking about logic, and doesn't this mean that we have to know our results anyway before we can prove them? We deal with this by distinguishing between formal logic, which is carried out in a precise formal language, and the informal logic we use to argue about the formal language; informal logic will be given in English. But if we look closely at the informal logic (and we could perform this look using yet another language, say French, to talk about it) we see that, although formal and informal logic parallel each other closely, informal logic is often much simpler. For instance, a formal result about all sentences may have a corresponding informal result, but the informal result may actually be used, not about all sentences, but only about (say) 64 specific sentences used in the specific results we wish to prove. Hilbert had hoped that it would be possible to reason about logic using informal methods that would be universally accepted as correct (in particular, using only finite, constructive reasoning), and that it might be possible to prove the consistency of mathematics by such methods. The Second Incompleteness Theorem of Gödel shows that this is not possible (at least, not if our reasoning is primarily about manipulation of formal symbols). For more about the relationship between formal and informal logic it is best to consult a book on logic with a philosophical approach, rather than a mathematical one. Such a book might well consider aspects of reasoning other than informal versions of the formal logic. For instance, it might discuss arguments of the forms 'A. So probably B' and 'A. So I believe B'. Arguments of this kind cannot be used to prove mathematical theorems, but probably every mathematician has used such arguments in discovering new results.

11.2 THE LANGUAGE OF PROPOSITIONAL LOGIC

Definition The **language P of propositional logic** is the set whose elements are

(1) the **propositional variables** p_n for all $n \in \mathbf{N}$,
(2) some of the **logical symbols** $\bot$, $\neg$, $\wedge$, $\rightarrow$, $\vee$, and $\leftrightarrow$,
(3) the **parentheses** (and).

This definition is not really satisfactory, because (2) is not precise enough. There are various choices possible in (2). We could take all the stated symbols, we could take only $\bot$, $\neg$, $\wedge$, and $\rightarrow$, we could take only $\bot$, $\wedge$, and $\rightarrow$, and other variations are possible (but we may not take only one of the symbols). The differences are unimportant, and are rather like the differences between dialects of a spoken language. For the sake of definiteness, in this section take the symbols as being, $\bot$, $\neg$, $\wedge$, and $\rightarrow$ (the reason for including $-$ is that it is used slightly differently from the other symbols). In later sections, we take the symbols as being only $\bot$, $\wedge$, and $\rightarrow$. Whichever symbols we choose, the other symbols can be regarded as auxiliary symbols, defined in terms of the given ones. This will be looked at further in the next section.

Intuitively, the symbols $\neg$, $\wedge$, $\vee$, $\rightarrow$, $\leftrightarrow$ are meant to correspond, respectively, to 'not', 'and', 'or', 'if . . . then . . .' (the word 'implies' is also used for this), and 'iff'; $\bot$ is referred to as 'lie' or 'absurdity', and it is meant to correspond to a sentence which must be false. We will not look at meanings until the next section, but the intended meanings will be a guide to the definitions we make.

For convenience, we may use p, q and r as propositional variables, rather than the more precise p_n for various n.

We will want to call certain strings **formulae.** It is obvious that $p\wedge$ should not be a formula (if we consider the intended meaning of $\wedge$ as 'and', this string doesn't correspond to a sentence); less obviously, but still fairly clearly, the string $p\wedge q\rightarrow r$ is not a formula, because there are two possible sentences it could mean. This latter example explains why parentheses are needed. It turns out that $p\wedge q$, which we might expect to be a formula, is not one; the reason is technical, as this happens because of the way it is convenient to use parentheses in formulae. We now proceed to give a definition of formulae.

Our first attempt at saying what a formula should be is the following.

(i) $\bot$ and every p_n are formulae,
(ii) if ϕ and ψ are formulae then so are $(\neg\phi)$, $(\phi\wedge\psi)$, and $(\phi\rightarrow\psi)$,
(iii) a string is a formula iff (i) and (ii) require it to be a formula.

The trouble is that, although in an intuitive sense this tells us what formulae are, condition (iii) is too imprecise to work with. So we make the following definition, which will require some justification before we can be sure it means anything.

Definition The set of **formulae** is the smallest set of strings such that

(1) $\bot$ and every p_n are in the set,

(2) if ϕ and ψ are in the set then so are $(\neg\phi)$, $(\phi\wedge\psi)$, and $(\phi\rightarrow\psi)$.

This definition would be satisfactory if we knew that there is a smallest set satisfying (1) and (2). We now proceed to prove this.

First observe that there are sets satisfying (1) and (2), for instance the set of all strings. Now suppose we have a collection of sets X_i (for i in some index set) each of which satisfies (1) and (2). It is easy to check that the intersection of the sets X_i also satisfies (1) and (2). In particular, the set W, which is defined to be the intersection of all sets satisfying (1) and (2), will itself satisfy (1) and (2). Since it is, by definition, contained in every set satisfying (1) and (2), it is the smallest set satisfying (1) and (2). We now simply define the set of formulae to be this set W.

It is easy to give examples of formulae. For instance, $(p_0\wedge p_1)$, $(\neg(p_0\wedge p_1))$, and $((\neg\ (p_0\wedge p_1))\rightarrow((\neg(p_4\rightarrow\bot))\rightarrow\bot))$ are all formulae. It is not clear from the definition how to tell whether or not a string is a formula. We look at this question shortly.

Plainly a string of length 1 is a formula iff it is either $\bot$ or p_n, while a string α of length greater than 1 is a formula iff α can be written either as $(\neg\ \beta)$ or as $(\beta\wedge\gamma)$ or $(\beta\rightarrow\gamma)$ where β is a formula and, in the two cases where it exists, γ is also a formula. Since β and γ (if it exists) are shorter than α, this seems at first sight to give an inductive definition of formulae which is simpler than the definition given. Informally, this process works (and is similar to the definition of abacus machines in Chapter 5). But if we attempted to give a formal definition along these lines, we would be defining a function from **N** into the set of all sets of strings (sending n to the set of all formulae of length $\leqslant n$). In order to justify the definition of such a function by induction, we would have to use the arguments of section 3.7. If this were done in detail, it would involve arguments similar in detail to those used in the justification of the current definition, but much more complicated.

Because the strings β and γ of the previous paragraphs are segments of α, the above can provide an inductive method of testing whether or not a string is a formula. We will obtain later extra conditions which all formulae satisfy and which shorten this search.

Many results about formulae are proved by induction. Ordinary induction (on the length of the formula) could be used, but it is usually more convenient to use the following result.

Theorem 11.1 (Principle of Induction for formulae of propositional logic) *A property of strings holds for all formulae iff it holds for* $\bot$ *and every* p_n *and holds for the strings* $(\neg\phi)$ *and* $(\phi\wedge\psi)$ *and* $(\phi\rightarrow\psi)$ *whenever it holds for* ϕ *and* ψ.

Proof Let X be the set of all strings having the property. Then X satisfies (1) and (2) in the definition of formulae. Since W, the set of all formulae, is the intersection of all sets satisfying (1) and (2), it follows that X contains W, as required. ■

It is obvious that every formula, except $\perp$ and p_n, must begin with (and end with).

Lemma 11.2 (*i*) *The number of occurrences of the left parenthesis (in a formula is the same as the number of occurrences of the right parenthesis) in the formula.*

(*ii*) *The number of occurrences of (in a proper initial segment of a formula is greater than the number of occurrences of) in this segment.*

Corollary *A proper initial segment of a formula is not a formula.*

Proof The corollary is obvious from the lemma.

Plainly $\perp$ and p_n have zero occurrences both of (and of). Suppose ϕ and ψ each have the same number of occurrences of (and), say m occurrences in ϕ and n in ψ. Then $(\neg\phi)$ has $m+1$ occurrences of each parenthesis, while $(\phi\wedge\psi)$ and $(\phi\rightarrow\psi)$ both have $m+n+1$ occurrences of each. By the Principle of Induction, all formulae have the same number of occurrences of (and).

The formulae $\perp$ and p_n have no proper initial segments. Hence all their proper initial segments have more occurrences of (than of). Now suppose ϕ and ψ have this property. The proper initial segments of $(\neg\phi)$ are plainly either (or $(\neg$ or $(\neg\alpha$, where α is a proper initial segment of ϕ, or $(\neg\phi$. Since we already know that (and) occur the same number of times in ϕ, we see that all these have more occurrences of (than of).

Now consider the proper initial segments of $(\phi\rightarrow\psi)$. They are (or $(\alpha$, where α is a proper initial segment of ϕ, or $(\phi$ or $(\phi\rightarrow$ or $(\phi\rightarrow\beta$, where β is a proper initial segment of ψ, or $(\phi\rightarrow\psi$. By the first part, ϕ and ψ have the same number of occurrences of (and), while by assumption α and β have more occurrences of (than of). Hence all these have more occurrences of (than of). A similar argument works for $(\phi\wedge\psi)$. The result follows, by the Principle of Induction. ■

In addition to proving results about formulae by induction, we will also want to define functions from the set of formulae by induction. We shall also define properties of formulae by induction; this can be regarded as defining a function (the characteristic function of the set given by the property) in order to fit it into the same framework. A formal justification of definitions by induction can be given along the lines of section 3.7 or the exercises to that section (if we follow the approach in those exercises an allowable function is a function satisfying the relevant conditions and defined either for all formulae or for all formulae of length $\leqslant n$ for some n).

There is, however, one point that needs to be cleared up before definitions by induction can be made. If it were possible to write a formula ϕ as, for instance, both $(\phi_1\wedge\phi_2)$ and $(\phi_3\rightarrow\phi_4)$, then the value of the function on ϕ would be given, on the one hand in terms of its values on ϕ_1 and ϕ_2, and on the other hand in terms of its values on ϕ_3 and ϕ_4, and these could give two incompatible expressions for the value of the function on ϕ. (Formally, if we use admissible sets, then Lemma 3.18 would go wrong if ϕ could be written in two ways. If we used allowable functions then it would not be possible to extend an allowable functio., Jefined on the set of formulae of length $\leqslant n$ to one defined on the set of formulae of length $\leqslant n+1$ if a formula could be written in two ways.) The Unique Reading Lemma below shows that this situation can never arise.

Lemma 11.3 (Unique Reading Lemma) *Any formula ϕ is of exactly one of the following forms.*
(1) $\perp$
(2) p_n *for some n which is determined by ϕ,*
(3) $(\neg\phi_1)$ *for some formula ϕ_1 which is uniquely determined by ϕ,*
(4) $(\phi_1 \wedge \phi_2)$ *for some formulae ϕ_1 and ϕ_2 which are uniquely determined by ϕ.*
(5) $(\phi_1 \to \phi_2)$ *for some formulae ϕ_1 and ϕ_2 which are uniquely determined by ϕ.*

Proof By definition every formula is of one of these forms. Plainly ϕ cannot be of form (1) or (2) and of another form. Since the propositional variables are distinct, n is given by ϕ if ϕ is of type (2).

If ϕ is of type (3) its second symbol is $\neg$. If it is of types (4) or (5) its second symbol is the first symbol of the formula ϕ_1, and this is either $\perp$ or p_n for some n or is (. Hence a formula of type (3) is not of types (4) or (5). Also, if ϕ is of type (3) the formula ϕ_1 is obtained from ϕ by removing the first two symbols of ϕ and the last symbol, and so ϕ_1 is uniquely determined by ϕ.

We have to show that a formula cannot be both of forms (4) and (5), and that the expression is unique. Both of these are done at once. Let $*$ and ! stand for either $\wedge$ or $\to$, and suppose that ϕ can be written as $(\phi_1 * \phi_2)$ and as $(\phi_3 ! \phi_4)$. Then both $(\phi_1$ and $(\phi_3$ are initial segments of ϕ. Hence one of ϕ_1 and ϕ_3 is an initial segment of the other. By the corollary to Lemma 11.2, neither can be a proper initial segment of the other, so they must coincide. Hence $*$ is the same as !, since both are the symbol immediately after the last symbol of ϕ_1. Then ϕ_2 and ϕ_4 are plainly also the same.■

Notice that, by Lemma 11.2(ii), if the formula ϕ is either $(\phi_1 \wedge \phi_2)$ or $(\phi_1 \to \phi_2)$ then the shortest string β such that (β is an initial segment of ϕ and β has the same number of left parentheses and right parentheses is ϕ_1. This leads to the following (inductive) test for whether or not a string is a formula. It follows (using Church's Thesis) that the set of formulae is recursive; more will be said about this in the next chapter.

Let α be a string. Step 1: if α is $\perp$ or p_n then α is a formula, otherwise go to step 2. Step 2: perform simple tests on α, such as 'does α begin with (and end with)' or 'does α have the same number of (as of)', and conclude that α is not a formula if it fails these tests (other simple tests can be used, for instance that of Exercise 11.1; also step 2 can be omitted if we wish, but it tends to shorten the procedure); if α passes these tests go to step 3. Step 3: if α can be written as $(\neg\beta)$ then α is a formula iff β is a formula; if α cannot be written in this way go to step 4. Step 4: find the shortest string β such that (β is an initial segment of α and β has the same number of (as of); if there is no such string, conclude that α is not a formula; if there is such a β but the symbol immediately following the last symbol of β is not $\wedge$ or $\rightarrow$, conclude that α is not a formula; finally, if this symbol is $\wedge$ or $\rightarrow$, then write α as $(\beta\wedge\gamma)$ or as $(\beta\rightarrow\gamma)$, or conclude that α is not a formula if this can't be done; then α is a formula iff both β and γ are formulae.

We can define by induction the **parsing tree** of a formula. This is a tree whose vertices are each labelled by a formula. For the precise definition of a tree see section 11.4. In fact, our inductive definition will automatically give a graph which is a tree.

Definition The **parsing tree** of a formula is given inductively as follows. If ϕ is either $\perp$ or p_n then the parsing tree of ϕ has one vertex labelled with ϕ and has no edges. If ϕ is $(\neg\phi_1)$ then the parsing tree has a vertex with label ϕ, an edge from ϕ to the vertex labelled ϕ_1 of the parsing tree of ϕ_1, the remaining edges and vertices with labels being those of the parsing tree of ϕ_1. If ϕ is $(\phi_1\wedge\phi_2)$ or $(\phi_1\rightarrow\phi_2)$ then the parsing tree of ϕ has a vertex labelled ϕ, and two edges starting at this vertex, one going to the vertex labelled ϕ_1 of the parsing tree of ϕ_1, the other to the vertex labelled ϕ_2 of the parsing tree of ϕ_2, the other edges and labelled vertices being those of the parsing trees of ϕ_1 and ϕ_2.

This definition looks complicated, but is quite easy when we draw pictures. Our pictures will be like family trees, with the start at the top, not the bottom. For convenience, when ϕ is $(\neg\phi_1)$ we draw the edge from ϕ vertically downwards, while if ϕ is $(\phi_1\wedge\phi_2)$ or $(\phi_1\rightarrow\phi_2)$ we draw the edges down and to the left and right; in the latter case, we mark the angle at the vertex with $\wedge$ or $\rightarrow$, respectively. Some examples are given in Fig. 1.

Reversing this process, if we have a tree, marked with $\wedge$ or $\rightarrow$ wherever there are two edges coming down from a vertex, and with the bottom vertices having labels which are either $\perp$ or p_n, for some n, we can build upwards from the bottom labels on the vertices until we reach at the top vertex a formula ϕ whose parsing tree this is. Examples are given in Fig. 2.

We can combine the process of finding the parsing tree and the process for determining whether or not a string is a formula into one process, which provides the parsing tree of α if α is a formula and tells us α is not a formula if it is not. Step 0: take a vertex and label it α. Perform steps 1 and 2 of the previous process. The third step becomes the following. Step 3: if α is of form $(\neg\beta)$ take the vertex labelled α and join it (by an edge going downwards) to a vertex labelled β; otherwise go to step 4. Step 4: try to find β and γ as in the previous step 4; if this cannot be done α is not a formula; if it can be done join the vertex labelled α by an edge going down and to the left to a vertex labelled β and by an edge going down and to the right to a vertex

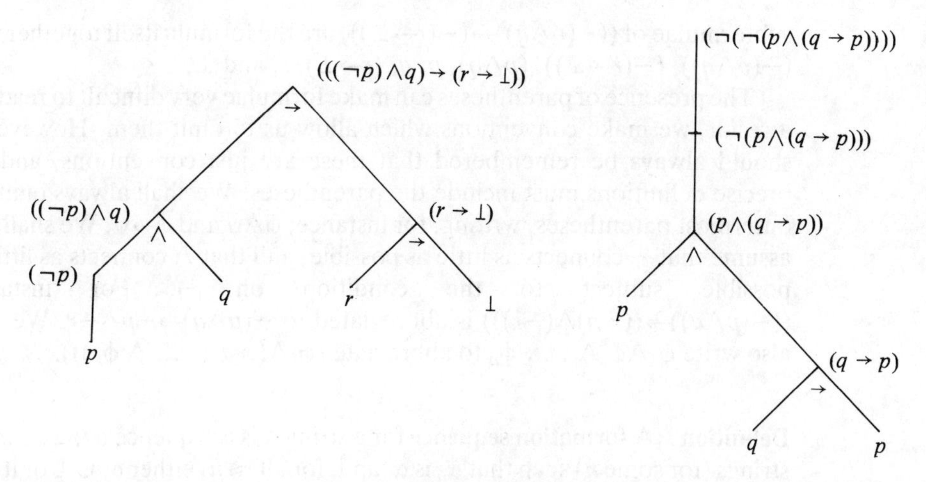

Fig. 1 — Examples of parsing trees.

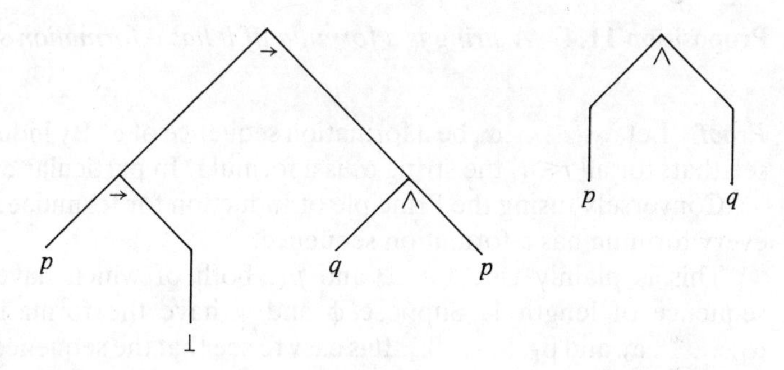

Fig. 2 — These correspond to $((p \rightarrow (\neg\bot)) \rightarrow (q \wedge p))$ and to $((\neg p) \wedge (\neg q))$, respectively.

labelled γ. Step 5: keep repeating these steps on the various strings β (and γ) obtained; if α is a formula the process will stop (when we have got strings of length 1 at every bottom vertex) with the parsing tree of α; if, at any stage in the process, we reach a string which is not a formula then the original string α is not a formula. One of these two cases must happen, since at each stage we introduce lower down the tree a shorter string than the previous one.

Definition The **subformulae** of a formula are defined inductively as follows. If ϕ is either $\bot$ or p_n then ϕ itself is the only subformula of ϕ. If ϕ is $(\neg\phi_1)$ then the subformulae of ϕ consists of ϕ itself and all subformulae of ϕ_1. If ϕ is either $(\phi_1\wedge\phi_2)$ or $(\phi_1\rightarrow\phi_2)$ then the subformulae of ϕ are ϕ itself and all subformulae of ϕ_1 or of ϕ_2.

For instance, the subformulae of $(p\wedge q)$ are $(p\wedge q)$, p, and q, while the subformulae of $((\neg(p\wedge q))\rightarrow(\neg(r\rightarrow\perp)))$ are the formula itself together with $(\neg(p\wedge q))$, $(\neg(r\rightarrow\perp))$, $(p\wedge q)$, p, q, $(r\rightarrow\perp)$, r, and $\perp$.

The presence of parentheses can make formulae very difficult to read. As a result, we make conventions which allow us to omit them. However, it should always be remembered that these are just conventions, and the precise definitions must include the parentheses. We shall always omit the outermost parentheses, writing, for instance, $\phi\wedge\psi$ and $\phi\rightarrow\psi$. We shall also assume that $\neg$ connects as little as possible, and that $\wedge$ connects as little as possible subject to the condition on $\neg$. For instance, $((\neg(p\wedge q))\rightarrow((\neg p)\wedge(\neg r)))$ is abbreviated to $\neg(p\wedge q)\rightarrow\neg p\wedge\neg r$. We shall also write $\phi_1\wedge\phi_2\wedge\ldots\phi_n$ to abbreviate $(\phi_1\wedge(\phi_2\wedge(\ldots\wedge\phi_n)))\ldots$.

Definition A **formation sequence** for a string α is a sequence $\alpha_1,\ldots,\alpha_n$ of strings (for some n) such that α_n is α, and, for all $r\leqslant n$, either α_r is $\perp$ or it is p_k for some k, or there is some $i<r$ such that α_r is $(\neg\alpha_i)$ or there are some i,j both less than r such that α_r is either $(\alpha_i\wedge\alpha_j)$ or $(\alpha_i\rightarrow\alpha_j)$.

Proposition 11.4 *A string is a formula iff it has a formation sequence.*

Proof Let $\alpha_1,\ldots,\alpha_n$ be a formation sequence of α. By induction on r, we see that, for all $r\leqslant n$, the string α_r is a formula. In particular α is a formula.

Conversely, using the Principle of Induction for formulae, we show that every formula has a formation sequence.

This is plainly true for $\perp$ and p_k, both of which have a formation sequence of length 1. Suppose ϕ and ψ have the formation sequences $\alpha_1,\ldots,\alpha_n$ and $\beta_1,\ldots,\beta_m$. It is easy to see that the sequence $\alpha_1,\ldots,\alpha_n$, $(\neg\phi)$ of length $n+1$ is a formation sequence for $(\neg\phi)$, because α_n is ϕ. Also the sequence $\alpha_1,\ldots,\alpha_n,\beta_1,\ldots,\beta_m,(\phi\rightarrow\psi)$ of length $m+n+1$ is a formation sequence for $(\phi\rightarrow\psi)$, and similarly for $(\phi\wedge\psi)$. The result follows by induction.■

Note that any formula will have infinitely many formation sequences, since we can always insert irrelevant entries in the sequence. Even if we only consider the shortest formation sequences they will not be unique, because we can alter the order of the entries.

Exercise 11.1 Show that any formula has the same number of left parentheses as it has (in total) of the symbols $\neg$, $\wedge$, and $\rightarrow$.

Exercise 11.2 Show that the subformulae of a formula ϕ are exactly those formulae which occur in the parsing tree of ϕ.

Exercise 11.3 Show that every subformula of ϕ occurs in every formation sequence of ϕ.

Exercise 11.4 Show that every formula ϕ has a formation sequence consisting entirely of subformulae of ϕ.

Exercise 11.5 Find the parsing tree and a formation sequence for $(\neg(\neg(\neg p)))$ and for $(((\neg p)\rightarrow(p\wedge q))\wedge(\neg(\bot\rightarrow q)))$.

Exercise 11.6 Find the formulae whose parsing trees have the patterns shown in Fig. 3.

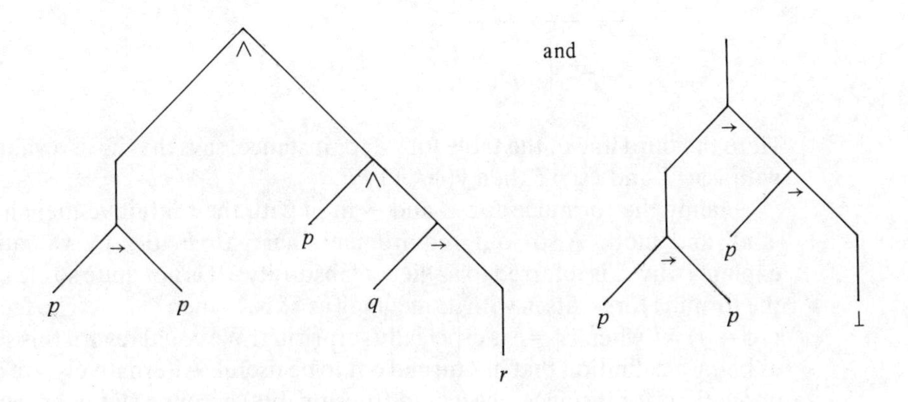

Fig. 3.

11.3 TRUTH

Up to now formulae have simply been regarded as strings, with no meaning attached to them. We now see how to give them meanings, and to talk of their truth or falsehood.

From now on, small Greek letters will denote formulae (we shall never need to look at arbitrary strings) and capital Greek letters will denote sets of formulae.

We take the two-element set $\{T,F\}$, where we think of T and F as standing for 'true' and 'false'. We then define a valuation as a function from W, the set of all formulae, to $\{T,F\}$ satisfying certain conditions.

Definition A **valuation** is a function v from W to $\{T,F\}$ such that

$$v\bot=F,\ v(\neg\phi)=T \text{ iff } v\phi=F,\ v(\phi\wedge\psi)=T \text{ iff } v\phi=T=v\psi, \text{ and}$$
$$v(\phi\rightarrow\psi)=F \text{ iff } v\phi=T \text{ and } v\psi=F.$$

We can express this information in a table, known as a **truth-table.** The truth-tables for $\wedge$ and $\rightarrow$ are

ϕ	ψ	$\phi \wedge \psi$
T	T	T
T	F	F
F	T	F
F	F	F

and

ϕ	ψ	$\phi \rightarrow \psi$
T	T	T
T	F	F
F	T	T
F	F	T

respectively,

while that for $-$ is

ϕ	$\neg\phi$
T	F
F	T

Here the third line of the table for $\wedge$, for instance, says that if v is a valuation with $v\phi = F$ and $v\psi = T$ then $v(\phi \wedge \psi) = F$.

Plainly the formulae for $\wedge$ and $\neg$ fit in with their intuitive meanings as 'and' and 'not'. Also, our requirement that $v\bot = F$ for any valuation v explains why $\bot$ is referred to as 'lie' or 'absurdity'. It is not quite so clear how the formula for $\rightarrow$ fits in with its meaning as 'if . . . then . . . '. The fact that $v(\phi \rightarrow \psi) = T$ when $v\phi = F$ is especially surprising. We could regard this simply as being a definition that has turned out to be useful. Alternatively, we could argue that, for instance, we would (presumably) accept a statement such as 'For all n in **N**, if 4 divides n, then 2 divides n' as being true. But then we would (probably) insist that this requires the truth of the specific instances 'if 4 divides 5, then 2 divides 5' and 'if 4 divides 6, then 2 divides 6'. Thus it does seem that, if we want $v(\phi \rightarrow \psi)$ to depend only on $v\phi$ and $v\psi$ and if we also want the symbol $\forall$, which is used in the predicate logic considered later, to fit in with its intended meaning as 'for all', we are forced into the given definition.

Now consider the truth-table for $\neg(\neg\phi \wedge \neg\psi)$. It is

ϕ	ψ	$\neg\phi$	$\neg\psi$	$(\neg\phi \wedge \neg\psi)$	$\neg(\neg\phi \wedge \neg\psi)$
T	T	F	F	F	T
T	F	F	T	F	T
F	T	T	F	F	T
F	F	T	T	T	F

Now, if the symbol $\vee$ is to correspond to its intuitive meaning of 'or', we would like $v(\phi \vee \psi)$ to be T except in the case when $v\phi = F = v\psi$. From the above truth-table, we see that this can be done by defining $\phi \vee \psi$ to be an abbreviation for $\neg(\neg\phi \wedge \neg\psi)$ (note that, in accordance with our conventions, we have omitted the outer pair of parentheses). When we use $\vee$ and omit parentheses, we require it to govern as little as possible subject to the rule that $\neg$ must govern as little as possible.

Similarly, we define $\phi\leftrightarrow\psi$ to be an abbreviation of $(\phi\rightarrow\psi)\wedge(\psi\leftrightarrow\phi)$, and we find that $v(\phi\leftrightarrow\psi)=T$ iff $v\phi=v\psi$.

We also find that $v(\phi\rightarrow\perp)=T$ iff $v\phi=F$, so that $v(\phi\rightarrow\perp)=v(\neg\phi)$.

Because of this, we now change the logical symbols used in our language to $\perp$, $\wedge$, and $\rightarrow$ only, regarding $\neg$ as an auxiliary symbol with $\neg\phi$ being simply an abbreviation of $\phi\rightarrow\perp$. It follows that, in inductive arguments, we have only to show that a property holds for $\phi\wedge\psi$ and $\phi\rightarrow\psi$ when it holds for ϕ and ψ.

We refer to the symbols $\wedge$, $\vee$ and $\neg$ as 'conjunction', 'disjunction', and 'negation', respectively.

Let S be any set of propositional variables. We define $W(S)$ to be the set of those formulae whose propositional variables are in S. A **partial valuation** is a function with values in $\{T,F\}$, whose domain is $W(S)$ for some S, which satisfies the conditions for a valuation when the formulae are in $W(S)$.

Let v and w be partial valuations defined on ϕ, and suppose that $vp_n=wp_n$ for every propositional variable p_n occurring in ϕ. It is easy to check (inductively) that $v\phi=w\phi$.

Let S be any set of propositional variables. Because we can make definitions by induction, any function v from S to $\{T,F\}$ extends to a unique partial valuation with domain $W(S)$.

The formula ϕ is called a **tautology** if $v\phi=T$ for all valuations v, and is called a **contradiction** if $v\phi=F$ for all v. If there is at least one v with $v\phi=T$, we say ϕ is **satisfiable.** When $v\phi=T$, we say that ϕ **satisfies** v (or, if it is more convenient, that v **satisfies** ϕ). If Γ is a set of formulae, we say v satisfies Γ (or Γ satisfies v) if v satisfies γ for every γ in Γ.

Plainly ϕ is a contradiction iff $\neg\phi$ is a tautology iff ϕ is not satisfiable. At first sight it is not clear how to tell whether or not ϕ is a tautology (or a contradiction or satisfiable) since there are infinitely many valuations to consider. But let S be the set of propositional variables occurring in ϕ. Let v be any valuation, and let w be its restriction to S. Then w extends uniquely to a partial valuation w defined on $W(S)$, and $v\phi=w\phi$. Hence we do not have to look at all valuations, but only at those partial valuations which extend a function from S to $\{T,F\}$. If S has n elements there are 2^n such functions, and we simply have to check all the corresponding partial valuations on ϕ.

There is no known method of finding whether or not ϕ is a tautology (or satisfiable) which is essentially faster than making this check. We can do this by making a truth-table for ϕ. However, there are methods which may be faster in particular cases.

Consider the formula $(\theta\rightarrow(\phi\rightarrow\psi))\rightarrow((\theta\rightarrow\phi)\rightarrow(\theta\rightarrow\psi))$, which we wish to show is a tautology. First observe that we do not need a truth-table with 2^n lines, where n is the number of propositional variables occurring in the formula. A truth-table of eight lines will be enough, corresponding to the eight possibilities for $v\theta$, $v\phi$, and $v\psi$. The reader should construct such a truth-table, and will find it is quite complicated because of its large number of columns (and even eight rows is a lot to deal with).

Now ask instead what conditions a valuation v must satisfy if v sends this formula to F. By the truth-table for $\rightarrow$, it must have $v(\theta\rightarrow(\phi\rightarrow\psi))=T$ and

$v((\theta\to\phi)\to(\theta\to\psi))=F$. The second of these requires $v(\theta\to\phi)=T$ and $v(\theta\to\psi)=F$, which in turn requires $v\theta=T$ and $v\psi=F$. Since $v\theta$ and also $v(\theta\to\phi)$ and $v(\theta\to(\phi\to\psi))$ are all T, we then get that $v\phi=T$ and $v(\phi\to\psi)=T$. But these contradict $v\psi=F$, so no such v can exist.

Let ϕ be a formula and Γ a set of formulae. We call ϕ a **semantic consequence** of Γ, and write $\Gamma\models\phi$ if $v\phi=T$ for every v which satisfies Γ. If we do not have $\Gamma\models\phi$ we write $\Gamma\not\models\phi$.

We can now explain what a correct argument (for propositional logic) is. Let Γ be a set of formulae, which we will assume uses any of $\neg$, $\vee$, $\wedge$, $\to$, and $\leftrightarrow$ (but, for convenience, does not use $\perp$). Associate, in some way, an English sentence to every subformula of every formula in Γ. Do this so that if A_1 and A_2 correspond to ϕ_1 and ϕ_2 then the sentence corresponding to $\phi_1\to\phi_2$ has the same meaning as 'if A_1 then A_2', the sentence corresponding to $\phi_1\wedge\phi_2$ has the same meaning as 'A_1 and A_2', and so on. (We must say 'has the same meaning as' and not 'is', since we can express these meanings in various ways. For instance, 'A_1 only if A_2' has the same meaning as 'if A_1 then A_2' and A_1 but A_2' has the same meaning as 'A_1 and A_2'.) We refer to this collection of sentences as an example of Γ.

Now let $A_1, \ldots, A_n$, and B be sentences. We say that 'A_1. A_2. . . . A_n. Therefore B' is a correct argument if there are formulae $\phi_1, \ldots, \phi_n$, and ψ such that $\{\phi_1, \ldots, \phi_n\}\models\psi$ and there is an example of $\{\phi_1, \ldots, \phi_n, \psi\}$ with A_i corresponding to ϕ_i and B to ψ. In our original examples in the previous section, the first two correct arguments come from $p\wedge q\models q$ and the third from $\{p\wedge\neg q\to r\vee s, p\wedge\neg r\}\models q\vee s$. The incorrect arguments can only (apart from renaming the propositional variables) be examples of $\{p\vee q, q\}$, and they are incorrect because $p\vee q\not\models q$.

Exercise 11.7 Show that the following are tautologies.

(a) $\phi\to(\theta\to\phi)$
(b) $(\theta\to\phi)\to(-\phi\to-\theta)$
(c) $\theta\wedge(\phi\vee\psi)\leftrightarrow(\theta\wedge\phi)\vee(\theta\wedge\psi)$
(d) $\neg\theta\to(\theta\to\phi)$
(e) $\neg\neg\phi\leftrightarrow\phi$.

Exercise 11.8 Show that $p\vee q\to p\wedge q$ and $(p\to q)\to(p\to\perp)$ are satisfiable but not tautologies.

11.4 PROOF

In this section we consider the notion of **proof** or **derivation.** (These are two words for the same notion. We will usually keep the word 'proof', though, for the informal notion, and call the formal notion a derivation; but we may say 'provable' rather than 'derivable'.) This will turn out to be a finite process of manipulation of sets of formulae which does not require any mention of truth. The connection with truth will be made later, and is, of course, important.

There are three standard definitions of a proof (or derivation) of a formula ϕ from a set of formulae Γ. Although these give different definitions for what a derivation is, it can be shown that the set of formulae derivable from a given set is the same under all three definitions. They are the **axiomatic method,** the **method of natural deduction**, and the **tableau method.** The tableau method is very similar to the method of natural deduction, but is particularly suited to automated theorem-proving. We shall not say anything further about it.

The axiomatic method is very easy to define. We begin by specifying a set Δ of **logical axioms**; this chosen set can be shown to be recursive. We then define a derivation of ϕ from Γ to be a sequence $\phi_1, \ldots, \phi_n$ of formulae such that ϕ_n is ϕ and, for all $r \leq n$, either ϕ_r is in Γ or it is in Δ or there are i and j less than r such that ϕ_j *is* $\phi_1 \rightarrow \phi_r$. This corresponds to the intuitive property that we should be able to derive ψ from θ and $\theta \rightarrow \psi$. This is a very straightforward definition. It is also easy to check that the set of formulae which can be derived from Γ is r.e. if Γ is recursive. However, it turns out that derivations by the axiomatic method are long and complicated (unless one fudges the issue by including so many logical axioms that any derivation one actually wants is obvious). Also there is no intuitive link between a derivable formula and its derivation. We will therefore concentrate on derivations by natural deduction.

The precise definition of a derivation by natural deduction is much more complicated than for the axiomatic method (in the axiomatic method, most of the complication goes into the choice of the set Δ of logical axioms). We shall have to begin by defining **trees.** It is reasonably clear intuitively what a tree is, and for much of the theory an intuitive idea is all we need. Some examples are given in Fig. 4, which also explains the name 'tree'.

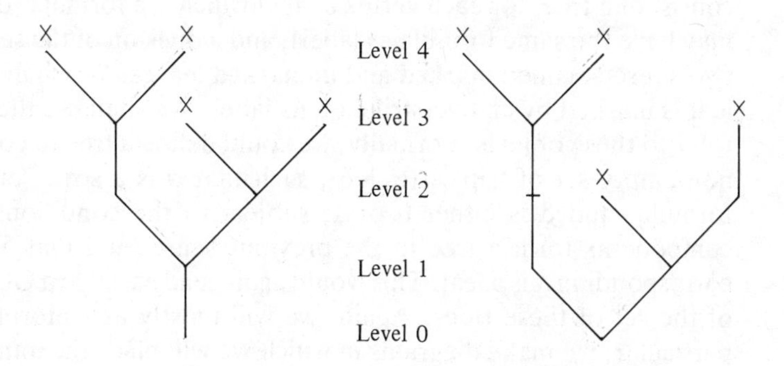

Fig. 4 — Some trees, whose leaves are indicated by ×.

Geometrically a tree can be defined as a graph (that is, a set of vertices and set of edges, with each edge having two associated vertices) such that any two vertices are joined by exactly one path. Because our trees have a starting point, it is convenient to add the following extra condition. To each vertex there corresponds a number (the level of the vertex) such that the levels of vertices joined by an edge differ by 1; there is exactly one vertex (the **root**) of level 0, and any vertex of level $n>0$ is joined to exactly one vertex of level $n-1$. A vertex is called a **leaf** if it is not joined to any vertex at higher level. It can be shown that any graph satisfying this condition is a tree.

The only trees we will be looking at have at most two immediate successors to any vertex; this restriction will be made without further comment. For such trees the following is a convenient formal definition, because it also leads very straightforwardly to a Gödel numbering of the set of trees. For most purposes this definition can be ignored as long as the reader has an intuitive understanding of what a tree is like.

Definition A **tree** is a non-empty finite set T of strings on $\{0,1\}$ satisfying the following conditions:

(1) no string in T begins with 0,
(2) if the string $\alpha 0$ is in T, then the string α is in T,
(3) if the string $\alpha 1$ is in T, then the string $\alpha 0$ is in T.

We refer to the elements of T as **vertices.** (No confusion should occur with the use of T in the previous section. They are very different objects, and we will not need to use both notations together.) If α is a vertex of T, we refer to those of the strings $\alpha 0$ and $\alpha 1$ which are in T as **immediate successors** of α. The string 1 is called the **root** of T. A string α with no immediate successor is called a **leaf** of T.

The objects we shall need to consider are more than just trees. They will consist of a tree, for each vertex a label which is a formula (distinct vertices may have the same formula as label), and a division of the set of leaves into two subsets, called **marked** and **unmarked** leaves. We shall indicate that a leaf is marked by an overstrike on its label. We shall use the word 'tree' to refer to these objects. Formally, we could define a tree to consist of a finite non-empty set of triples (α,ϕ,δ), such that α is a string on $\{0,1\}$, ϕ is a formula, and δ is either 0 or 1, subject to the conditions that the first components form a tree in the previous sense, and that $\delta=1$ only if the corresponding α is a leaf. This would again lead easily to a Gödel numbering of the set of these trees. Again, we will mostly use informal pictures. In particular, we make diagrams in which we will place the immediate successors of a vertex α above that vertex, directly above if $\alpha 0$ is the only immediate successor of α, while if α has immediate successors $\alpha 0$ and $\alpha 1$, we put $\alpha 0$ above and to the left of α, and $\alpha 1$ above and to the right of α.

We shall need various ways of building new trees from old ones, and of representing these processes diagrammatically. Thus we shall write $\begin{matrix} T \\ \phi \end{matrix}$ to

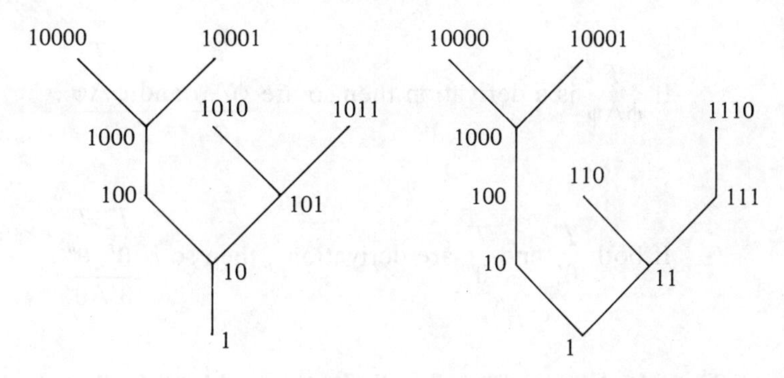

Fig. 5 — The trees of Fig. 4 with the strings corresponding to their vertices.

denote a tree whose root is labelled ϕ, $\dfrac{\overset{T}{\theta}}{\phi}$ to denote a tree whose root is labelled ϕ and for which the root has one immediate successor such that the tree $\overset{T}{\theta}$ is left when the root is removed, and $\dfrac{\overset{T'}{\theta'}\ \overset{T''}{\theta''}}{\phi}$ for the tree with root ϕ where the root has two immediate successors and such that the two trees $\overset{T'}{\theta'}$ and $\overset{T''}{\theta''}$ remain when the root is removed. Finally, the notation $\overset{\not\theta}{T}$ is the tree obtained from T by requiring all leaves labelled θ (if any) to be marked, the originally marked leaves remaining marked.

We now call certain trees **derivations,** the definition being inductive. As with the definition of formulae, there are certain difficulties associated with this inductive definition. We could be more precise and define the set of derivations as the smallest set of trees satisfying certain conditions, showing that there is a smallest such set. We can then show that a tree is a derivation iff it is built up from smaller derivations according to the relevant rules.

Definition A tree with only one vertex is a derivation iff that vertex is unmarked. A tree with more than one vertex is a derivation iff it is made from smaller derivations by one of the six following rules.

The rules E$\wedge$ and I$\wedge$ **for elimination and introduction** of $\wedge$.

If $\begin{matrix} T \\ \phi\wedge\psi \end{matrix}$ is a derivation then so are $\dfrac{\begin{matrix} T \\ \phi\wedge\psi \end{matrix}}{\phi}$ and $\dfrac{\begin{matrix} T \\ \phi\wedge\psi \end{matrix}}{\psi}$.

If both $\begin{matrix} T' \\ \theta' \end{matrix}$ and $\begin{matrix} T'' \\ \theta'' \end{matrix}$ are derivations, then so is $\dfrac{\begin{matrix} T' & T'' \\ \theta' & \theta'' \end{matrix}}{\theta'\wedge\theta''}$.

The rules E→ and I→ **for elimination and introduction** of →.

If both $\begin{matrix} T' \\ \phi \end{matrix}$ and $\begin{matrix} T'' \\ \phi\rightarrow\psi \end{matrix}$ are derivations, then so is $\dfrac{\begin{matrix} T' & T'' \\ \phi & \phi\rightarrow\psi \end{matrix}}{\psi}$.

If $\begin{matrix} T \\ \psi \end{matrix}$ is a derivation, then so is $\dfrac{\begin{matrix} \not{\phi} \\ T \\ \psi \end{matrix}}{\phi\rightarrow\psi}$.

Notice that in using the rule for introduction of →, we do not require any leaves to have the label ϕ. In particular, $\dfrac{\psi}{\phi\rightarrow\psi}$ is a derivation.

The **rules** ⊥ and **RAA**.

If $\begin{matrix} T \\ \bot \end{matrix}$ is a derivation, then so is $\dfrac{\begin{matrix} T \\ \bot \end{matrix}}{\phi}$, for any ϕ.

If $\begin{matrix} T \\ \bot \end{matrix}$ is a derivation, then so is $\dfrac{\begin{matrix} \not{\neg\phi} \\ T \\ \bot \end{matrix}}{\phi}$, for any ϕ.

Notice the difference between these two cases. In the second rule (which is called RAA, because of its connection with proof by contradiction, also called Reductio Ad Absurdum) in the new derivation additional leaves are marked, namely all those labelled with $\neg\phi$, whereas in the rule $\perp$, the new derivation has the same marked leaves as the original one.

These ways of building new derivations from old ones are very similar to the ways we construct informal proofs, which is why this approach is called natural deduction. For instance, the rule for introduction of $\rightarrow$ corresponds to the fact that if we want to prove 'if P, then Q' from given hypotheses, we usually add P to the hypotheses and then try to prove Q. Similarly, the rule RAA mimics the informal proof by contradiction.

The formula which is the label at the root of a derivation D is called the **conclusion** of D, and the labels on the unmarked leaves are called the **hypotheses** of D. We write $\Gamma \vdash \phi$ (and say ϕ can be **derived from** Γ) if there is a derivation with conclusion ϕ and hypotheses in Γ. In particular, we write $\vdash \phi$, and call ϕ a **theorem,** if $\Gamma \vdash \phi$ with Γ empty. We also write $\Gamma \vdash \phi$ when we do not have $\Gamma \vdash \phi$.

Since $\neg\phi$ is short for $\phi \rightarrow \perp$, we have derivations $\dfrac{\phi \quad \neg\phi}{\perp}$ and $\dfrac{\neg\phi \quad \neg\neg\phi}{\perp}$. Applying the rule for introduction of $\rightarrow$ to the first of these, and the rule RAA to the second, we find that $\phi \vdash \neg\neg\phi$ and $\neg\neg\phi \vdash \phi$.

The rules for introduction and elimination of $\rightarrow$ show at once that $\Gamma \cup \{\phi\} \vdash \psi$ iff $\Gamma \vdash \phi \rightarrow \psi$. In the axiomatic approach, this is a deep theorem.

In order to keep track of the stage at which various leaves become marked, it is convenient to number each marking of a leaf and put the corresponding number on the formula which requires this marking. These numbers are not part of the formal definition, but are simply aids to comprehension.

Lemma 11.5 *(i) If* $\Gamma \vdash \phi$ *and* $\Gamma \subset \Gamma'$ *then* $\Gamma' \vdash \phi$. *(ii) If* $\Gamma \vdash \phi$ *then* $\Gamma_0 \vdash \phi$, *for some finite subset* Γ_0 *of* Γ. *(iii) If* $\Gamma \vdash \phi$ *and* $\Delta \vdash \gamma$, *for every* γ *in* Γ *then* $\Delta \vdash \phi$.

Proof (i) is obvious, by definition. (ii) is also obvious, since a derivation is finite.

Suppose the assumptions of (iii) hold. By (ii), we may assume Γ is finite, and we can use induction on its size. If Γ is empty, then $\Gamma \subset \Delta$, and the result follows by (i). Suppose Γ is $\{\gamma_1, \ldots, \gamma_n\}$. As $\Gamma \vdash \phi$, we know that $\{\gamma_1, \ldots, \gamma_{n-1}\} \vdash \gamma_n \rightarrow \phi$. Inductively, $\Delta \vdash \gamma_n \rightarrow \phi$. Since $\Delta \vdash \gamma_n$, by assumption, the rule for eliminating $\rightarrow$ tells us that $\Delta \vdash \phi$.■

Lemma 11.6 *(i)* $\Gamma \cup \{\phi\} \vdash \psi$ *iff* $\Gamma \cup \{\neg\psi\} \vdash \neg\phi$. *(ii)* $\Gamma \cup \{\neg\phi\} \vdash \psi$ *iff* $\Gamma \cup \{\neg\psi\} \vdash \phi$. *(iii)* $\Gamma \cup \{\phi\} \vdash \neg\psi$ *iff* $\Gamma \cup \{\psi\} \vdash \neg\phi$.

Proof (i) Let D be a derivation of ψ from $\Gamma \cup \{\phi\}$. Then $\dfrac{\begin{matrix} D \\ \psi \end{matrix} \quad \neg\psi}{\bot}$ is a derivation, and so

$$\dfrac{\begin{matrix} \not\phi \ (1) \\ D \\ \dfrac{\phi \quad \neg\psi}{\bot} \end{matrix}}{\phi \to \bot \ (1)}$$

is also a derivation; its conclusion is $\neg\phi$, and its hypotheses are obtained from the hypotheses of D by adding $\neg\psi$ and removing ϕ. Thus it is a derivation of $\neg\phi$ from $\Gamma \cup \{\neg\psi\}$. Conversely, let D be a derivation of $\neg\phi$ from $\Gamma \cup \{\neg\psi\}$. Then $\dfrac{\begin{matrix} D \\ \neg\phi \end{matrix} \quad \phi}{\bot}$ is a derivation. Applying RAA to this, we get a derivation of ψ whose hypotheses are those of D with ϕ added and $\neg\psi$ removed, as required.

(ii) Let D be a derivation of ψ from $\Gamma \cup \{\neg\phi\}$. Then we have a derivation

$$\begin{matrix} \not\neg\phi \\ D \\ \dfrac{\psi \quad \neg\psi}{\bot} \\ \hline \phi \end{matrix}$$

of ϕ from $\Gamma \cup \{\neg\psi\}$. The last line of this derivation comes by RAA. By symmetry, the converse is also true.

(iii) Finally, let D be a derivation of $\neg\psi$ from $\Gamma \cup \{\phi\}$. Then we have a derivation

$$\begin{matrix} \not\phi \\ D \\ \dfrac{\psi \quad \neg\psi}{\bot} \\ \hline \neg\phi \end{matrix}$$

of $\neg\phi$ from $\Gamma \cup \{\psi\}$, the last line coming by introduction of $\to$. (It is frequently necessary to remember that $\neg\phi$ is $\phi \to \bot$.)■

These two lemmas will be used frequently without specific mention. It is often easier to use them rather than to construct derivations in detail.

I leave to the reader the proof that $\alpha \vdash \alpha \vee \beta$ and $\beta \vdash \alpha \vee \beta$, from which it follows that $\Gamma \vdash \alpha \vee \beta$ if either $\Gamma \vdash \alpha$ or $\Gamma \vdash \beta$. These will be referred to as the rules for **right-introduction** of $\vee$.

Now suppose that $\Gamma \cup \{\alpha\} \vdash \phi$ and $\Gamma \cup \{\beta\} \vdash \phi$. Then $\Gamma \cup \{\neg\phi\} \vdash \neg\alpha$ and $\Gamma \cup \{\neg\phi\} \vdash \neg\beta$. Hence $\Gamma \cup \{\neg\phi\} \vdash \neg\alpha \wedge \neg\beta$. Since $\alpha \vee \beta$ is defined to be $\neg(\neg\alpha \wedge \neg\beta)$, we find that $\Gamma \cup \{\alpha \vee \beta\} \vdash \phi$. We call this the rule for **left-introduction** of $\vee$. As a particular case, observe that $\vdash \theta \vee \neg\theta$, from which it follows that $\Gamma \vdash \phi$ if both $\Gamma \cup \{\theta\} \vdash \phi$ and $\Gamma \cup \{\neg\theta\} \mid \phi$. Also, if $\phi \vdash \psi$ then $\theta \vee \phi \mid \theta \vee \psi$.

We have seen that $\psi \vdash \phi \to \psi$. It follows from Lemma 11.6 that $\neg\psi$ can be derived from $\neg(\phi \to \psi)$. Also, ϕ can be derived from $\neg(\phi \to \psi)$ by the derivation below.

$$\dfrac{\dfrac{\dfrac{\dfrac{\dfrac{{}^{(2)}\,\cancel{\neg\phi}\quad \cancel{\phi}\,{}^{(1)}}{\bot}}{\psi}}{(1)\ \phi\to\psi} \qquad \neg\,(\phi\to\psi)}{\bot}}{\phi\ (2)}$$

Here we first use introduction of $\to$ to get the derivation of $\phi\to\psi$ from $\neg\,\phi$, and then apply RAA to get the derivation we need.

As a further example, we show that $\vdash\ (\theta\to(\phi\to\psi))\to((\theta\to\phi)\to(\theta\to\psi))$; we have already seen that this formula is a tautology. The derivation is the following.

$$\dfrac{\dfrac{\dfrac{\dfrac{\dfrac{{}^{(1)}\,\cancel{\theta}\quad \cancel{\theta\to\phi}\,{}^{(2)}}{\phi} \qquad \dfrac{{}^{(1)}\,\cancel{\theta}\quad \cancel{\theta\to(\phi\to\psi)}\,{}^{(3)}}{\phi\to\psi}}{\psi}}{\theta\to\psi\quad (1)}}{(\theta\to\phi)\to(\theta\to\psi)\quad (2)}}{(\theta\to(\phi\to\psi))\to((\theta\to\phi)\to(\theta\to\psi))\quad (3)}$$

As another example, we show that $\vdash\ ((\phi\to\psi)\to\phi)\to\phi$. The derivation is

$$\dfrac{\dfrac{\dfrac{\dfrac{\dfrac{\dfrac{{}^{(2)}\,\cancel{\neg\phi}\quad \cancel{\phi}\,{}^{(1)}}{\bot}}{\psi}}{(1)\quad \phi\to\psi} \quad (\phi\to\psi)\to\phi}{\phi} \qquad \cancel{\neg\phi}\,{}^{(2)}}{\dfrac{\bot}{\phi}\ (2)}}{((\phi\to\psi)\to\phi)\to\phi}$$

Here the next to last line comes by RAA. These are the first examples where two leaves are marked at the same time.

While it is easy to check that the claimed derivations are indeed derivations, the reader may well be wondering how such derivations can be obtained. The following techniques can be used to construct derivations of ϕ from a finite set Γ (assuming ϕ can be derived from Γ). It can be shown that, with slight modifications, these techniques are certain to give a derivation if it exists.

(1) If ϕ is $\phi_1 \wedge \phi_2$ look for derivations of ϕ_1 and ϕ_2 from Γ, and apply the rule for introduction of $\wedge$.

(2) If ϕ is $\phi_1 \rightarrow \phi_2$ look for a derivation of ϕ_2 from $\Gamma \cup \{\phi_1\}$, and apply the rule for introduction of $\rightarrow$.
(3) If Γ is $\Delta \cup \{\alpha_1 \wedge \alpha_2\}$ look for a derivation of ϕ from $\Delta \cup \{\alpha_1, \alpha_2\}$. This can be modified to a derivation of ϕ from Γ, using the rule for elimination of $\wedge$.
(4) If Γ is $\Delta \cup \{\alpha_1, \alpha_1 \rightarrow \alpha_2\}$ look for a derivation of ϕ from $\Delta \cup \{\alpha_1, \alpha_2\}$, and modify it using the rule for elimination of $\rightarrow$.
(5) If $\Gamma \vdash \beta$ look for a derivation of ϕ from $\Gamma \cup \{\beta\}$. By Lemma 11.5(iii) this can be modified to a derivation of ϕ from Γ.
(6) If Γ is $\Delta \cup \{\alpha_1 \rightarrow \alpha_2\}$ we may combine the previous two possibilities. That is, we can look for a derivation of α_1 from Δ. If we find one and we also find a derivation of ϕ from $\Delta \cup \{\alpha_1, \alpha_2\}$ we can combine the two to get a derivation of ϕ from Γ.
(7) If none of the above work we may try to use RAA, and then attempt to apply the other techniques.

The following theorems will be proved for predicate logic in section 12.3. The proof for propositional logic is a simplification of the later proofs, simply omitting those aspects which are not relevant.

Theorem 11.7 *The set of derivations is recursive.*

Theorem 11.8 *Let Γ be recursive. Then $\{\phi; \Gamma \vdash \phi\}$ is r.e.*

Lemma 11.9 *Let Γ be r.e. Then there is a recursive set Γ' such that $\Gamma \vdash \phi$ iff $\Gamma' \vdash \phi$.*

Theorem 11.10 *Let Γ be r.e. Then $\{\phi; \Gamma \vdash \phi\}$ is r.e.*

By contrast, the following theorem is true only for propositional logic.

Theorem 11.11 *Let Γ be finite. Then $\{\phi; \Gamma \vdash \phi\}$ is recursive.*

Proof Let Γ be $\{\gamma_1, \ldots, \gamma_n\}$, and define α to be $\gamma_1 \wedge \ldots \wedge \gamma_n$. It is easy to check that $\Gamma \vdash \phi$ iff $\alpha \vdash \phi$ iff $\vdash \alpha \rightarrow \phi$. Since the map sending ϕ to $\alpha \rightarrow \phi$ is recursive (by Church's Thesis), it is enough to prove that the set of theorems is recursive.

Now we know that the set of tautologies is recursive. By the Completeness Theorem below, tautologies and theorems are the same, so the result follows.■

In the following exercises, you may either construct detailed derivations or use results such as Lemmas 11.5 and 11.6 to show that derivations exist. You are recommended to look at both methods.

Exercise 11.9 Prove that $\alpha \vdash \alpha \vee \beta$ and $\beta \vdash \alpha \vee \beta$. Deduce the rule for right-introduction of $\vee$.

Exercise 11.10 Show that $\vdash ((\theta \to \phi) \to (\theta \to \psi)) \to (\theta \to (\phi \to \psi))$.

Exercise 11.11 Show that $\vdash (\theta \to (\phi \to \psi)) \leftrightarrow ((\theta \wedge \phi) \to \psi)$.

Exercise 11.12 Show that $\{-\theta, \theta \vee \phi\} \vdash \phi$ and that $\vdash \theta \vee (\phi \wedge \psi) \leftrightarrow (\theta \vee \phi) \wedge (\theta \vee \psi)$.

Exercise 11.13 Suppose that $\phi_i \vdash \psi_i$ and $\psi_i \vdash \phi_i$ for $i = 1,2$. Show that $\phi_1 \wedge \phi_2 \vdash \psi_1 \wedge \psi_2$ and that $\phi_1 \to \phi_2 \vdash \psi_1 \to \psi_2$.

11.5 SOUNDNESS

Whenever we have notions of truth and proof (in the current situation the relevant notions are tautology and theorem) it is natural to ask how these concepts are related. We would like the notion of proof to be **sound**; that is, anything that can be proved should be true. We would also like the notion to be **adequate**; that is, anything that is true should be provable. It seems fairly clear that an unsound theory would be extremely troublesome; however, an inadequate theory might be worth considering if it had other advantages (for instance, if it gave very simple proofs). Fortunately our notions are both sould and adequate. This was proved by Gödel in 1930.

Theorem 11.12 (Gödel's Completeness Theorem for propositional logic) $\Gamma \vdash \phi$ *iff* $\Gamma \models \phi$.

This theorem divides into two parts The Soundness Theorem, which we now prove, is fairly easy. The Adequacy Theorem is much harder; its proof will take most of the next section.

Theorem 11.13 (Soundness Theorem) *If* $\Gamma \vdash \phi$ *then* $\Gamma \models \phi$.

Proof We shall show, by induction on the number of vertices, that if D is a derivation, then the conclusion of D satisfies any valuation which satisfies all the hypotheses of D. Write Hyp$D$ for the set of hypotheses of D, and let v be a valuation which satisfies Hyp$D$.

If D contains only one vertex, the result is obvious, as the conclusion and hypotheses of the derivation D are the same.

Suppose D is $\dfrac{\overset{D'}{\phi \wedge \psi}}{\phi}$ or $\dfrac{\overset{D'}{\phi \wedge \psi}}{\psi}$. As Hyp$D'$ = Hyp$D$, we have, inductively, that $v(\phi\wedge\psi) = T$ (or $v(\psi\wedge\phi) = T$), from which we get $v\phi = T$, as required.

If D is $\dfrac{\overset{D'}{\phi} \quad \overset{D'}{\psi}}{\phi\wedge\psi}$, then $\mathrm{Hyp}D = \mathrm{Hyp}D' \cup \mathrm{Hyp}D''$. Inductively both $v\phi$ and $v\psi$ are T, and so $v(\phi\wedge\psi) = T$, as required. A similar argument works for elimination of $\rightarrow$.

Suppose D is $\dfrac{\overset{D'}{\bot}}{\phi}$. Then $\mathrm{Hyp}D = \mathrm{Hyp}D'$, so, inductively, there can be no v satisfying the hypotheses of D, as $v\bot$ is never T. Thus the results holds by default.

Suppose D is $\dfrac{\begin{matrix}\not\phi \\ D' \\ \psi\end{matrix}}{\phi\rightarrow\psi}$. Then $\mathrm{Hyp}D' \subset \mathrm{Hyp}D \cup \{\phi\}$. So, by induction, if $v\phi = T$ then also $v\psi = T$, since all the hypotheses of D' are true. In this case, we know that $v(\phi\rightarrow\psi) = T$. If $v\phi = F$ then, by definition, $v(\phi\rightarrow\psi) = T$.

If D is $\dfrac{\begin{matrix}\not\neg\phi \\ D' \\ \bot\end{matrix}}{\phi}$, then $\mathrm{Hyp}D' \subset \mathrm{Hyp}D \cup \{\neg\phi\}$. No valuation can satisfy the hypotheses of D'. Hence, if v satisfies $\mathrm{Hyp}D$ we must have $v(\neg\phi) = F$, and so $v\phi = T$, as required.

11.6 ADEQUACY

In this section we show that our notion of derivation is adequate; that is $\Gamma \vdash \phi$ if $\Gamma \models \phi$. Several other properties will have to be discussed first.

Lemma 11.14 *The following are equivalent*: *(i)* $\Gamma \vdash \bot$, *(ii)* $\Gamma \vdash \phi$ *for all* ϕ, *(iii) for some* θ *we have both* $\Gamma \vdash \theta$ *and* $\Gamma \vdash \neg\theta$.

Proof If (i) holds then (ii) holds by the rule $\bot$. If (ii) holds then (iii) obviously holds. If (iii) holds the rule for elimination of $\rightarrow$ tells us that (i) holds.

A set Γ of formulae satisfying these conditions is called **inconsistent.** If Γ is not inconsistent it is called **consistent.**

If Γ is inconsistent some finite subset must be inconsistent (by Lemma 11.5), and conversely. Equivalently, Γ is consistent iff every finite subset of Γ is consistent.

Lemma 11.15 *If* $\Gamma \cup \{\phi\}$ *is inconsistent, then* $\Gamma \vdash \neg\phi$. *If* $\Gamma \cup \{\neg\phi\}$ *is inconsistent, then* $\Gamma \vdash \phi$.

Corollary *If* $\Gamma \cup \{\neg\phi\}$ *and* $\Gamma \cup \{\phi\}$ *are inconsistent, then* Γ *is inconsistent.*

Proof We get a derivation of $\neg\phi$ from Γ by applying the rule for introduction of $\rightarrow$ to a derivation of $\perp$ from $\Gamma \cup \{\phi\}$. Similarly, we get a derivation of ϕ from Γ by applying the rule RAA to a derivation of $\perp$ from $\Gamma \cup \{\neg\phi\}$. This proves the lemma. The corollary follows, using one of the other equivalent definitions of inconsistency.■

If $\Gamma \vdash \phi$ then every member of $\Gamma \cup \{\phi\}$ can be derived from Γ. It follows by Lemma 11.5(iii), that if $\Gamma \vdash \phi$ and $\Gamma \cup \{\phi\}$ is inconsistent, so is Γ.

A consistent set of formulae Γ is **maximal consistent** if any Γ' strictly containing Γ is inconsistent. By the previous remark, if Γ is maximal consistent and $\Gamma \vdash \phi$, then ϕ is in Γ.

Γ is **complete** if for every ϕ at least one of $\neg\phi$ or ϕ is in Γ. Plainly, if Γ is complete consistent then for every ϕ exactly one of ϕ and $\neg\phi$ is in Γ.

Proposition 11.16 *Let* Γ *be consistent. Then* Γ *is maximal consistent iff it is complete.*

Proof Suppose Γ is maximal consistent, and that ϕ is not in Γ. Then $\Gamma \cup \{\phi\}$ is inconsistent. Hence $\Gamma \vdash \neg\phi$, and so $\neg\phi$ is in Γ. This shows that Γ is complete.

Suppose Γ is complete consistent, and let Γ' strictly contain Γ. Take ϕ in $\Gamma' - \Gamma$. Then $\neg\phi$ is in Γ. As both ϕ and $\neg\phi$ are in Γ', Γ' is inconsistent. Hence Γ is maximal consistent.■

Theorem 11.17 *Let* Γ *be consistent. Then there is a maximal consistent set* Γ' *with* $\Gamma \subset \Gamma'$.

Proof There are only countably many formulae. Let them be $\phi_0, \phi_1, \ldots$. Let Γ_0 be Γ. Define Γ_n inductively by $\Gamma_{n+1} = \Gamma_n \cup \{\phi_n\}$ if this is consistent, and $\Gamma_{n+1} = \Gamma_n \cup \{\neg\phi_n\}$ otherwise. Let Γ' be $\cup\, \Gamma_n$.

Inductively, every Γ_n is consistent. For Γ_0 is consistent, and if Γ_n is consistent at least one of $\Gamma_n \cup \{\phi_n\}$ and $\Gamma_n \cup \{\neg\phi_n\}$ is consistent, by the corollary to Lemma 11.15. By our definition, if $\Gamma_n \cup \{\phi_n\}$ is consistent, this set is Γ_{n+1}, while if $\Gamma_n \cup \{\phi_n\}$ is inconsistent, then Γ_{n+1} is $\Gamma_n \cup \{\neg\phi_n\}$, which is then consistent.

Suppose Γ' were inconsistent. Then some finite subset of Γ' would be inconsistent, and this finite subset would be in Γ_n for some n. This would make Γ_n inconsistent. Hence Γ' is consistent.

By construction Γ' is complete, so it is maximal consistent.■

Theorem 11.18 *A set Γ of formulae is consistent iff it satisfies some valuation.*

Proof If Γ satisfies v, by Soundness, $\bot$ cannot be derived from Γ, as $v\bot = F$.

Let v be the unique valuation such that $vp_n = T$ iff $p_n \in \Gamma$. We now show, inductively, that a formula ϕ is in Γ iff $v\phi = T$. This holds for $\bot$ which cannot be in Γ because Γ is consistent. It holds for all propositional variables by definition.

Suppose this property holds for ϕ and ψ. First, suppose that $v(\phi \wedge \psi) = T$. Then $v\phi = T = v\psi$, and ϕ and ψ are in Γ, inductively. Hence $\Gamma \vdash \phi \wedge \psi$. As Γ is maximal consistent, $\phi \wedge \psi$ is in Γ. Conversely, suppose $\phi \wedge \psi$ is in Γ. Then $\Gamma \vdash \phi$ and $\Gamma \vdash \psi$. As Γ is maximal consistent, ϕ and ψ are in Γ. Hence $v\phi$ and $v\psi$ are T, and so $v(\phi \wedge \psi) = T$.

Suppose $v(\phi \rightarrow \psi) = T$. Then either $v\phi = F$ or $v\psi = T$. In the first case, ϕ is not in Γ. As Γ is complete, $\neg\phi$ will be in Γ. Since $\neg\phi \vdash \phi \rightarrow \psi$, as before $\phi \rightarrow \psi$ will be in Γ. Similarly, as $\psi \vdash \phi \rightarrow \psi$, if $v\psi = T$ then $\phi \rightarrow \psi$ is in Γ. Conversely, suppose $\phi \rightarrow \psi$ is in Γ. If ϕ is not in Γ, then, inductively, $v\phi = F$, and so $v(\phi \rightarrow \psi) = T$. If ϕ is in Γ, then $\Gamma \vdash \psi$. As before, this shows that ψ is in Γ and so, inductively, $v\psi = T$, and so $v(\phi \rightarrow \psi) = T$.■

Theorem 11.19 (Adequacy Theorem for Propositional Logic) *If $\Gamma \models \phi$ then $\Gamma \vdash \phi$.*

Proof Suppose ϕ cannot be derived from Γ. Then we know that $\Gamma \cup \{\neg\phi\}$ is consistent. Hence there is a valuation v which satisfies $\neg\phi$ and all the members of Γ. As $v(\neg\phi) = T$ we have $v\phi = F$. By definition this means that we do not have $\Gamma \models \phi$.

Exercise 11.14 Show that Γ is maximal consistent iff it satisfies exactly one valuation.

11.7 EQUIVALENCE

The results of this section will only be used in the corresponding section of the next chapter and in Chapter 14.

Definition The formulae ϕ and ϕ' are called **equivalent** if $\models \phi \leftrightarrow \phi'$. This holds iff $v\phi = v\phi'$ for any valuation v, which shows that we do have an equivalence relation.

Let ϕ be equivalent to ϕ', and let ψ be equivalent to ψ'. It is easy to check that $\neg\phi$, $\phi \wedge \psi$, $\phi \vee \psi$, $\phi \rightarrow \psi$, and $\phi \leftrightarrow \psi$ are equivalent, respectively, to $\neg\phi'$, $\phi' \wedge \psi'$, $\phi' \vee \psi'$, $\phi' \rightarrow \psi'$, and $\phi' \leftrightarrow \psi'$.

Also $\neg\neg\phi$ is equivalent to ϕ, and $\phi \rightarrow \psi$ to $\neg\phi \vee \psi$ and to $-(\phi \wedge -\psi)$. Further we have the following rules, known as the associative, commutative, and distributive laws. $\theta \wedge (\phi \wedge \psi)$ and $\theta \vee (\phi \vee \psi)$ are equivalent to $(\theta \wedge \phi) \wedge \psi$ and $(\theta \vee \phi) \vee \psi$, $\phi \wedge \psi$ and $\phi \vee \psi$ are equivalent to $\psi \wedge \phi$ and $\psi \vee \phi$, and $(\theta \wedge \phi) \vee \psi$ and $(\theta \vee \phi) \wedge \psi$ are equivalent to $(\theta \vee \psi) \wedge (\phi \vee \psi)$ and $(\theta \wedge \psi) \vee (\phi \wedge \psi)$. By the Completeness Theorem, if two formulae are equivalent each can be derived from the other. It is not difficult to give directly derivations of each of the formulae above from the corresponding formula which is equivalent to it. This is sometimes convenient.

Definition A formula is said to be in **disjunctive normal form** if it is $\phi_1 \vee \ldots \vee \phi_n$ for some formulae ϕ_r, each of which is a conjunction of formulae ϕ_{rs} where s runs from 1 to n_r for some n_r, and finally each ϕ_{rs} is either $\perp$ or a propositional variable p_k or the negation of one of these.

Notice that we do not bother to bracket the disjunctions (or the conjunctions), since, by the associative law, all bracketings are equivalent.

If $\perp$ occurs in ϕ_{rs} we know that ϕ_r is equivalent to $\perp$. In this case we obtain an equivalent formula by omitting ϕ_r unless $n=1$, when ϕ is equivalent to $\perp$. If $\neg\perp$ occurs in ϕ_{rs} we obtain an equivalent formula by omitting it unless $n_r = 1$; in this case, ϕ_r is $\neg\perp$ and ϕ is equivalent to $\neg\perp$, and so is also equivalent to $p_1 \vee \neg p_1$. We can also ensure that each propositional variable occurs at most once (whether negated or not) in each ϕ_r. For if p occurs twice or $\neg p$ occurs twice we obtain an equivalent formula by omitting the second occurrence. If both p and $\neg p$ occur in ϕ_r then ϕ_r may be replaced by the equivalent formula $\perp$ (which can be omitted unless $n=1$).

Proposition 11.20 *Any formula is equivalent to one in disjunctive normal form.*

Proof Let α be equivalent to a formula ϕ in disjunctive normal form. Then $\neg\alpha$ is equivalent to $\neg\phi_1 \wedge \ldots \wedge \neg\phi_n$, and each $\neg\phi_r$ is equivalent to the disjunction of the corresponding $\neg\phi_{rs}$. By the distributive laws, we find that we may change the order of the conjunctions and disjunctions to get $\neg\alpha$ equivalent to a disjunction of conjunctions of the $\neg\phi_{rs}$. Each of these is the negation or the double negation of $\perp$ or some p_k and the double negation of a formula is equivalent to the formula itself. Thus $\neg\alpha$ is equivalent to a formula in disjunctive normal form.

Plainly, $\alpha \vee \beta$ is equivalent to a formula in disjunctive normal form if both α and β are. As $\alpha \rightarrow \beta$ and $\alpha \wedge \beta$ are equivalent to $\neg\alpha \vee \beta$ and $\neg(\neg\alpha \vee \neg\beta)$, respectively, both $\alpha \rightarrow \beta$ and $\alpha \wedge \beta$ will be equivalent to formulae in disjunctive normal form, and the result follows by induction. ■

It is very easy to check whether or not a formula in disjunctive normal form is satisfiable. (This does not provide a simpler method than truth-tables for determining whether or not an arbitrary formula is satisfiable. The difficulty just gets moved to constructing the disjunctive normal form). Let ϕ be in disjunctive normal form. Then ϕ is satisfiable iff some ϕ_r is satisfiable. And ϕ_r is satisfiable iff the corresponding ϕ_{rs} are simultaneously satisfiable. Since each ϕ_{rs} is (unless ϕ is just $\perp$) either p_i or $\neg p_i$ for some i, these can be simultaneously satisfied unless both p_i and $\neg p_i$ both occur in ϕ_r for some i.

There is an alternative proof of Proposition 11.20, which is also interesting. Let V_n denote the set of all partial valuations defined on those formulae whose only propositional variables are p_k for $k \leq n$. Thus V_n can be regarded as the set of functions from $\{p_0, \ldots, p_n\}$ to $\{T,F\}$. Any such formula ϕ defines a function from V_n to $\{T,F\}$, by sending v to $v\phi$. Two formulae ϕ and ψ are plainly equivalent iff they define the same function. Hence the previous result follows from the next one.

Proposition 11.21 *To any function f from V_n to $\{T,F\}$ there is a formula ϕ in disjunctive normal form such that $fv = v\phi$ for all v in V_n.*

Proof Suppose this is true for $n-1$, and let f be a function from V_n to $\{T,F\}$. Define two functions g_1 and g_2 from V_{n-1} to $\{T,F\}$ as follows. Let w be in V_{n-1}. Then define g_1 and g_2 by $g_1w = fv_1$ and $g_2w = fv_2$, where $v_1p_k = wp_k = v_2p_k$ for $k < n$, and $v_1p_n = T$, $v_2p_n = F$. Let ϕ_1 and ϕ_2 be the formulae corresponding to g_1 and g_2 by the inductive hypothesis. Let ϕ be $(\phi_1 \wedge p_n) \vee (\phi_2 \wedge \neg p_n)$. Then ϕ is in disjunctive normal form. Let v be in V_n and let w be the restriction of v to V_{n-1}. Then, with the above notation, if $vp_n = T$ then v is v_1 and $fv = g_1w = w\phi_1 = v\phi_1$. Since $vp_n = T$, we have $v(\phi_2 \wedge \neg p_n) = F$, and then $v\phi = v\phi_1$. So $fv = v\phi$ in this case. Similarly, we find $fv = v\phi$ when $vp_n = F$.

The induction starts when $n = 0$. In this case we can choose ϕ_1 and ϕ_2 to be either $\perp$ or $\neg\perp$, and with the right choice we will still have $fv = v\phi$ for all v (there are only two possibilities for v in this case.) ■

11.8 SUBSTITUTION

Let α be a formula and let $\Phi = \{\phi_0, \phi_1, \ldots\}$ be a sequence of formulae. We denote by $\alpha[\Phi]$ the result of **substituting** ϕ_n **for** p_n **in** α for all n. It is intuitively clear what this means and that the result is a formula. A precise definition is, as usual, by induction. We define $\perp[\Phi]$ to be $\perp$ and $p_n[\Phi]$ to be ϕ_n. We then define $(\alpha \wedge \beta)[\Phi]$ and $(\alpha \to \beta)[\Phi]$ to be $\alpha[\Phi] \wedge \beta[\Phi]$ and $\alpha[\Phi] \to \beta[\Phi]$, respectively.

An inductive proof now shows that $\alpha[\Phi]$ is a formula for all α. The following lemma is immediate, by induction.

Lemma 11.22 *Let α be a formula, Φ a sequence of formulae, and v a* *valuation. Then* $v\alpha[\Phi] = w\alpha$, *where* w *is the valuation such that* $wp_n = v\phi_n$ *for all* n.

Lemma 11.23 (*i*) *Let* α *and* β *be equivalent formulae, and let* Φ *be a sequence of formulae. Then* $\alpha[\Phi]$ *is equivalent to* $\beta[\Phi]$.

(*ii*) *Let* α *be a formula, and let* Φ *and* Ψ *be sequences of formulae such that* ϕ_n *is equivalent to* ψ_n *for all* n. *Then* $\alpha[\Phi]$ *is equivalent to* $\alpha[\Psi]$.

Proof Let v be any valuation, and let w be the valuation such that $wp_n = v\phi_n$ for all n.

(i) By the previous lemma, we have $v\alpha[\Phi] = w\alpha = w\beta$ (since α and β are equivalent), and $w\beta = v\beta[\Phi]$, as required.

(ii) As before, we have $v\alpha[\Phi] = w\alpha$. Since ϕ_n is equivalent to ψ_n for all n, we have also that $wp_n = v\psi_n$, and hence $v\alpha[\psi] = w\alpha$. Hence $v\alpha[\Phi] = v\alpha[\Psi]$, as required.■

With Φ as before, and Γ a set of formulae, we define $\Gamma[\Phi]$ be be $\{\gamma[\Phi]$; all γ in $\Gamma\}$. Lemma 11.22 shows at once that if $\Gamma \models \alpha$ then $\Gamma[\Phi] \models \alpha[\Phi]$. In particular, $\alpha[\Phi]$ is a tautology if α is a tautology.

If D is a derivation we let $D[\Phi]$ be the tree obtained from D by replacing each label α at a vertex by the corresponding $\alpha[\Phi]$. It is easy to check, inductively, that $D[\Phi]$ is also a derivation. It follows (without using the Completeness Theorem) that if $\Gamma \vdash \alpha$ then $\Gamma[\Phi] \vdash \alpha[\Phi]$.

Both these properties have in effect been used earlier (without specific mention). For instance, the comment that we only need a truth-table of eight lines to show that $(\theta \to (\phi \to \psi)) \to ((\theta \to \phi) \to (\theta \to \psi))$ is a tautology amounts to showing that $((p_0 \to (p_1 \to p_2)) \to ((p_0 \to p_1) \to (p_0 \to p_2)))$ is a tautology and then substituting.

12

Predicate logic

In this chapter we look at predicate logic, which adds to the connectives of propositional logic the quantifiers 'for all' and 'for some'. This adds significantly to the complexity of the theory, especially in the definitions. However, many results hold both for propositional and for predicate logic; where the proofs for predicate logic are not significantly different from the earlier ones they will be omitted.

12.1 LANGUAGES OF FIRST-ORDER PREDICATE LOGIC

A **language** L of **first-order predicate logic** (the words 'first-order' will usually be omitted) is a set consisting of the following elements (no element can be of more than one type).

(1) some elements (possibly none) called **constant symbols**,
(2) some elements (possibly none) called **function symbols**. To each function symbol f there is an associated positive integer called the **arity** of f,
(3) some elements (at least one) called **predicate symbols**. To each predicate symbol P there is an associated positive integer called the **arity** of P,
(4) countably many symbols $x_0, x_1, \ldots$ called **variables**,
(5) the **logical symbols** $\wedge, \rightarrow$, and $\perp$,
(6) the **universal quantifier** $\forall$,
(7) the **parentheses** (and).

Notice that there are many languages of predicate logic, depending on what elements we take in (1), (2), and (3). The choice of these will depend on what objects we want to talk about. Most of our results apply to all languages, but there are some deeper aspects where different languages have significantly different properties (for instance, the decidability results of the next chapter).

There are variations possible on what symbols are chosen in (5) and (6). One might wish to include the existential quantifier $\exists$ or some of the symbols $\neg$, $\vee$, and $\leftrightarrow$. We will regard these as auxiliary symbols, which will be defined later in terms of the given symbols.

For convenience we often use x, y, and z as variables, rather than the strict notation given in (4).

If a function symbol or predicate symbol has arity n, we usually refer to it as an n-ary symbol. We also refer to unary, binary, or ternary symbols if the arity is 1, 2, or 3. Most languages used in practice have a binary predicate symbol called equality, which plays a special role. We will include this at a later stage. If there is such a symbol, it may be the only predicate symbol.

Certain strings on L are called **terms** or **formulae**. These are the strings we are interested in; intuitively, terms are the things we can talk about, and formulae are what we can say about them. As with formulae of propositional logic, we begin with an informal look, in which the final condition is too vague to work with properly. Thus we begin by considering the set of terms to be given by

(i) the constant symbols and the variables are terms,
(ii) if f is an n-ary function symbol for some n and $t_1, \ldots, t_n$ are terms then $ft_1 . t_n$ is a term,
(iii) no string is a term unless (i) and (ii) require it to be a term.

In order to make this precise we make the following definition. As before, to show that there is a smallest such set we show that the intersection of all sets satisfying the conditions (1) and (2) below is itself a set satisfying (1) and (2). It must then be the smallest such set.

Definition The set of **terms** is the smallest set such that

(1) the constant symbols and the variables are in the set,
(2) for every n and every n-ary function symbol f, if $t_1, \ldots, t_n$ are in the set then so is $ft_1 \ldots t_n$.

A term is called **closed** if it does not involve any variables.

Symbols such as s and t (possibly with subscripts) will always stand for terms.

We can then see that a string of length >1 is a term iff it can be written as $ft_1 \ldots t_n$, where f is an n-ary function symbol (for some n) and $t_1, \ldots, t_n$ are terms. We have the following, which is proved as in Theorem 11.1.

Theorem 12.1 (Principle of induction for terms) *Let S be a property of strings which is true for all constant symbols and for all variables. Suppose S is true for $ft_1 \ldots t_n$, where f is an n-ary function symbol, whenever it is true for each of $t_1, \ldots, t_n$. Then S is true for all terms.*

Definition A string is an **atomic formula** iff it is $\perp$ or is $Pt_1 \ldots t_n$ for some n-ary predicate symbol P and some terms $t_1, \ldots, t_n$.

Definition The set of **formulae** is the smallest set of strings such that

(1) any atomic formula is in the set,
(2) if ϕ and ψ are in the set then $(\phi \wedge \psi)$, $(\phi \rightarrow \psi)$, and $\forall x \phi$, where x is any variable, are in the set.

Symbols such as θ, ϕ, and ψ (possibly subscripted) will always stand for formulae; after this section (when we no longer have to look at arbitrary strings) any small Greek letter will stand for a formula and any capital Greek letter for a set of formulae.

As before, we can show that there is a smallest such set, and we can also show that a string containing $\wedge$, $\rightarrow$, or $\forall$ is a formula iff it can be written as $(\phi \wedge \psi)$, $(\phi \rightarrow \psi)$, or $\forall x \phi$, respectively, where ϕ is a formula and so is ψ (when it occurs) — which must be shorter than the original string, of course — and x is any variable. Again, we have induction for formulae.

Theorem 12.2 (Principle of induction for formulae of predicate logic) *Let S be a property of strings which holds for all atomic formulae. Suppose that S holds for* $(\phi \wedge \psi)$, $(\phi \rightarrow \psi)$, *and for* $\forall x \phi$, *for any variable x, whenever it holds for* ϕ *and* ψ. *Then S holds for all formulae.*

If we wish to discuss questions of recursiveness it is essential that our language L is countable; this assumption will be taken for granted when such topics are mentioned. In practice we usually have only finitely many constant, function, and predicate symbols. If this happens we can obtain a bijection from L to $\mathbf{N}$ sending x_n to $n+r$ for some fixed r, the remaining symbols (constant, function, and predicate symbols, logical symbols, quantifier and parentheses) being sent to $0, \ldots, r-1$. If there are countably many constant symbols, but only finitely many function and predicate symbols, we could map x_n to $2n+r$ and the constant symbol c_n to $2n+1+r$. If there are countably many function or predicate symbols, the situation is rather more complicated, as we have to allow for countably many symbols of each arity; however, we could (for instance) map the ith function symbol of arity n onto $4J(n,i)+r$. For all countable languages L we have in this way a Gödel numbering of L, which leads to a Gödel numbering of the strings on L. It is this Gödel numbering that is used to refer to certain sets as recursive. Readers are advised, for convenience, to consider only languages with finitely many constant, function, and predicate symbols; the general case is messier in detail, but essentially the same.

The set of terms is recursive. For, given any string not containing function symbols, we can plainly tell whether or not it is a term. If the string α contains function symbols it is not a term unless it begins with a function symbol. Suppose α begins with a function symbol f whose arity is n. Then we can find all ways of writing α as $f\alpha_1 \ldots \alpha_n$, where each α_i is a string. For each such way we can check inductively whether or not $\alpha_1, \ldots, \alpha_n$ are terms. Then α will be a term iff there is a way of writing α as $f\alpha_1 \ldots \alpha_n$ with each α_i being a term.

It is possible to form the parsing trees for terms in a way similar to the parsing trees for formulae of propositional logic, but here there can be n vertices following a given vertex, not just 1 or 2. Further, the definition of weight in Lemma 12.3 and the proof of that lemma show that α_1 can be identified as the smallest string of weight 1 such that $f\alpha_1$ is an initial segment of α, and then α_2 is the smallest segment of weight 1 such that $f\alpha_1\alpha_2$ is an initial segment of α, and so on. As with formulae of propositional logic, we can in one construction determine whether or not α is a term and find its parsing tree if it is a term.

The set of atomic formulae is recursive. For if a string α other than $\perp$ is an atomic formula it must begin with a predicate symbol. If it begins with a n-ary predicate symbol P it is an atomic formula iff it can be written as $P\alpha_1 \ldots \alpha_n$ for some terms $\alpha_1, \ldots, \alpha_n$. Since we can tell whether or not a string is a term, we can tell whether or not α can be written in the required form.

The set of all formulae is recursive. A string not involving $\wedge$, $\rightarrow$, or $\forall$ is a formula iff it is an atomic formula. A string involving one of $\wedge$, $\rightarrow$, and $\forall$ is a formula iff it can be built from smaller strings in certain ways (this is a property which we can check) and if these smaller strings are themselves formulae (which can be checked inductively).

We can construct a parsing tree for formulae, and in one construction determine whether or not a string is a formula and find its parsing tree if it is a formula. The parsing tree of a formula has a kind of hybrid nature. At first it is constructed in a similar way to the parsing tree of a formula of propositional logic. This procedure continues until there are no occurrences of $\rightarrow$, $\wedge$, or $\forall$. After this, the atomic formulae give rise to terms, and we must continue using the parsing trees of these terms. This process is not very convenient. Readers may feel that this argument, even allowing for the standard appeal to Church's Thesis, is too vague to be convincing. We therefore give a more precise proof that the set of all formulae is recursive, assuming that the set of all atomic formulae is recursive.

We begin by defining three functions $F_\wedge$, $F_\rightarrow$, and $F_\forall$ from $\mathbf{N}^2$ to $\mathbf{N}$. If m and n are the Gödel numbers of the strings α and β we define $F_\wedge(m,n)$ and $F_\rightarrow(m,n)$ to be the Gödel numbers of $(\alpha\wedge\beta)$ and $(\alpha\rightarrow\beta)$, respectively. If m is the Gödel number of α and n is arbitrary we define $F_\forall(m,n)$ to be the Gödel number of $\forall x_n\alpha$. in all other cases we define the values of the functions to be $m+n+1$. It is then clear that the value of each function on (m,n) is greater than $\max(m,n)$. It is also clear, by Church's Thesis, that these functions are recursive.

The definition of a formula tells us that n is the Gödel number of a formula iff either it is the Gödel number of an atomic formula or it is either $F_\wedge(r,s)$ or $F_\rightarrow(r,s)$ where r and s (which must be less than n) are Gödel numbers of formulae or it is $F_\forall(r,s)$ where r is the Gödel number of a formula (again r and s are less than n). We have discussed exactly this situation (but using only one function in place of the two functions $F_\wedge$ and

$F_{\rightarrow}$) as an example on course-of-values recursion. That example was given largely for its application here. The same techniques as in that example can be used to give a formal proof that the set of all formulae is recursive assuming that the set of atomic formulae is recursive.

If there are only finitely many function symbols a similar approach, using one function from $\mathbf{N}^n$ to $\mathbf{N}$ for each n-ary function symbol, gives a formal proof that the set of terms is recursive. It is then easy, assuming that there are only finitely many predicate symbols, to show that the set of atomic formulae is recursive.

If there are countably many function symbols the situation is trickier. It is not enough to define one function for each function symbol. Instead, it is necessary to define one function from $\mathbf{N}$ to $\mathbf{N}$ such that the Gödel number (as constructed in Chapter 8) of the finite sequence $(n_0, n_1, \ldots, n_k)$ maps to the Gödel number of $f\alpha_1 \ldots \alpha_k$ when n_0 is the Gödel number of a k-ary function symbol f and n_i is the Gödel number of a string α_i. The details will be omitted, as they are messy and not very illuminating, and the finite case is all that we need. The proof above is the only one we shall give in detail. Other results will be justified by an apeal to Church's Thesis. Full proofs can be given along the above lines.

Most results about terms and formulae are proved by induction. In proving a result for formulae, the start of the induction is the case of an atomic formula. Sometimes the result for atomic formulae is obvious. More usually, one has to find a proof for this case. The typical way of doing this is to obtain, inductively, a result for terms related to the result we want for formulae, and to obtain the result for atomic formulae from the result for terms. We shall see several examples of this later.

As in propositional logic, we will want to define by induction various functions (or properties) of terms and formulae. In order to ensure that there are no problems with such definitions, we need the Unique Reading Lemmas below.

Lemma 12.3 *A proper initial segment of a term is not a term. A proper initial segment of a formula is not a formula.*

Proof We begin by associating with each symbol of L an integer, called the weight of the symbol. For any constant symbol c and any variable x, we define wt$c$ and wt$x$ to be 1. For any n-ary function symbol f and any n-ary predicate symbol P, we define wt$f$ and wt$P$ to be $1-n$. We define wt) and wt $\bot$ to be 1, and wt (, wt $\forall$, wt$\wedge$, and wt$\rightarrow$ are all -1. We define the weight of a string to be the sum of the weights of the elements of the string.

Any term has weight 1. This is proved by induction. By definition, any constant symbol or variable has weight 1. Also, if f is an n-ary function symbol and $t_1, \ldots, t_n$ have weight 1, then the weight of $ft_1 \ldots t_n$ is $1-n+1+\ldots+1$, with n entries being 1, and so this weight is 1, as needed.

Any proper initial segment of a term has weight less than 1. This is true for constant symbols and variables, because they do not have any proper initial segments. It is easy to see that the proper initial segments of $ft_1 \ldots t_n$

are f, $ft_1 \ldots t_r$ for $r<n$, and $ft_1 \ldots t_ru$, where u is a proper initial segment of t_{r+1} and $r<n$. Using the first part, the result follows by induction.

Having proved these results on terms, we see, in the same way, that any atomic formula has weight 1, while any proper initial segment of an atomic formula has weight less than 1.

The weight of $(\phi \rightarrow \psi)$ and of $(\phi \wedge \psi)$ is $-1+\text{wt}\phi+(-1)+\text{wt}\psi+(-1)$, while the weight of $\forall x\phi$ is $-1+1+\text{wt}\phi$. It follows by induction that every formula has weight 1. Another inductive proof now shows that any proper initial segment of a formula has weight less than 1. The lemma follows.■

Lemma 12.4 (Unique Reading Lemma for Terms) *Let t be a term. Then t is of exactly one of the follwing forms.*

(1) *a constant symbol,*
(2) *a variable,*
(3) $ft_1 \ldots t_n$, *where f is an n-ary function symbol and $t_1, \ldots, t_n$ are terms.*

In this case f, n, and each of $t_1, \ldots, t_n$ are determined by t.

Proof By definition any term is of one of these three types. It cannot be of two of these types, since a term of type (3) begins with a function symbol, while terms of types (1) and (2) do not.

Let t be a term of type (3). Then f must be its first symbol, and n is the arity of f. Suppose we can write t as $fs_1 \ldots s_n$, where each s_i is a term. Suppose that t_i coincides with s_i for all $i<r$. We show t_r is the same as s_r, proving the result by induction. Plainly t_r is an initial segment of the string obtained from t by deleting its initial segment $ft_1 \ldots t_{r-1}$, and s_r is an initial segment of the string obtained from t by deleting its initial segment $fs_1 \ldots s_{r-1}$. By hypothesis, the two deleted segments are the same, and so t_r and s_r are both initial segments of the same string. Hence one of t_r and s_r is an initial segment of the other. By the previous lemma, neither can be a proper initial segment of the other, and so they are the same.■

Lemma 12.5 (Unique Reading Lemma for Formulae) *Any formula ϕ is of exactly one of the following forms.*

(1) $\perp$.
(2) $Pt_1 \ldots t_n$, *where P is an n-ary predicate symbol and each t_i is a term. In this case P, n, and each t_i are determined by ϕ.*
(3) $\forall x\psi$, *for some variable x and formula ψ which are determined by ϕ.*
(4) $(\phi_1 \wedge \phi_2)$ *for some formulae ϕ_1 and ϕ_2 which are uniquely determined by ϕ.*
(5) $(\phi_1 \rightarrow \phi_2)$ *for some formulae ϕ_1 and ϕ_2 which are uniquely determined by ϕ.*

Proof By definition every formula is of one of these forms. Plainly ϕ cannot be of form (1) and of another form. Since formulae of form (2) begin with a predicate symbol, of form (3) begin with $\forall$, and of forms (4) and (5) begin with (, a formula of form (2) or of form (3) cannot be of any other form.

Uniqueness in form (2) follows the same proof as for the previous lemma. Uniqueness in form (3) is obvious, since x must be the second symbol and ψ is obtained from ϕ by deleting the first two symbols.

The proof that a formula cannot be both of forms (4) and (5), and that the expression is unique, will be omitted, since it is exactly the same as the proof in Lemma 11.3.■

The reason we need parentheses in formulae but not in terms is that the connectives $\wedge$ and $\rightarrow$ are placed between the formulae, whereas the function symbols are placed before their corresponding block of terms. For the same reason, we can write $\forall x\phi$ without parentheses. If we used the notations $\wedge\phi\psi$ and $\rightarrow\phi\psi$ no parentheses would be needed. Similarly, no parentheses would be needed if we placed these connectives and all function symbols at the end of the relevant strings. This latter notation is often used in programming, and is called Reverse Polish notation.

We use the same rules of abbreviation by omitting parentheses as for propositional logic. Notice that $\forall x\phi\rightarrow\psi$ abbreviates $(\forall x\phi\rightarrow\psi)$, while $\forall x(\phi\rightarrow\psi)$ must be written in that form.

We can now make some definitions by induction.

Definition The **subformulae** of the formula ϕ are given by: if ϕ is atomic, then ϕ is the only subformula of ϕ; if ϕ is either $(\phi_1\wedge\phi_2)$ or $(\phi_1\rightarrow\phi_2)$ then the subformulae of ϕ are ϕ and all subformulae of either ϕ_1 or ϕ_2; if ϕ is $\forall x\psi$ then the subformulae of ϕ are ϕ and all subformulae of ψ.

Definition Every occurrence of a variable in a formula is either **free** or **bound** (but not both). This is given inductively by: if ϕ is atomic, every occurrence of any variable is free; if ϕ is either $(\phi_1\wedge\phi_2)$ or $(\phi_1\rightarrow\phi_2)$ then an occurrence of x in ϕ is free (or bound) if the corresponding occurrence of x in ϕ_1 or ϕ_2 is free (or bound); every occurrence of x in $\forall x\psi$ is bound; an occurrence of x in $\forall y\psi$ is free (or bound) if the corresponding occurrence in ψ is free (or bound).

A **sentence** is a formula in which no variable occurs free.

There are two uses of the symbol x in informal language. If we say '$x>2$' we are referring to some definite x. This statement does not mean the same as '$y>2$'. However, 'for all x, $x^2\geqslant 0$' does not refer to any specific x, and it has the same meaning as 'for all y, $y^2\geqslant 0$'. These correspond to the free and bound occurrences, respectively, of a variable in a formula of the formal language.

As an example, take a language with two unary predicate symbols P and Q and one binary predicate symbol R. Let ϕ be $(\forall xPx\wedge(Rxy\rightarrow Qx))$. Then x occurs four times, the first two occurrences being bound and the last two being free. Also, ϕ is $(\phi_1\wedge\phi_2)$, where ϕ_1 is $\forall xPx$ and ϕ_2 is $(Rxy\rightarrow Qx)$. The third and fourth occurrences of x in ϕ correspond to the first and second occurrences in ϕ_2.

We can plainly tell, inductively, whether or not an occurrence of a variable in a formula is free. It follows that the set of sentences is recursive. It appears at first sight that we need to determine for every n whether or not the variable x_n occurrs free in ϕ, which would cause trouble. However x_n can only occur at all in ϕ if n is at most the Gödel number of ϕ. Thus we only have a finite number of variables to check, and this can be done.

We might wonder whether this result means 'The set of sentences is a recursive subset of the set of all strings' or whether it means 'The set of sentences is a recursive subset of the set of all formulae'. However Proposition 8.2 shows that these two statements are equivalent.

Let t be a term and x a variable. We want to define the **result of substituting** t **for** x (we may also say that x is **replaced by** t) in the formula ϕ, which we shall denote by $\phi(t/x)$. Intuitively, we simply replace each free occurrence of x by t to get the new string (but bound occurrences are not affected). Formally, this will be done by induction, but first we must define by induction the result $s(t/x)$ of substituting t for x in the term s. In these expressions the parentheses are simply punctuation marks, and are not to be regarded as symbols of the language.

If s is a constant symbol or a variable other than x we define $s(t/x)$ to be s, while $x(t/x)$ is to be t. If s is $fs_1 \ldots s_n$ we define $s(t/x)$ to be $ft_1 \ldots t_n$, where t_i is $s_i(t/x)$.

If ϕ is $Ps_1 \ldots s_n$ we define $\phi(t/x)$ to be $Pt_1 \ldots t_n$, where each t_i is $s_i(t/x)$, while $\perp(t/x)$ is just $\perp$. If $\phi_i(t/x)$ is ψ_i for $i=1,2$, and ϕ is either $(\phi_1 \wedge \phi_2)$ or $(\phi_1 \rightarrow \phi_2)$ we define $\phi(t/x)$ to be $(\psi_1 \wedge \psi_2)$ or $(\psi_1 \rightarrow \psi_2)$. If ϕ is $\forall x \theta$ we define $\phi(t/x)$ to be ϕ (since in this case all occurrences of x are bound), while if ϕ is $\forall y \theta$ we define $\phi(t/x)$ to be $\forall y \psi$ where ψ is $\theta(t/x)$. We see easily that this map is recursive, as a function of the three variables t, x, and ϕ.

More generally we could simultaneously substitute for a number of variables. In the case when we are substituting constant symbols this can be done by substituting successively. In general the two results are different. For instance, if t_1 contains x_2, the result of simultaneously substituting t_1 for x_1 and t_2 for x_2 will not be the same as first substituting t_1 for x_1 and then substituting t_2 for x_2 in the result.

While substitutions can always be made, when we come to consider meanings in the next section, we find that certain substitutions turn out to have the wrong meanings. An example of a similar situation from analysis is the following. We know that

$$\int x \cos xy \, \mathrm{d}y = \sin x.$$

If we substitute any expression not involving y for x we get a true result. If, however, we substitute y for x on both sides of this equation the result is not true.

We define inductively what is meant by saying that t is **free for** x in ϕ. If ϕ is atomic, then t is free for x in ϕ. If ϕ is either $(\phi_1 \wedge \phi_2)$ or $(\phi_1 \rightarrow \phi_2)$ then t is free for x in ϕ iff t is free for x in both of ϕ_1 and ϕ_2. If ϕ is $\forall x \psi$ then t is free for x in ϕ. Finally, if ϕ is $\forall y \psi$, then t is free for x in ϕ if t is free for x in ψ and either y does not occur in t or x does not occur free in ψ. It can be shown that t is free for x in ϕ iff there is no subformula of ϕ of form $\forall y \theta$ with y occurring in t and x occurring free in θ. The set of triples (t,x,ϕ) for which t is free for x in ϕ is easily seen to be recursive.

12.2 TRUTH

Up to now formulae have simply been regarded as strings, with no meaning attached to them. We now see how to give them meanings, and to talk of their truth or falsehood.

Definition An L-**structure A** consists of a non-empty set A, an element $c_{\mathbf{A}}$ of A corresponding to each constant symbol c, a function $f_{\mathbf{A}}: A^n \to A$ for each n-ary function symbol f, and a subset $P_{\mathbf{A}}$ of A^n (or, equivalently, a property or relation of n variables) to each n-ary predicate symbol P.

This definition suggests that we start with a language L and then look at the relevant structures. This is the most convenient definition, but, in practice, the situation is reversed. We start with a class of objects we wish to consider, and then take an appropriate language. For instance, if we wish to talk about groups, our language might have a binary predicate symbol corresponding to equality, a constant symbol, and one unary and one binary function symbol (so that we can talk about the identity element, and about the inverse function and multiplication); alternatively, we might omit the constant symbol and the unary function symbol. Similarly, if we wanted to talk about ordered sets, we might require our language to contain two binary predicate symbols (one for equality, and the other for the order relation). Note that if we took the above-mentioned language L for group theory, then L-structures are not groups. They are simply objects of which we can meaningfully ask whether or not they are groups (whereas it would not make sense to ask of a set with an order relation whether it was a group).

Let **A** be an L-structure. Let t be a closed term of L. We define inductively the corresponding element $t_{\mathbf{A}}$ of A. We have already defined $c_{\mathbf{A}}$. If $t = ft_1 \ldots t_n$, we define $t_{\mathbf{A}}$ to be $f_{\mathbf{A}}(t_{1\mathbf{A}}, \ldots, t_{n\mathbf{A}})$.

We wish to define a function $v_{\mathbf{A}}$ from the set of sentences of L to the two-element set $\{T,F\}$. To do this, we have to define a larger language $L(\mathbf{A})$, and define $v_{\mathbf{A}}$ on the set of sentences of $L(\mathbf{A})$. For convenience, we shall usually omit the subscript **A** of v, and will only use it when we need to look at several L-structures.

Let **A** be any L-structure. We define the language $L(\mathbf{A})$ to be obtained from L by adding a new constant symbol **a** for each a in A. (In the remaining chapters, bold type will be used for certain constant symbols or closed terms of a language, and not to denote tuples of elements.) Plainly, A can be regarded as an $L(\mathbf{A})$-structure, by defining $\mathbf{a}_{\mathbf{A}}$ to be a itself. As before, we have an element $t_{\mathbf{A}}$ of A to each closed term t of $L(\mathbf{A})$.

We now define the function v (more strictly, $v_{\mathbf{A}}$) on all sentences of $L(\mathbf{A})$ by induction. If ϕ is the atomic sentence $Pt_1 \ldots t_n$, we define $v\phi$ to be T iff $(t_{1\mathbf{A}}, \ldots, t_{n\mathbf{A}})$ is in the set $P_{\mathbf{A}}$, while $v\bot$ is F. When ϕ and ψ are sentences, we define, as in propositional logic, $v(\phi\wedge\psi)$ to be T iff both $v\phi$ and $v\psi$ are T, and $v(\phi\rightarrow\psi)$ to be F iff $v\phi=T$ and $v\psi=F$. If the sentence ϕ is $\forall x\psi$, we define $v\phi$ to be T iff $v\psi(\mathbf{a}/x)=T$ for every a in A. (Notice that ψ must have no free variables other than x, since ϕ is a sentence. Hence $\psi(\mathbf{a}/x)$ is a sentence; also, this sentence is a sentence of $L(\mathbf{A})$ even if ψ is a sentence of L. It is for this reason that we must work in $L(\mathbf{A})$ rather than L.) We say the sentence ϕ is **true** or **valid** in $\mathbf{A}$ if $v_{\mathbf{A}}\phi=T$; otherwise ϕ is **false** or **invalid** in $\mathbf{A}$.

Plainly the definition of $v\forall x\phi$ fits in with the intuitive meaning of $\forall$ as 'for all'.

The name 'first-order predicate logic' can now be explained. It is 'predicate logic' because it concerns properties or predicates of elements. It is 'first-order' because our use of $\forall$ lets us talk about all elements of a structure, but we are not permitted to talk about all subsets or all properties. For instance, induction in the way we normally use it is not a first-order concept, although there is a closely related first-order concept.

We define the auxiliary symbols $\neg$, $\vee$, and $\leftrightarrow$ as in propositional logic. The auxiliary symbol $\exists$ is defined by requiring $\exists x\phi$ to be an abbreviation for $\neg\forall x\neg\phi$. It then follows that $v_{\mathbf{A}}\exists x\phi=T$ iff $v_{\mathbf{A}}\phi(\mathbf{a}/x)=T$ for some a in A, which fits in with the intended meaning of $\exists$ as 'for some'.

We write $\mathbf{A}\models\phi$ if ϕ is true in $\mathbf{A}$, and $\models\phi$ if $\mathbf{A}\models\phi$ for all $\mathbf{A}$. We call ϕ **valid** if $\models\phi$. If Γ is a set of sentences of L, we call $\mathbf{A}$ a **model** of Γ if every member of Γ is true in $\mathbf{A}$. Finally, we write $\Gamma\models\phi$ (and call ϕ a **semantic consequence** of Γ) if ϕ is true in every model of Γ. In particular, this last notion makes sense when Γ is empty; in this case it coincides with the previous meaning of $\models\phi$.

Observe that the notion of $\models$ is essentially infinite. That is, $\models\phi$ holds iff $\mathbf{A}\models\phi$ holds of every L-structure $\mathbf{A}$, and there are infinitely many structures to consider. Even if we look at only one structure $\mathbf{A}$, and a formula ϕ containing quantifiers, to find out whether or not $\mathbf{A}\models\phi$ we will have to look at all elements of A, and there may be infinitely many of them. Thus, at this point in our work, we have no reason to expect that $\{\phi; \phi \text{ is valid}\}$ is even listable. This does hold, and will be shown much later. If we are concerned only with finite structures, the position is simpler.

Theorem 12.6 *The set of finite sets Γ of sentences such that Γ has a finite model is r.e.*

Proof If Γ has a finite model it has a model defined on the set of natural numbers $\leqslant n$ for some n. For any finite set can be mapped bijectively to such a set, and we can plainly translate the constants, functions, and predicates of the original structure to make this set an L-structure which will also be a model of Γ. There can only be finitely many L-structures for a given n. (At least, if L does not have infinitely many constant, function, or predicate symbols. But Γ is finite, and only the symbols occurring in some formula of Γ

can be relevant, so we may forget about any others.) We simply look at all these structures for each n, checking each member of Γ to see whether or not it is true in the given structure. As the structure is finite, we have only a finite amount of checking to do for each element of the finite set Γ. Thus we can look through all finite structures until, if ever, we find a model of Γ. In other words, the set of all such finite sets is the domain of a computable function (assigning to each finite Γ, for instance, the smallest n for which Γ has a model with $n+1$ elements), and so is r.e.■

We need to extend matters of truth and falsehood to arbitrary formulae, which may contain free variables, not just sentences. To do this, we have to assign a member of A to each variable. We define an **interpretation i in $\mathbf{A}$** to be a function i from the set of free variables into A. To each formula ϕ of $L(\mathbf{A})$, we obtain a sentence $i\phi$ of $L(\mathbf{A})$ by simultaneously replacing in ϕ the free occurrences of the variables x_r for all r by the corresponding constant symbols $\mathbf{a}_r$, where a_r is ix_r. In particular, if ϕ is a sentence, and so has no free variables, $i\phi$ is just ϕ. Combining this with the previously defined $v_{\mathbf{A}}$, we have a function $v_{\mathbf{A}}i$ from the set of all formulae into $\{T,F\}$. If this function sends ϕ to T we say ϕ is **true under the interpretation** i. Notice that if i and j are interpretations such that $ix=jx$ for every variable x which occurs free in ϕ, then $i\phi$ and $j\phi$ coincide.

Lemma 12.7 *$vi\forall x\phi=T$ iff $vj\phi=T$ for every interpretation j such that $jy=iy$ for every variable y other than x.*

Proof Let θ be the formula obtained from ϕ by replacing the free occurrences of the variables x_r other than x by the corresponding $\mathbf{a}_r$. Then $i\forall x\phi$ is $\forall x\theta$, by definition. So $vi\forall x\phi=T$ iff $v\theta(\mathbf{b}/x)=T$ for all b in A. Let j be the interpretation with $jx=b$ and $jy=iy$ for all other variables y. Then $j\phi$ is just $\theta(\mathbf{b}/x)$. The result follows.■

We now write $\mathbf{A}\models\phi$ if ϕ is true under every intepretation in $\mathbf{A}$, and $\models\phi$ if $\mathbf{A}\models\phi$ for every L-structure $\mathbf{A}$. We write $\Gamma\models\phi$, where Γ is a set of formulae, if all those interpretations (into all structures) which make all the members of Γ true also make ϕ true. When ϕ is a sentence, and Γ a set of sentences, these definitions coincide with the previous ones. However, it should be noted that $\Gamma\models\phi$ does not mean that $\mathbf{A}\models\phi$ for every $\mathbf{A}$ such that $\mathbf{A}\models\gamma$ for every γ in Γ. The latter would require us only to look at those $\mathbf{A}$ such that $v_{\mathbf{A}}i\gamma=T$ for all interpretations i into $\mathbf{A}$, whereas we want to look at those particular i (and corresponding $\mathbf{A}$) for which $v_{\mathbf{A}}i\gamma=T$.

We will leave to a later section further properties of interpretations.

Exercise 12.1 Show that the following hold.

(a) $\forall x\phi\models\exists x\phi$.
(b) $\exists x\forall y\phi\models\forall y\exists x\phi$.
(c) $\forall x\forall y\phi\models\forall y\forall x\phi$.

(d) $\exists x(\phi \rightarrow \forall x\phi)$.

Exercise 12.2 Is it true that $\forall y \exists x\phi \models \exists x \forall y \phi$? Justify your answer.

Exercise 12.3 Let x be a variable which does not occur free in ϕ. Show that $\models \phi \leftrightarrow \forall x\phi$ and that $\models \phi \leftrightarrow \exists x\phi$.

Exercise 12.4 Show that $\models \forall x(\phi \wedge \psi) \leftrightarrow \forall x\phi \wedge \forall x\psi$. Show that when x is a variable which does not occur free in ϕ we also have $\models \forall x(\phi \vee \psi) \leftrightarrow \phi \vee \forall x\psi$.

Exercise 12.5 Show that $\forall x\phi \vee \forall x\psi \models \forall x(\phi \vee \psi)$. Is it true that $\forall x(\phi \vee \psi) \models \forall x\phi \vee \forall x\psi$?

12.3 PROOF

Just as in propositional logic, we wish to define **derivations**, and we will use the method of natural deduction rather than the axiomatic to tableau methods. As before, the definition is inductive.

Definition A tree with only one vertex is a derivation iff that vertex is unmarked. A tree with more than one vertex is a derivation iff it is made from smaller derivations by one of the eight following rules.

The rules E$\wedge$ and I$\wedge$ **for elimination and introduction of** $\wedge$.

If $\begin{array}{c}T\\ \phi\wedge\psi\end{array}$ is a derivation then so are $\dfrac{\begin{array}{c}T\\ \phi\wedge\psi\end{array}}{\phi}$ and $\dfrac{\begin{array}{c}T\\ \phi\wedge\psi\end{array}}{\psi}$.

If both $\begin{array}{c}T'\\ \theta'\end{array}$ and $\begin{array}{c}T''\\ \theta''\end{array}$ are derivations, then so is $\dfrac{\begin{array}{cc}T' & T''\\ \theta' & \theta''\end{array}}{\theta'\wedge\theta''}$.

The rules E$\rightarrow$ and I$\rightarrow$ **for elimination and introduction of** $\rightarrow$.

If both $\begin{array}{c}T'\\ \phi\end{array}$ and $\begin{array}{c}T''\\ \phi\rightarrow\psi\end{array}$ are derivations, then so is $\dfrac{\begin{array}{cc}T' & T''\\ \phi & \phi\rightarrow\psi\end{array}}{\psi}$.

If $\begin{array}{c}T\\ \psi\end{array}$ is a derivation, then so is $\dfrac{\begin{array}{c}\not\phi\\ T\end{array}}{\phi\rightarrow\psi}$.

The rules $\bot$ and **RAA**.

If $\begin{matrix} T \\ \bot \end{matrix}$ is a derivation, then so is $\dfrac{\begin{matrix} T \\ \bot \end{matrix}}{\phi}$, for any ϕ.

If $\begin{matrix} T \\ \bot \end{matrix}$ is a derivation, then so is $\dfrac{\begin{matrix} \cancel{\neg\phi} \\ T \\ \bot \end{matrix}}{\phi}$, for any ϕ.

The rules E$\forall$ and I$\forall$ **for elimination and introduction of** $\forall$.

If $\begin{matrix} T \\ \forall x\phi \end{matrix}$ is a derivation, then so is $\dfrac{\begin{matrix} T \\ \forall x\phi \end{matrix}}{\phi(t/x)}$, provided that t is free for x in ϕ.

Notice that x is free for x in any formula, so, as a special case of this, if $\begin{matrix} T \\ \forall x\phi \end{matrix}$ is a derivation, so is $\dfrac{\begin{matrix} T \\ \forall x\phi \end{matrix}}{\phi}$.

If $\begin{matrix} T \\ \phi \end{matrix}$ is a derivation, then so is $\dfrac{\begin{matrix} T \\ \phi \end{matrix}}{\forall x\phi}$, provided x does not occur free in any label of an unmarked leaf of T.

The first six rules are the ones already used in propositional logic. Consequently all the derivations obtained in the previous chapter are also derivations of predicate logic. In particular, Lemmas 11.5 and 11.6 hold.

Notice that the rules of $\forall$ follow the usual way of working with 'for all x'. For instance, the rule for introduction of $\forall$ corresponds to an informal argument where we prove a result for x, and then say that x was arbitrary and so the result holds for all x.

As previously, the formula which is the label at the root of a derivation D is called the **conclusion** of D, and the labels on the unmarked leaves are called the **hypotheses** of D. We write $\Gamma \vdash \phi$ (and say ϕ can be **derived from** Γ) if there is a derivation with conclusion ϕ and hypotheses in Γ. When Γ is empty, we write $\vdash \phi$ instead of $\Gamma \vdash \phi$, and we call ϕ a **theorem**.

We can now show that $\forall x\phi \vdash \forall x\psi$ if $\phi \vdash \psi$. For (by elimination of $\forall$ with t being x) we know that $\forall x\phi \vdash \phi$. So Lemma 11.5(iii) tells us that $\forall x\phi \vdash \psi$. As x does not occur free in $\forall x\phi$ we can apply introduction of $\forall$ to this derivation to get a derivation of $\forall x\psi$ from $\forall x\phi$.

As particular cases of the last result, we see that $\forall x\phi \vdash \forall x\neg\neg\phi$ and $\forall x\neg\neg\phi \vdash \forall x\phi$. It follows that $\neg\forall x\neg\neg\phi \vdash \neg\forall x\phi$ and $\neg\forall x\phi \vdash \neg\forall x\neg\neg\phi$. By definition of $\exists$, these say that $\exists x\neg\phi \vdash \neg\forall x\phi$ and $\neg\forall x\phi \vdash \exists x\neg\phi$. The fact that $\forall x\neg\phi \vdash \neg\exists x\phi$ and $\neg\exists x\phi \vdash \forall x\neg\phi$ comes from the propositional rules only, using the definition of $\exists$.

There is a rule for **left-introduction of $\exists$**, which is obtained by a similar method to the rule for left-introduction of $\vee$ in propositional logic. It states that if $\Gamma\cup\{\phi\}\vdash\psi$ then $\Gamma\cup\{\exists x\phi\}\vdash\psi$ provided that x does not occur free in $\Gamma\cup\{\psi\}$. The rule for **right-introduction** of $\exists$ comes from the rule for elimination of $\forall$. It says that if t is free for x in ϕ then $\Gamma\vdash\exists x\phi$ provided that $\Gamma\vdash\phi(t/x)$.

For suppose $\Gamma\cup\{\phi\}\vdash\psi$, and that x does not occur free in $\Gamma\cup\{\psi\}$. Then $\Gamma\cup\{\neg\psi\}\vdash\neg\phi$. By the rule for introduction of $\forall$, we then have $\Gamma\cup\{\neg\psi\}\vdash\forall x\neg\phi$. From this we get, as needed, $\Gamma\cup\{\neg\forall x\neg\phi\}\vdash\psi$.

We now look at some examples using the rules for $\forall$.

$\vdash\forall x(\phi\wedge\psi)\leftrightarrow\forall x\phi\wedge\forall x\psi$. Known properties of $\leftrightarrow$, of $\wedge$, and of $\rightarrow$ tell us that we need only show that $\forall x(\phi\wedge\psi)\vdash\forall x\phi\wedge\forall x\psi$ and $\forall x\phi\wedge\forall x\psi\vdash\forall x(\phi\wedge\psi)$. The derivations are as follows.

$$\dfrac{\dfrac{\dfrac{\dfrac{\forall x(\phi\wedge\psi)}{\phi\wedge\psi}}{\phi}}{\forall x\phi}\qquad\dfrac{\dfrac{\dfrac{\forall x(\phi\wedge\psi)}{\phi\wedge\psi}}{\psi}}{\forall x\psi}}{\forall x\phi\wedge\forall x\psi}\quad\text{and}\quad\dfrac{\dfrac{\dfrac{\dfrac{\forall x\phi\wedge\forall x\psi}{\forall x\phi}}{\phi}\qquad\dfrac{\dfrac{\forall x\phi\wedge\forall x\psi}{\forall x\psi}}{\psi}}{\phi\wedge\psi}}{\forall x(\phi\wedge\psi)}$$

Now suppose x does not occur free in ϕ. We will show that $\vdash\forall x(\phi\vee\psi)\leftrightarrow\phi\vee\forall x\psi$. As before, we need only show that $\forall x(\phi\vee\psi)\vdash\phi\vee\forall x\psi$ and that $\phi\vee\forall x\psi\vdash\forall x(\phi\vee\psi)$. For the first of these, we begin by observing that $\{\neg\phi,\phi\vee\psi\}\vdash\psi$ (this is Exercise 11.12). Since $\forall x(\phi\vee\psi)\vdash\phi\vee\psi$, we have $\{\neg\phi,\forall x(\phi\vee\psi)\}\vdash\psi$. Because x does not occur free in either $\forall x(\phi\vee\psi)$ (by definition) or in ϕ (given), we find that $\{\neg\phi,\forall x(\phi\vee\psi)\}\vdash\forall x\psi$. The result now follows by propositional reasoning. For the other property, note first that we need only show that $\phi\vdash\forall x(\phi\vee\psi)$ and that $\forall x\psi\vdash\forall x(\phi\vee\psi)$, by the rule for left-introduction of $\vee$. Now $\phi\vdash\phi\vee\psi$, and as x does not occur free in ϕ we obtain $\phi\vdash\forall x(\phi\vee\psi)$. Similarly, as $\psi\vdash\phi\vee\psi$ and $\forall x\psi\vdash\psi$ we see first $\forall x\psi\vdash\phi\vee\psi$ and then that $\forall x\psi\vdash\forall x(\phi\vee\psi)$. Because of the complicated use of $\neg$ in the definition of $\vee$, it is usually easier to use various rules and known derivations rather than directly constructing the derivation we need.

Theorem 12.8 *The set of derivations is recursive.*

Proof A tree with one vertex is a derivation iff the vertex is unmarked. A tree with more than one vertex is a derivation iff it is built from smaller trees by one of eight methods, and these smaller trees are derivations. Thus we can tell inductively whether or not a tree is a derivation, provided we can tell whether or not it is built from smaller trees by one of the eight ways.

We can plainly tell whether or not a tree is of one of the forms

$$\dfrac{\overset{T}{\phi\wedge\psi}}{\phi},\quad\dfrac{\overset{T}{\psi\wedge\phi}}{\phi},\quad\dfrac{\overset{T'}{\phi}\quad\overset{T''}{\psi}}{\phi\wedge\psi},\quad\dfrac{\overset{T'}{\bot}}{\phi},\quad\text{or}\quad\dfrac{\overset{T'}{\phi}\quad\overset{T''}{\phi\rightarrow\psi}}{\psi}.$$

If we ask whether or not T is of one of the forms $\dfrac{\overset{\not\phi}{T'}\atop\psi}{\phi\to\psi}$ or $\dfrac{\overset{\not\neg\phi}{T'}\atop\bot}{\phi}$, we find that T' is not uniquely given. This is because T' is obtained from T by removing the root, and then removing the marks on some leaves. We can choose which, if any, of the leaves labelled ϕ (or $\neg\phi$, in the RAA case) to remove the marks from. However, this provides finitely many possibilities for T', all of which we can find, and we can then continue the checking for each of these possibilities.

We can check whether or not T is $\dfrac{\overset{T'}{\phi}}{\forall x\phi}$. We can find the labels on the unmarked leaves of T', and we can then see whether or not x occurs free in any of these labels.

The final case we have to check is whether or not T is $\dfrac{\overset{T'}{\forall x\phi}}{\psi}$, where we require that ψ is $\phi(t/x)$ for some term t free for x in ϕ. We can certainly check whether or not T is of this form for some ϕ and ψ which do not satisfy the extra condition. The problem is that we appear to have infinitely many t to consider. However, we can first check whether or not x occurs free in ϕ. If not, then $\phi(t/x)$ always coincides with ϕ, so in this case we have only to check whether or not ψ is ϕ. If x does occur free in ϕ, it is easy to see that the string t occurs as a segment of $\phi(t/x)$. So, in this case, we look at all segments of ψ, see which of them are terms, and for each such term t decide whether or not t is free for x in ϕ and ψ is $\phi(t/x)$. Thus we have reduced the apparently infinitely many terms t we need to look at to only finitely many. ■

Theorem 12.9 *Let Γ be recursive. Then both $\{\phi;\ \phi$ is a formula with $\Gamma\vdash\phi\}$ and $\{\phi;\ \phi$ is a sentence with $\Gamma\vdash\phi\}$ are r.e.*

Proof As Γ is recursive, the set of trees whose unmarked leaves have labels in Γ is recursive. Since the set of derivations is recursive, the intersection of these two sets, which is the set of derivations with hypotheses in Γ, is recursive.

The function sending each tree to the label on its root is recursive (by Church's Thesis, of course). So the image by this function of the set of derivations with hypotheses in Γ is r.e. But this set is, by definition, exactly $\{\phi;\Gamma\vdash\phi\}$.

As the set of all sentences is recursive, its intersection with $\{\phi;\Gamma\vdash\phi\}$ will also be r.e., as required. ■

Lemma 12.10 *Let Γ be r.e. Then there is a recursive set Γ' such that $\Gamma\vdash\phi$ iff*

$\Gamma' \vdash \phi$. *Further* Γ' *may be chosen to consist of sentences if* Γ *consists of sentences.*

Theorem 12.11 *Let* Γ *be r.e. Then both* $\{\phi;\ \phi$ *is a formula with* $\Gamma \vdash \phi\}$ *and* $\{\phi;\ \phi$ *is a sentence with* $\Gamma \vdash \phi\}$ *are r.e.*

Proof The theorem follows at once from the lemma and the previous theorem.

When referring to an r.e. set of formulae, we are implicitly considering a bijection from **N** to the set of all formulae. So we shall assume the formulae are given as $\phi_0, \phi_1, \ldots$.

If Γ is empty, it is itself recursive. Hence we may assume there is a recursive function f such that Γ is $\{\phi_{fn};\ \text{all } n\}$. Let Γ' consist of all formulae of form $\phi_{fn} \wedge (\phi_n \rightarrow \phi_n)$. Now we can tell for a given formula whether or not it is $\phi_m \wedge (\phi_n \rightarrow \phi_n)$ for some m and n. If not, it is not in Γ'. If so, it is in Γ' iff $m = fn$, which can be checked. Hence Γ' is recursive.

Now $\phi_{fn} \wedge (\phi_n \rightarrow \phi_n) \vdash \phi_{fn}$, obviously. Also, $\vdash \phi_n \rightarrow \phi_n$, and so $\phi_{fn} \vdash \phi_{fn} \wedge (\phi_n \rightarrow \phi_n)$. So by Lemma 11.5(iii), $\Gamma \vdash \phi$ iff $\Gamma' \vdash \phi$.

If Γ consists only of sentences, we similarly take Γ' to be the set of all $\sigma_{fn} \wedge (\sigma_n \rightarrow \sigma_n)$, where $\sigma_0, \sigma_1, \ldots$ are all the sentences.■

Observe that, both in propositional logic and in predicate logic, the set of theorems is r.e. In propositional logic, the set of tautologies was easily seen to be recursive; in contrast, in predicate logic, it is not even obvious that the set of valid formulae is r.e. The Completeness Theorem tells us that in propositional logic theorems and tautologies are the same, and so the set of theorems is recursive. In predicate logic the Completeness Theorem (proved later) tells us that theorems and valid formulae are the same, and so the set of valid formulae is r.e. In predicate logic the set of theorems need not be recursive. As we shall see in the next chapter, for some languages the set of theorems is recursive, while for others it is not.

Exercise 12.6 Show that $\vdash \exists x(\phi \vee \psi) \leftrightarrow \exists x\phi \vee \exists x\psi$. Show also that if x does not occur free in ϕ then $\vdash \exists x(\phi \wedge \psi) \leftrightarrow \phi \wedge \exists x\psi$ and that $\vdash (\phi \rightarrow \forall x\psi) \leftrightarrow \forall x(\phi \rightarrow \psi)$.

Exercise 12.7 Show that $\forall x \forall y \phi \vdash \forall y \forall x \phi$ and that $\exists x \forall y \phi \vdash \forall y \exists x \phi$.

Exercise 12.8 Show that if $\phi \vdash \psi$ then $\exists x\phi \vdash \exists x\psi$.

12.4 SOUNDNESS

Just as in propositional logic, we can ask about soundness and adequacy, and we find that the corresponding results hold for predicate logic. The proof of soundness is only slightly more difficult than in the propositional case, but the proof of adequacy is considerably harder. As before, the Completeness

Theorem, which we now state, divides into two parts, the Soundness Theorem and the Adequacy Theorem.

Theorem 12.12 (Gödel's Completeness Theorem) $\Gamma \vdash \phi$ *iff* $\Gamma \models \phi$.

We now prove the Soundness Theorem for Predicate Logic. The Adequacy Theorem will be proved in the next section.

Theorem 12.13 (Soundness Theorem) *If* $\Gamma \vdash \phi$ *then* $\Gamma \models \phi$.

Proof We shall show, by induction on the number of vertices, that if D is a derivation, then the conclusion of D is true under any interpretation which makes all the hypotheses of D true. Write Hyp$D$ for the set of hypotheses of D, and let i be an interpretation under which all members of Hyp$D$ are true.

The start of the induction and the six propositional cases follow exactly as in section 11.5, except that we must replace v by vi.

If D is $\dfrac{\begin{matrix} D' \\ \phi \end{matrix}}{\forall x \phi}$, we need, by Lemma 12.7, to show that ϕ is true under any interpretation j for which $jy = iy$ for all variables other than x. However, x does not occur free in any formula γ of Hyp$D$ (= HypD'). Hence the hypotheses of D' are also true under j, so that ϕ is true under j.

We are left with the case when D is $\dfrac{\begin{matrix} D' \\ \forall x \phi \end{matrix}}{\phi(t/x)}$. To deal with this case, we must show that $\forall x \phi \models \phi(t/x)$ when t is free for x in ϕ. This will take the remainder of the section. ■

Before looking at this, we observe that $\phi(t/x)$ need not be a semantic consequence of $\forall x \phi$ when t is not free for x in ϕ. This is why this concept has to be defined, and required in the rule for elimination of $\forall$. As an example, let L have no constant and function symbols, and have exactly one predicate symbol P whose arity is 2. Then $\mathbf{N}$ is an L-structure with (m,n) in $P_{\mathbf{N}}$ iff $m<n$. Plainly $\forall x \phi$ is true in $\mathbf{N}$ if ϕ is $\exists y Pxy$. But $\phi(y/x)$ is not true in $\mathbf{N}$.

Similarly $\forall x \phi$ is not a semantic consequence of ϕ if x occurs free in ϕ. Let L be as before, and let ϕ be Pxy. Then ϕ is true in the interpretation which sends x and y to 0 and 1, respectively. But $\forall x \phi$ is plainly not true in this interpretation.

Let i be any interpretation. We can extend i to a map, which we still denote by i, from the set of all terms of $L(\mathbf{A})$ into $\mathbf{A}$. We need only define i on a constant symbol c to be the corresponding element $c_{\mathbf{A}}$ of $\mathbf{A}$, and extend inductively.

Now let t be a term free for x in the formula ϕ, and let ψ be $\phi(t/x)$. Suppose $\forall x \phi$ is true under i. In particular, ϕ will be true under the interpretation j for which $jx = it$ and $jy = iy$ for all other y. Hence, to show that $\forall x \phi \models \psi$, it is enough to prove the lemma below.

Lemma 12.14 *Let t be free for x in ϕ, and let ψ be $\phi(t/x)$. Let i be an interpretation, and let j be the interpretation with $jx=it$, $jy=iy$ for all other y. Then $vi\psi=vj\phi$.*

Proof As usual, by induction. If ϕ is either $\phi_1 \wedge \phi_2$ or $\phi_1 \rightarrow \phi_2$, the result is plainly true for ϕ if it is true for ϕ_1 and ϕ_2. If ϕ is $\forall x \theta$, then, as x does not occur free in ϕ, we find that ψ is ϕ and $j\phi$ is $i\phi$, as required.

Let ϕ be $\forall y \theta$, where y is different from x. Then ψ is $\forall y \alpha$, where α is $\theta(t/x)$. Since t is free for x in ϕ, y cannot occur in t and t is free for x in θ. Now $vi\psi=T$ iff $vi'\alpha=T$ for every interpretation i' such that $i'z=iz$ for all z different from y. Inductively, $vi'\alpha=vj'\theta$, where $j'z=i'z$ if z is not x and $j'x=i't$. Since i' is i except on y, and t does not contain y, $i't=it$. Thus j' differs from j only on y, and j' can be any such interpretation. So $vj\phi=T$ iff $vj'\theta=T$ for all such j', and we have $vi\psi=vj\phi$, as required. To start our induction, we need some auxiliary results, which will also be useful later.

Let i be an interpretation, and let ϕ be an atomic formula $Pt_1 \ldots t_n$. Then ϕ is true under i iff $(s_{1\mathbf{A}}, \ldots, s_{n\mathbf{A}})$ is in $P_{\mathbf{A}}$, where s_r comes from t_r by replacing each variable x by the corresponding constant symbol $\mathbf{a}$, where $a=ix$. The usual induction shows that $s_{r\mathbf{A}}$ is just it_r. So ϕ is true under i iff $(it_1, \ldots, it_n)$ is in $P_{\mathbf{A}}$.

Let ψ be $\phi(t/x)$, with ϕ as in the previous paragraph. Then ψ is true under i iff $(iu_1, \ldots, iu_n)$ is in $P_{\mathbf{A}}$, where u_r is $t_r(t/x)$, while ϕ is true under j iff $(jt_1, \ldots, jt_n)$ is in $P_{\mathbf{A}}$. So the result we want for atomic formulae holds provided $iu_r=jt_r$. As usual, this is proved by a straightforward induction.■

Lemma 12.15 *Let t be free for x in ϕ. Let i be an interpretation, and let $it=a$. Then $vi\phi(t/x)=vi\phi(\mathbf{a}/x)$.*

Proof By definition, we also have $i\mathbf{a}=a$. Thus, by the previous lemma, both sides equal $vj\phi$, where $jx=a$, and $jy=iy$ for all other y.■

Lemma 12.16 *Let $\mathbf{A}$ be an L-structure, and suppose there is a set S of closed terms of L such that every element of A is $t_{\mathbf{A}}$ for some t in S. Then $vi\forall x\phi=T$ iff $vi\phi(t/x)=T$ for all t in S.*

Proof By definition, $vi\forall x\phi=T$ iff $vi\phi(\mathbf{a}/x)=T$ for all a in A. By the previous lemma and our assumption on S, this holds iff $vi\phi(t/x)=T$ for all t in S, as required.■

Exercise 12.9 Complete the details in the proof of Lemma 12.14.

12.5 ADEQUACY

In this section we show that our notion of derivation is adequate; that is, $\Gamma \vdash \phi$ if $\Gamma \models \phi$. Several other properties will have to be discussed first.

12.5.1 Consistency

As in propositional logic we call a set Γ of formulae of predicate logic inconsistent if $\Gamma \vdash \bot$. A consistent set of formulae Γ is maximal consistent if any Γ' strictly containing Γ is inconsistent. Γ is complete if for every ϕ either ϕ or $\neg\phi$ is in Γ. We are often interested only in sentences, rather than formulae in general. We therefore call a consistent set Σ of sentences maximal consistent if any set Σ' of sentences which strictly contains Σ is inconsistent, and we call Σ complete if for every sentence σ either σ or $\neg\sigma$ is in Σ. It follows that a set of sentences which is maximal consistent or complete as a set of sentences is not maximal consistent or complete as a set of formulae. This should not cause any difficulties. Most of our results are true both for sets of formulae and for sets of sentences; the proofs for sentences come from the proofs for formulae by just restricting attention to sentences. All the results of section 11.6 preceding Theorem 11.17 hold for predicate logic (whether for formulae or sentences) as well as for propositional logic, the details of the proofs being identical.

Theorem 12.17 *Let Γ be a consistent set (of formulae or sentences). Then there is a maximal consistent set Γ' (of formulae or sentences, respectively) with $\Gamma \subset \Gamma'$.*

Proof If our language L is countable, the theorem is proved in exactly the same way as Theorem 11.17.

Readers may wish to restrict attention to the countable case. For the general case, a certain amount of familiarity with the Axiom of Choice or its equivalents is needed. One method is to follow the above approach, replacing finite induction by transfinite induction. Alternatively, we could use Zorn's Lemma to find a maximal consistent Γ' containing Γ. The best proof is to use a result known as Tukey's Lemma to get the result. Tukey's Lemma, which is equivalent to Zorn's Lemma, says that for any property P such that a set has P iff all its finite subset have P, then any set having P is contained in a maximal set having P; Tukey's Lemma applies at once to the property of being consistent.■

12.5.2 Witnesses

Let ϕ be a formula, c a constant symbol, and y a variable. Then $\phi(y/c)$ denotes the result of replacing each occurrence of c in ϕ by y. If T is a tree, $T(y/c)$ is the tree obtained from T by replacing each label ϕ by the corresponding $\phi(y/c)$.

Lemma 12.18 *If D is a derivation so is $D(y/c)$ provided y does not occur in D.*

Proof In the usual induction, seven of the ways of building D from smaller derivations immediately give $D(y/c)$ as built in the same way from smaller derivations. The difficult case is when D is built from D' by eliminating $\forall$. To deal with this case, take a formula ϕ and term t free for x in ϕ, and let ψ be

$\phi(t/x)$. Let y be a variable not occurring in ϕ or t, and let θ be $\phi(y/c)$. We need to prove that there is a term s free for x in θ such that $\theta(s/x)$ is $\psi(y/c)$.

It is clear that s should be $t(y/c)$. The proof that this is as required is, as ever, an induction.

We show first that if u is a term not involving y and v and w are the terms $u(t/x)$ and $u(y/c)$ then $v(y/c)$ is $w(s/x)$. When u is a constant symbol other than c or a variable other than x this is obvious (remembering that u cannot be y). If u is c then v is c and w is y, and so $w(s/x)$ is y, which is also $v(y/c)$. If u is x then v is t and w is x; in this case $v(y/c)$ is $t(y/c)$, which is s, and $w(s/x)$ is also s. The inductive step is obvious.

It is now immediate that we have the result we want when the formula ϕ is atomic. It clearly follows inductively if ϕ is either $(\phi_1 \wedge \phi_2)$ or $(\phi_1 \rightarrow \phi_2)$.

Now suppose that ϕ is $\forall z\alpha$, where z is neither x nor y. As t is free for x in ϕ, it follows that t cannot involve z, and so neither can s. Now θ is $\forall z\beta$, where β is $\alpha(y/c)$. Hence s is free for x in θ. Also ψ is $\forall x\gamma$, where γ is $\alpha(t/x)$. Hence $\psi(y/c)$ is the same as $\theta(s/x)$, because $\gamma(y/c)$ is $\beta(s/x)$, inductively.

Finally, let ϕ be $\forall x\alpha$. Then ψ is ϕ, since there are no free occurrences of x in ϕ. Also θ will be $\forall x\beta$, and so, similarly, s is free for x in θ and $\theta(s/x)$ is just θ. Again we have $\psi(y/c)$ being $\theta(s/x)$, completing the induction (since y does not occur in ϕ).■

Corollary *Let c be a constant symbol not occurring in* Γ *or* ϕ. *If* $\Gamma \vdash \phi(c/x)$ *then* $\Gamma \vdash \forall x\phi$.

Proof If D is the derivation, and y a variable other than x which does not occur in D, we know that $D(y/c)$ is a derivation of $\phi(y/x)$ from the same hypotheses as D. Since y does not occur in these hypotheses, we can introduce $\forall$, and we find that $\Gamma \vdash \forall y\phi(y/x)$. The corollary follows from the next lemma.■

Lemma 12.19 *Let y be a variable not occurring in* ϕ, and let ψ *be* $\phi(y/x)$. *Then x is free for y in* ψ, ϕ *is* $\psi(x/y)$, *and* $\forall y\psi \vdash \forall x\phi$ *and* $\forall x\phi \vdash \forall y\psi$.

Proof Since y does not occur in ϕ, it is certainly free for x in ϕ. Hence $\forall x\phi \vdash \psi$. As y does not occur in $\forall x\phi$, this gives $\forall x\phi \vdash \forall y\psi$. A similar argument shows that $\forall y\psi \vdash \forall x\phi$, once we have shown that x is free for y in ψ and that ϕ is $\psi(x/y)$. This requires the usual induction, which will be left as an exercise.■

Let L' be a language containing L, and let ϕ be a formula of L, and Γ a set of formulae of L. Every L'-structure is an L-structure (just forget about the extra symbols), and every L-structure can be made into an L'-structure (define the elements, functions, and subsets corresponding to the new symbols arbitrarily). Hence $\Gamma \models \phi$ for L iff $\Gamma \models \phi$ for L'. From the Completeness Theorem, it follows that $\Gamma \vdash \phi$ for L iff $\Gamma \vdash \phi$ for L'. We shall need to prove a special case of this as a step in proving the Completeness Theorem.

Lemma 12.20 *Let L' be a language containing L, such that $L'-L$ consists only of constant symbols. Let ϕ be a formula of L, and Γ a set of formulae of L. Then $\Gamma \vdash \phi$ for L iff $\Gamma \vdash \phi$ for L'.*

Proof Plainly a derivation of ϕ from Γ in L is also a derivation in L'. Conversely, any derivation in L' which does not use any of the symbols of $L'-L$ will be a derivation in L.

So it is enough to show that if there is a derivation D of ϕ from Γ in L', and D uses $n+1$ new symbols, then there is another derivation using only n new symbols. Let c be one of the new constant symbols which occurs in D, and let y be a variable which does not occur in D. By the previous lemma $D(y/c)$ is a derivation, and it only uses n of the new symbols. Because ϕ and Γ, being in L, do not involve c, the conclusion $\phi(y/c)$ of $D(y/c)$ is just ϕ, and similarly the hypotheses of $D(y/c)$ are in Γ. ■

A formula $\exists x\phi \rightarrow \phi(c/x)$, where c is a constant symbol not occurring in ϕ, is called a **Henkin formula** (for $\exists x\phi$), and c is called a **witness** for $\exists x\phi$.

Lemma 12.21 *Let Γ be consistent, and let Φ be a set of Henkin formulae such that each witness occurs in only one formula of $\Gamma \cup \Phi$. Then $\Gamma \cup \Phi$ is consistent.*

Proof Suppose $\Gamma \cup \Phi$ is inconsistent. Then there is a finite subset Φ_0 of Φ such that $\Gamma \cup \Phi_0$ is inconsistent. We show that if we remove one member from Φ_0 the result is still inconsistent. Thus, by induction, Γ will be inconsistent.

So it is enough to show that Γ is inconsistent assuming $\Gamma \cup \{\exists x\phi \rightarrow \phi(c/x)\}$ is inconsistent, where c is a constant symbol not occurring in $\Gamma \cup \{\phi\}$. As $\Gamma \cup \{\exists x\phi \rightarrow \phi(c/x)\} \vdash \bot$, we know that $\Gamma \vdash \neg(\exists x\phi \rightarrow \phi(c/x))$. But we have shown (immediately after Lemma 11.6) that $\neg(\alpha \rightarrow \beta) \vdash \alpha$ and $\neg(\alpha \rightarrow \beta) \vdash \neg\beta$. Hence $\Gamma \vdash \exists x\phi$ and $\Gamma \vdash \neg\phi(c/x)$. Since c does not occur in Γ or ϕ, the corollary to Lemma 12.18 tells us that $\Gamma \vdash \forall x \neg\phi$. As $\exists x\phi$ is, by definition, $\neg\forall x\neg\phi$, we see that Γ is inconsistent. ■

Lemma 12.22 *Let Γ be a consistent set (of formulae or sentences) of L. Then there is a language $L*$, with $L \subset L*$ and $L*-L$ consisting only of constant symbols, and a consistent set $\Gamma*$ (of formulae or sentences, respectively) such that $\Gamma \subset \Gamma*$ and $\Gamma*$ contains a Henkin formula (sentence) for each formula (sentence) $\exists x\phi$ of $L*$.*

Proof First take a language L' containing L, such that $L'-L$ consists only of constant symbols $c_{x\phi}$, one for each pair (x, ϕ) with ϕ a formula of L. Let Γ' be $\Gamma \cup \{\exists x\phi \rightarrow \phi(c_{x\phi}/x)\}$. By the previous lemma, Γ' is consistent.

This is not yet what we want, since Γ' only contains Henkin formulae for each $\exists x\phi$ with ϕ a formula of L, and we now have to look at all formulae of

L'. We define, inductively, L^0 to be L, Γ^0 to be Γ, and $L^{n+1}=(L^n)'$, and $\Gamma^{n+1}=(\Gamma^n)'$. Finally, we define $L*$ and $\Gamma*$ to be $\cup L^n$ and $\cup\Gamma^n$.

Plainly $L*$ satisfies the relevant conditions. For any formula ϕ of $L*$, there is some integer n such that ϕ is a formula of L^n. Then Γ^{n+1}, and so $\Gamma*$ also, contains a Henkin formula for $\exists x\phi$. If $\Gamma*$ were inconsistent, there would be a finite inconsistent subset, which would be contained in Γ^n for some n. Hence Γ^n would be inconsistent in $L*$, and hence, by Lemma 12.20, would also be inconsistent in L^n. But we know, inductively, that each Γ^n is consistent.

If we are only interested in sentences, we can simply cut $\Gamma*$ down to the set of sentences which it contains.■

12.5.3 Adequacy

Theorem 12.23 *A set Γ of sentences is consistent iff it has a model.*

Proof If Γ has a model, by Soundness, $\perp$ cannot be derived from Γ, as it is not true in the model.

Let Γ be consistent. Suppose we can find a language containing L and a set of sentences of this language containing Γ which has a model. Then this can be regarded as an L-structure (by ignoring the new symbols) and, as such, is a model of Γ. Hence, by Lemma 12.22 and Theorem 12.17 we may assume that Γ contains Henkin sentences for every sentence $\exists x\phi$, and that Γ is maximal consistent.

Let A be the set of all closed terms of L (note that A is not empty, since it includes all the witnesses). We can define an L-structure $\mathbf{A}$ on A as follows. For each constant symbol c, $c_{\mathbf{A}}$ is just c itself. The function $f_{\mathbf{A}}$ is given the natural definition, namely, $f_{\mathbf{A}}(t_1, \ldots, t_n)$ is $ft_1 \ldots t_n$. Finally, $P_{\mathbf{A}}$ consists of all $(t_1, \ldots, t_n)$ such that $Pt_1 \ldots t_n$ is in Γ. Plainly, for any closed term t, the element $t_{\mathbf{A}}$ of A is just t.

We now show, inductively, that a sentence ϕ of L is in Γ iff $v\phi = T$ (where, as usual, v is short for $v_{\mathbf{A}}$). This holds for $\perp$, which cannot be in Γ because Γ is consistent. It holds for all other atomic sentences by the definition of $P_{\mathbf{A}}$ and the remark about $t_{\mathbf{A}}$.

Suppose this property holds for ϕ and ψ. Then it holds for $\phi\wedge\psi$ and for $\phi\rightarrow\psi$, exactly as in Theorem 11.18.

Now let θ be $\forall x\phi$. If θ is in Γ, then $\phi(t/x)$ is in Γ for all closed terms t. Thus, inductively, $v\phi(t/x)=T$ for all closed terms t. By Lemma 12.16, this tells us that $v\phi(\mathbf{a}/x)=T$ for every element a of A. Hence $v\theta=T$.

Conversely, suppose θ is not in Γ. As Γ is complete, $\neg\theta$ will be in Γ. Now we know that $\neg\forall x\phi \vdash \exists x-\phi$. Hence $\exists x\neg\phi$ is in Γ. Let c be the witness to $\exists x\neg\phi$. By hypothesis, Γ contains the Henkin sentence $\exists x\neg\phi\rightarrow\neg\phi(c/x)$. Hence $\Gamma\vdash\neg\phi(c/x)$, and so $\neg\phi(c/x)$ is in Γ. Its value is then T, by induction, so $v\phi(c/x)=F$. This guarantees that $v\forall x\phi$ is F, completing the inductive proof.■

If we want to prove a similar result for formulae, a technical lemma is needed.

Lemma 12.24 *Let Y be any infinite set of variables. To any formula ϕ, there is a formula ψ such that ψ comes from ϕ by replacing all the bound variables by variables in Y, and such that $\phi \vdash \psi$ and $\psi \vdash \phi$.*

Proof It is easy to check that if $\phi_r \vdash \psi_r$ and $\psi_r \vdash \phi_r$ for $r = 1, 2$, then each of $\phi_1 \wedge \phi_2$ and $\psi_1 \wedge \psi_2$ can be derived from the other, and similarly for $\phi_1 \rightarrow \phi_2$ and $\psi_1 \rightarrow \psi_2$ (this was Exercise 11.13).

Suppose the result holds for ϕ. Then $\phi \vdash \psi$ and $\psi \vdash \phi$, from which we know that $\forall x\phi \vdash \forall x\psi$ and $\forall x\psi \vdash \forall x\phi$. Let y be a variable in Y and not in ψ. Let θ be $\psi(y/x)$. Then, by Lemma 12.19, $\forall x\psi \vdash \forall y\theta$ and $\forall y\theta \vdash \forall x\psi$. Thus $\forall y\theta$ is the relevant formula for $\forall x\phi$, and the result follows by induction.

Theorem 12.25 *Let Γ be a consistent set of formulae. Then there is an L-structure A and an interpretation i in A under which all the members of Γ are true.*

Proof As before, we may assume Γ contains Henkin formulae for all $\exists x\phi$, and that Γ is maximal consistent. We take A to consist of all terms (not just the closed ones), and make the obvious definitions, such as requiring ix to be x. The proof is almost as before, but there is one difficulty.

Suppose $\forall x\phi$ is in Γ. We want to show that $vi\phi(\mathbf{a}/x) = T$ for all a in A. Now a is just a term t of L (but it is convenient to make distinctions in the notation). We cannot claim that $vi\phi(t/x) = T$, because t may not be free for x in ϕ. However, by the previous lemma, we can find ψ such that no variable in t occurs bound in ψ, and with $\phi \vdash \psi$. Then $\forall x\phi \vdash \forall x\psi$, so $\forall x\psi$ is in Γ. Since t is free for x in ψ, we find that $\psi(t/x)$ is in Γ. As ψ is obtained from ϕ by changing the bound variables, it will contain the same number of $\wedge$, $\rightarrow$, and $\forall$ as ϕ does. So, in our inductive proof, we are entitled to assume that $vi\psi(t/x)$ is T. Thus, by Lemma 12.15, $vi\psi(\mathbf{a}/x) = T$. Let j be the interpretation with $jx = a$ and $jy = iy$ for all other y. Then $vi\phi(\mathbf{a}/x) = vj\phi$, and similarly for ψ. Since $\psi \vdash \phi$, we know that $\psi \models \phi$, and hence $T = vj\psi = vj\phi$, so we obtain $vi\phi(\mathbf{a}/x) = T$, as needed.■

Theorem 12.26 (Adequacy Theorem) *If $\Gamma \models \phi$ then $\Gamma \vdash \phi$. This holds both for sentences and formulae.*

Proof Suppose ϕ cannot be derived from Γ. Then we know that $\Gamma \cup \{\neg\phi\}$ is consistent. If we are dealing only with sentences, Theorem 12.23 tells us that $\Gamma \cup \{\neg\phi\}$ has a model, and this shows that we cannot have $\Gamma \models \phi$. If we are dealing with formulae, Theorem 12.25 shows there is an interpretation i under which $\neg\phi$ and all the members of Γ are true. As ϕ is false under this interpretation, by definition this means that we do not have $\Gamma \models \phi$.■

Exercise 12.10 Complete the proof of Lemma 12.19. That is, given a variable y not occurring in ϕ, and defining ψ to be $\phi(y/x)$, show that x is free for y in ψ and that ϕ is $\psi(x/y)$.

Exercise 12.11 Give an example of a situation where ψ is $\phi(y/x)$ with y occurring in ϕ and such that ϕ is not $\psi(x/y)$.

12.6 EQUALITY

Most objects that we wish to consider are sets, and so have the relation of equality. The languages we use to talk about such objects should then have a binary predicate symbol to correspond to this.

We define a **language with equality** to consist of a language L in the previous sense together with a specified binary predicate symbol of L. We call this symbol 'equals', and write it as =. Then = can mean either equality in a set, or the predicate symbol; this should not lead to any confusion.

We shall need to define structures and derivations for languages with equality, and to compare these with the previous notions. We will call a language in which no binary predicate symbol has been specified a **pure language**. Thus any language with equality can also be regarded as a pure language.

Let L be a language with equality, and **A** an L-structure when L is regarded as a pure language. To any binary predicate symbol of L there is a corresponding subset of A^2, and this subset can be regarded as a binary relation on A. If the relation corresponding to = is the relation of equality on A, we say **A** is a structure of L as language with equality.

We could define derivations for languages with equality by adding further rules covering the use of =. But it is rather easier to use a hybrid method involving **axioms of equality**. We shall write $s=t$, where s and t are terms, instead of the more formal $=st$, to make easier reading.

Definition The **axioms of equality** are the following set E of sentences:

(1) $\forall x(x=x)$;
(2) $\forall x \forall y(x=y \rightarrow y=x)$;
(3) $\forall x \forall y \forall z(x=y \wedge y=z \rightarrow x=z)$;
(4) for all n, to every n-ary function symbol f the sentence

$$\forall x_1 \ldots \forall x_{2n}(x_1=x_{n+1} \wedge \ldots \wedge x_n=x_{2n} \rightarrow fx_1 \ldots x_n=fx_{n+1} \ldots x_{2n})$$

(5) for all n, to every n-ary predicate symbol P the sentence

$$\forall x_1 \ldots .\ x_{2n}(x_1=x_{n+1} \wedge \ldots \wedge x_n=x_{2n} \rightarrow (Px_1 \ldots x_n \rightarrow Px_{n+1} \ldots x_{2n})).$$

We now say that $\Gamma \vdash \phi$ for L as language with equality if $\Gamma \cup E \vdash \phi$ for L as pure language. Since every structure for L as language with equality is plainly a model of E, the Soundness Theorem for languages with equality follows at once from the Soundness Theorem for pure languages. We need a lemma before we can prove the Adequacy Theorem (and hence the Completeness Theorem).

Lemma 12.27 *Let* **A** *and* **B** *be structures for a pure language. Let* $*$ *be a map from A onto B satisfying the following conditions*:

(1) $c_{\mathbf{A}}* = c_{\mathbf{B}}$, *for every constant symbol c*,
(2) *if* $f_{\mathbf{A}}(a_1, \ldots, a_n) = a$ *then* $f_{\mathbf{B}}(a_1*, \ldots, a_n*) = a*$, *for every n-ary function symbol f*,
(3) $(a_1*, \ldots, a_n*)$ *is in* $P_{\mathbf{B}}$ *iff* $(a_1, \ldots, a_n)$ *is in* $P_{\mathbf{A}}$, *for every n-ary predicate symbol P*.

Then, for any sentence ϕ, $v_{\mathbf{A}}\phi = v_{\mathbf{B}}\phi$. *More generally, let i be any interpretation in A, and let i* be the interpretation in B obtained by following i with* $*$. *Then, for any formula* ϕ, $v_{\mathbf{A}}i\phi = v_{\mathbf{B}}i*\phi$.

Remark Suppose L happens to be a language with equality, and let **A** and **B** be structures for the language with equality. Condition (3) applied to the predicate symbol = tells us that $a_1* = a_2*$ iff $a_1 = a_2$. Thus $*$ is one–one in this case, as well as being onto. When this happens, we call $*$ an **isomorphism**.

Proof We can regard **B** as an $L(\mathbf{A})$-structure, defining $\mathbf{a}_{\mathbf{B}}$ to be $a*$. Inductively, for any closed term t of $L(\mathbf{A})$, we have $t_{\mathbf{A}}* = t_{\mathbf{B}}$; more generally, for any term t of $L(\mathbf{A})$, we have $(it)* = i*t$.

The last part of the proof of Lemma 12.14 tells us that $v_{\mathbf{B}}iPt_1 \ldots t_n = T$ iff $\{it_1, \ldots, it_n\} \in P_{\mathbf{A}}$. The corresponding property for $i*$ and **B**, together with condition (3), shows that the result holds when ϕ is atomic. The general case follows inductively, using Lemma 12.16 to show that $v_{\mathbf{B}}i*\forall x\phi = T$ iff $v_{\mathbf{B}}i*\phi(\mathbf{a}/x) = T$ for all a in A.■

Theorem 12.28 *Let* Γ *be a consistent set of sentences for a language with equality. Then* Γ *has a model. More generally, if* Γ *is a consistant set of formulae there is an interpretation under which all the members of* Γ *are true.*

Proof Since $\Gamma \cup E$ is consistent for L as pure language, it has a model **A**. Corresponding to the symbol = of L is a binary relation in A, which we denote by $\equiv$.

As **A** is a model of E, we see that for any a_1, a_2, and a_3 in A we have $a_1 \equiv a_1$, if $a_1 \equiv a_2$ then $a_2 \equiv a_1$, and if $a_1 \equiv a_2$ and $a_2 \equiv a_3$ then $a_1 \equiv a_3$. Thus $\equiv$ is an equivalence relation on A, and we let B be the set of equivalence classes. For any constant symbol c, let $c_{\mathbf{B}}$ be the equivalence class of $c_{\mathbf{A}}$.

Also, if $a_r \equiv a_{n+r}$ for $r = 1, \ldots, n$ we know that $f_{\mathbf{A}}a_1 \ldots a_n \equiv f_{\mathbf{A}}a_{n+1} \ldots a_{2n}$. It follows that we can define $f_{\mathbf{B}}$ by $f_{\mathbf{B}}(b_1, \ldots, b_n) = a*$, where $a = f_{\mathbf{A}}(a_1, \ldots, a_n)$ and a_r is any element in the equivalence class of b_r.

With the same notation, we define $P_{\mathbf{B}}$ to be the set of those $(b_1, \ldots, b_n)$ such that $(a_1, \ldots, a_n)$ is in $P_{\mathbf{A}}$. Condition (5) of the axioms of equality and the rest of the construction show that the conditions of the previous lemma hold, and so give the result. The more general result that a consistent set of formulae in a language with equality satisfies some interpretation can be obtained similarly.■

The Adequacy Theorem for languages with equality follows from this, as before.

We note some consequences of E. First, for any terms $t_1, \ldots, t_{2n}$, the formula $t_1 = t_{n+1} \wedge \ldots \wedge t_n = t_{2n} \rightarrow ft_1 \ldots t_n = ft_{n+1} \ldots t_{2n}$ can be derived from E, and there is a similar result for predicate symbols. This cannot be derived directly from (4) of E by eliminating $\forall$, both because t_r may not be free for x_r in the formula (4), and because, since we must work one step at a time, we would run into trouble if, for instance, t_1 contained x_2. Instead, we first show that, for any $m > 2n$, we can derive from E the formula $\forall x_{m+1} \ldots \forall x_{m+2n} (x_{m+1} = x_{m+n+1} \wedge \ldots \rightarrow fx_{m+1} \ldots x_{m+n} = fx_{m+n+1} \ldots x_{m+2n})$. This is obtained directly, by $2n$ eliminations of $\forall$ followed by $2n$ introductions of $\forall$. If we take m such that the only variables occurring in $t_1, \ldots, t_{2n}$ are x_k for some $k < m$, a further $2n$ eliminations of $\forall$ from this will give the result we want.

Now let t be any term, and let t' be obtained from t by replacing, for some m, each variable x_r by the variable x_{m+r}. Inductively we see that $\vdash \forall x_1 \ldots \forall x_{2m}(x_1 = x_{m+1} \wedge \ldots \rightarrow t = t')$ in the language with equality. For we need only show that $x_1 = x_{m+1} \wedge \ldots \vdash t = t'$. This is obvious if t is a constant symbol or a variable. Suppose t is $ft_1 \ldots t_n$. Inductively the hypotheses let us derive $t_r = t_r'$ for $r = 1, \ldots, n$, and the result follows from the previous paragraph.

Finally, let ϕ be any formula, and let ϕ' be obtained from ϕ by replacing simultaneously the free occurrences of x_k, for $k = 1, \ldots, m$ by x_{m+k}. If the new variables do not occur in ϕ, we have $x_1 = x_{m+1} \wedge \ldots \vdash \phi \rightarrow \phi'$, and, as usual, we may move the hypothesis to the right and then introduce $\forall$. This is, as usual, proved inductively. The start of the induction follows from (5) of E and the result just proved for terms.

Exercise 12.12 Fill in the details in the proof of Lemma 12.27.

Exercise 12.13 Prove the result stated in the last paragraph. Deduce that if ϕ is any formula and s and t any terms free for x in ϕ then $s = t \vdash \phi(s/x) \rightarrow \phi(t/x)$. In particular, $s = x \vdash \phi(s/x) \rightarrow \phi$.

12.7 COMPACTNESS AND THE LOWENHEIM–SKOLEM THEOREMS

This section will only be needed to prove, in the next chapter, that certain theories are decidable. We shall consider only languages with equality. The next theorem is very easy in our approach; it is sometimes proved before the Adequacy Theorem, and used to obtain the latter.

Theorem 12.29 (Compactness Theorem) *A set Γ of sentences has a model iff every finite subset of Γ has a model.*

Proof We know that a set has a model iff it is consistent, and that a set is consistent iff all its finite subsets are consistent. The result is immediate.■

Let λ be any infinite cardinal number. For the next results, we need to know that the union of countably many sets of cardinality λ has cardinality λ, and that the set of strings on a set of cardinality λ also has cardinality λ. When λ is the cardinality of the natural numbers this is easy. It is also straightforward when λ is the cardinality of the real numbers; these two cases suffice for the examples in the next chapter.

Theorem 12.30 (Downward Lowenheim–Skolem Theorem) *Let L be a language of cardinality λ, and Γ a set of sentences of L. If Γ has a model it has a model of cardinality at most λ.*

Proof Since Γ has a model, it is consistent. Then we have explicitly constructed, in the proofs of Theorems 12.25 and 12.28, a model for Γ, and we show this model has cardinality at most λ.

First observe that any language of cardinality λ has at most (in fact, exactly) λ pairs (x,ϕ). Hence, if we add witnesses for all $\exists x\phi$, the new language still has cardinality λ.

It follows that, in the construction, all the languages L^n will have cardinality λ, and hence so will $L*$. Now the L-structure constructed in the proof of Theorem 12.25 as a model of $\Gamma \cup E$ regarding L as a pure language has as its elements the closed terms of $L*$. There are at most (in fact, exactly) λ of these. Finally, the structure which is a model for Γ when L is regarded as a language with equality is a set of equivalence classes of this set. Hence its cardinality is at most λ.■

Theorem 12.31 (Upward Lowenheim–Skolem Theorem) *Let L have cardinality λ. If Γ has arbitrarily large finite models it has an infinite model. If Γ has an infinite model, it has models of any cardinality $\mu \geqslant \lambda$.*

Proof Suppose Γ has arbitrarily large finite models. Enlarge L to a language L' by adding countably many new constant symbols. Let Δ' be the set of sentences $\{\neg(c=c')\}$ for all distinct new constant symbols c and c'. It is plain that any L'-model of $\Gamma \cup \Delta'$ is an infinite L-model of Γ. Thus it is enough to show that $\Gamma \cup \Delta'$ has a model.

By compactness, it is enough to show that $\Gamma \cup \Delta$ has a model, where Δ is any finite subset of Δ'. Now Δ contains at most n constant symbols for some integer n. Take an L-model of Γ with at least n elements. Make this into an L'-structure by requiring the elements of the structure corresponding to the constant symbols occurring in Δ to be distinct, which is possible, and letting the elements corresponding to the other constant symbols be arbitrary. This is plainly a model of $\Gamma \cup \Delta$.

Now supose we have a cardinal number μ which is at least λ. This time enlarge L to L' by adding μ new constant symbols. Then L' has cardinality

μ. Let Δ' be as before. Again let Δ be any finite subset of Δ'. The given infinite L-model of Γ can be regarded as an L'-model of $\Gamma \cup \Delta$, as before. Hence $\Gamma \cup \Delta'$ has an L'-model.

By the previous theorem, $\Gamma \cup \Delta'$ has an L'-model of cardinality at most μ. Plainly any model of Δ' has cardinality at least μ, so this model has cardinality exactly μ, as required.■

12.8 EQUIVALENCE

The results of this section will only be used in Chapter 14. We return to considering pure languages. The formulae ϕ and ϕ' are called **equivalent** if $\models \phi \leftrightarrow \phi'$. This holds iff $vi\phi = vi\phi'$ for any interpretation i; this shows that we do have an equivalence relation.

The simple equivalences of the previous chapter also hold for predicate logic. In addition, let ϕ be equivalent to ϕ'. It is easy to check that $\forall x \phi$ and $\exists x \phi$ are equivalent, respectively, to $\forall x \phi'$ and $\exists x \phi'$. Also $\forall x \phi$ is equivalent to $\neg \exists x \neg \phi$.

The formulae $\forall x(\phi \wedge \psi)$ and $\exists x(\phi \vee \psi)$ are equivalent to $\forall x \phi \wedge \forall x \psi$ and $\exists x \phi \vee \exists x \psi$. We do not have any equivalence for $\forall x(\phi \vee \psi)$ and $\exists x(\phi \wedge \psi)$ in general. However, if x does not occur free in ϕ, they are equivalent to $\phi \vee \forall x \psi$ and $\phi \wedge \exists x \psi$. These may be checked directly. Alternatively, we may show that each formula can be derived from the corresponding one, and then use Soundness.

Lemma 12.32 *Let Y be any infinite set of variables. Then to any formula ϕ there is an equivalent formula ψ obtained from ϕ by replacing the distinct bound variables in ϕ by distinct variables in Y.*

Proof This follows immediately from Lemma 12.24, using Soundness. Alternatively, it can be proved directly along the same lines.■

A formula is said to be in **prenex normal form** if it is $Q_1 x_{i(1)} \ldots Q_n x_{i(n)} \phi$, where each Q_r is one of the quantifiers $\forall$ and $\exists$, the variables $x_{i(r)}$ are distinct, and ϕ contains no quantifiers.

Proposition 12.33 *Every formula is equivalent to one in prenex normal form.*

Proof Any atomic formula is in prenex normal form. Suppose ϕ is equivalent to ψ where ψ is in prenex normal form. Let y be a variable not occurring in ψ. Then $\forall x \phi$ is equivalent to $\forall x \psi$, which in turn is equivalent to $\forall y \psi(y/x)$, and the latter is in prenex normal form.

Suppose ϕ and ψ are equivalent to the formulae α and β which are in prenex normal form. Replacing β by an equivalent formula according to the previous lemma, we may assume that no variable which is bound in β occurs in α. Making a similar replacement for α, we can also assume that no variable which is bound in α occurs in β. Then $\phi \wedge \psi$ is equivalent to $\alpha \wedge \beta$.

Writing β as $Qx\gamma$, where Q is either $\forall$ or $\exists$, we find that $\alpha\wedge\beta$ is equivalent to $Qx(\alpha\wedge\gamma)$, since x does not occur in α. Repeating this procedure as necessary, we obtain a formula in prenex normal form equivalent to $\alpha\wedge\beta$, as required.

Also $-\phi$ is equivalent to α', where α' is obtained from α by replacing each $\forall$ and $\exists$ by $\exists$ and $\forall$ respectively, and then taking the negation of the quantifier-free formula. Finally $\phi\rightarrow\psi$, being equivalent to $\neg(\phi\wedge\neg\psi)$ has an equivalent prenex normal form, using the previous two cases. ■

Let α be a formula of propositional logic, and let Φ be a sequence of formulae of predicate logic. Exactly as in the previous chapter, we can define a formula $\alpha[\Phi]$ of predicate logic (notice that α must be a formula of propositional logic, not a formula of predicate logic). Let i be any interpretation, and let w be the valuation given by $wp_n = vi\phi_n$ for all n. As before, we find that $vi\alpha[\Phi] = w\alpha$, and the analogue of Lemma 11.23 holds, the proof being almost identical to the previous one.

We now look at quantifier-free formulae. In predicate logic a formula without quantifiers is said to be in disjunctive normal form if it is $\phi_1 \vee \ldots \vee \phi_n$ for some formulae ϕ_r, each of which is a conjunction of formulae ϕ_{rs} where s runs from 1 to n_r for some n_r, and finally each ϕ_{rs} is either atomic or the negation of an atomic formula. Note that we do not bother to bracket the disjunctions (or the conjunctions), since, by the associative law, all bracketings are equivalent.

Proposition 12.34 *Any formula without quantifiers is equivalent to one in disjunctive normal form.*

Proof It is easy to check by induction that any quantifier-free formula can be written as $\alpha[\Phi]$, where α is a formula of propositional logic and each ϕ_n in Φ is atomic. Now α is equivalent to a formula β of propositional logic such that β is in disjunctive normal form. Then $\alpha[\Phi]$ is equivalent to $\beta[\Phi]$, and this latter is in disjunctive normal form for predicate logic, because each ϕ_n is atomic. ■

13

Undecidability and Incompleteness

In the later sections of this chapter we prove various versions of Gödel's Incompleteness Theorem and the Undecidability of Number Theory (and of Logic). In the first section we give some decidable theories.

13.1 SOME DECIDABLE THEORIES

A **theory** is a set Γ of sentences such that any sentence σ for which $\Gamma \vdash \sigma$ is in Γ. Let $\mathbf{A}$ be a structure; then the **theory of A**, written Th$\mathbf{A}$, is $\{\sigma; \mathbf{A} \models \sigma\}$. This is a theory, by the Soundness Theorem. The theory of the language L, written ThL, is $\{\sigma; \models \sigma\}$; by completeness, this is also $\{\sigma; \vdash \sigma\}$. More generally, we could choose any set of L-structures, and the set of sentences true in these structures would be a theory (for instance, group theory).

For any set Δ of sentences, the set of **consequences** of Δ, written $C(\Delta)$, is $\{\sigma; \Delta \vdash \sigma\}$; this is plainly a theory. We call Δ a **set of axioms** for Γ if $\Gamma = C(\Delta)$. If Γ has a recursive set of axioms we say it is **recursively axiomatisable**. By Theorem 12.9 and Lemma 12.10, this holds iff Γ is r.e.

For the remainder of this section all languages considered will be countable. If a theory is recursive, we shall usually call it a 'decidable theory' rather than saying 'recursive theory'.

Proposition 13.1 *Let L be a pure language with no constant or function symbols whose only predicate symbols are finitely many unary symbols. Then* ThL *is decidable.*

Proof Let the predicate symbols be $P_1, \ldots, P_k$. Let X be the set of all k-tuples $(\beta_1, \ldots, \beta_k)$ where each β_i is either 0 or 1. Let B be any non-empty subset of X. Then B can be made into an L-structure $\mathbf{B}$ by defining $P_{i\mathbf{B}}$ to be $\{(\beta_1, \ldots, \beta_k) \in B; \beta_i = 0\}$.

Let $\mathbf{A}$ be any L-structure. Then A can be mapped into X by sending a to $(\alpha_1, \ldots, \alpha_k)$, where $\alpha_i = 0$ if $a \in P_{i\mathbf{A}}$ and $\alpha_i = 1$ otherwise. Let B be the image of A under this map. By construction, the assumptions of Lemma 12.27 are satisfied, and so for any sentence σ, $v_{\mathbf{A}}\sigma = v_{\mathbf{B}}\sigma$. Hence $\models \sigma$ iff $\mathbf{B} \models \sigma$ for every $\mathbf{B}$ defined from a subset of X. Since each such B is finite, we can decide for each $\mathbf{B}$ whether or not $\mathbf{B} \models \sigma$. Since there are only finitely many $\mathbf{B}$, we can decide whether or not $\mathbf{B} \models \sigma$ for all $\mathbf{B}$.■

Recall that a theory is **complete** iff for every sentence σ either σ or $-\sigma$ is in the theory.

Proposition 13.2 *For any* $\mathbf{A}$, Th$\mathbf{A}$ *is complete. Conversely, if* $\mathbf{A}$ *is a model of the complete theory* Γ *then* Γ = Th$\mathbf{A}$.

Proof Plainly Th$\mathbf{A}$ is complete. Let $\mathbf{A}$ be a model of the complete theory Γ. By soundness, if σ is in Γ it is in Th$\mathbf{A}$. If σ is not in Γ then $\neg\sigma$ is in Γ. In this case, $\neg\sigma$ is in Th$\mathbf{A}$, so, as required, σ is not in Th$\mathbf{A}$.■

Proposition 13.3 *A complete recursively axiomatisable theory is decidable.*

Proof Let Γ be a complete recursively axiomatisable theory. If Γ is inconsistent it is the set of all sentences, and so it is decidable.

Suppose Γ is consistent. Then, for any σ, exactly one of σ and $\neg\sigma$ is in Γ. Hence $\{\sigma;\ \sigma \text{ not in } \Gamma\}$ is $\{\sigma;\ \neg\sigma \text{ in } \Gamma\}$. As Γ is recursively axiomatisable it is r.e. But then $\{\sigma;\ \neg\sigma \text{ in } \Gamma\}$ is also r.e. Since Γ and its complement are both r.e., Γ is recursive.■

So we want to have conditions which ensure a theory is recursive. The following criterion uses the Lowenheim–Skolem theorems in its proof.

Proposition 13.4 *Let* Γ *be a theory in a language with equality. Let* Γ *have no finite models, and, for some infinite cardinality* λ, *let all models of* Γ *with cardinality* λ *be isomorphic. Then* Γ *is complete.*

Proof Suppose not. Then there is a sentence σ such that neither σ nor $\neg\sigma$ is in Γ. As Γ is a theory, neither σ nor $\neg\sigma$ can be derived from Γ. Hence both $\Gamma \cup \{\sigma\}$ and $\Gamma \cup \{\neg\sigma\}$ are consistent. Therefore they both have models. These models will be models of Γ, and so must be infinite. By the Upward Lowenheim–Skolem Theorem, they both have models of cardinality λ. Let these models be $\mathbf{A}$ and $\mathbf{B}$. Then $\mathbf{A} \models \sigma$ and $\mathbf{B} \models \neg\sigma$. But $\mathbf{A}$ and $\mathbf{B}$ are models of Γ whose cardinality is λ, so they are isomorphic. Hence, by Lemma 12.27, a sentence true in $\mathbf{A}$ is also true in $\mathbf{B}$, and we have a contradiction.■

We now give examples of such theories.

Let L have binary predicate symbols $=$ and $<$, and no other constant, function, or predicate symbols. For convenience, we write $x < y$ rather than $<xy$. Let DLO be the theory with the following axioms.

(1) $\forall x \forall y \forall z(x < y \wedge y < z \rightarrow x < z)$
(2) $\forall x \forall y(x = y \rightarrow \neg(x < y))$
(3) $\forall x \forall y(x < y \rightarrow \neg(y < x))$

(4) $\forall x \forall y(x = y \lor x < y \lor y < x)$
(5) $\forall x \forall y(x < y \rightarrow \exists z(x < z \land z < y))$
(6) $\forall x \exists y(x < y)$
(7) $\forall x \exists y(y < x)$.

Any model of DLO is a set with a relation which we will still denote by $<$ (rather than $<_{\mathbf{A}}$). Conditions (1), (2), and (3) tell us this is a partial order, and (4) ensures it is a linear order (also called a simple order). (5) says that the order is dense; that is, that there is an element between any two distinct elements. (6) and (7) say there is no last or first elements. The abbreviation DLO stands for Dense Linear Order. Plainly both the rational numbers and the real numbers are models of DLO.

Proposition 13.5 *DLO is complete.*

Proof We will show that any two countable models of DLO are isomorphic. As DLO plainly has no finite models, the result follows by 13.4.

Let A and B be any two countable models of DLO. Let A' be any finite subset of A, and a any element of A. We first show that any order-preserving map f from A' to B can be extended to an order-preserving map from $A' \cup \{a\}$ into B.

Let the elements of A' be $a_1, \ldots, a_n$ with $a_1 < a_2 < \ldots < a_n$. If a is in A' there is nothing to prove. If $a_n < a$, map a to any element of B which is larger than fa_n. If $a < a_1$ map a to any element smaller than fa_1. Otherwise there is some i such that $a_i < a < a_{i+1}$. In this case we map a to any element lying between fa_i and fa_{i+1}.

Now let the elements of A and B be (arranged in some way, not in order of increasing size, which is impossible) $a_0, a_1, \ldots$ and $b_0, b_1, \ldots$. By the previous paragraph, it is easy to define (by induction) an order-preserving map from A into B. We are looking for such a map which is onto B, and we need to use what is called a 'back-and-forth' argument.

Let A_0 and B_0 be $\{a_0\}$ and $\{b_0\}$. Let f_0 be the only possible map from A_0 to B_0. Suppose we have defined, for some n, subsets A_n and B_n of A and B such that A_n and B_n contain a_i and b_i (respectively) for all $i \leqslant n$, and an order-preserving map f_n from A_n to B_n. We will define inductively similar subset A_{n+1} and B_{n+1} and an order-preserving map f_{n+1} which equals f_n on A_n. If this is done, we have, by construction, $\cup A_n = A$ and $\cup B_n = B$, and we may define a map f, which is certainly order-preserving, from A onto B by requiring fa to be $f_n a$ for any n with a in A_n.

By the previous discussion we can extend f_n to an order-preserving map g_n from $A_n \cup \{a_{n+1}\}$ into B. Let C_n be the image of g_n. Then g_n^{-1} is an order-preserving map from C_n into A. Define B_{n+1} to be $C_n \cup \{b_{n+1}\}$. Then g_n^{-1} can be extended, for the same reason, to an order-preserving map h_n from B_{n+1} into A. We need only define A_{n+1} to be the image of h_n, and define f_{n+1} to be h_n^{-1}.■

We now look at a theory in the language with equality L which has exactly one constant symbol **0**, exactly one unary function symbol **S** and no other function symbols, and whose only predicate symbol is $=$. The theory SUC is defined by the following axioms.

(1) $\forall x \neg (\mathbf{S}x = \mathbf{0})$
(2) $\forall x(x = \mathbf{0} \vee \exists y(\mathbf{S}y = x))$
(3) $\forall x \forall y(\mathbf{S}x = \mathbf{S}y \rightarrow x = y)$.
(4) $\forall x \neg (\mathbf{S}^n x = x)$ for every positive integer n. Here $\mathbf{S}^n$ is an abbreviation for $\mathbf{SS} \ldots \mathbf{S}$, with $\mathbf{S}$ occurring n times. This infinite set of axioms is obviously recursive.

Plainly **N** is a model of SUC, with $\mathbf{0_N}$ and $\mathbf{S_N}$ being 0 and the successor function. It is easy to see that the union of **N** and any number of copies of **Z** is also a model of SUC, with the corresponding function sending any n in a copy of **Z** to $n+1$ in the same copy. Thus there are infinitely many non-isomorphic countable models of SUC, since we may choose any finite number of copies of **Z** or a countable number.

Proposition 13.6 *SUC is complete.*

Proof Because of axioms (3) and (4), any model **A** of SUC must be infinite, since the elements $\mathbf{S}^n_{\mathbf{A}}\mathbf{0_A}$ will all be distinct.

We shall show that any model of SUC is isomorphic to the union of a copy of **N** and some copies of **Z**. If there are λ copies of **Z**, where λ is an infinite cardinal, then the cardinality of **A** is λ. Hence, for any uncountable cardinal λ, all models of cardinality λ are isomorphic.

So take a model of SUC consisting of a set A together with a function S and an element a_0. We define a relation $\equiv$ on A by $a \equiv a'$ iff there is $n \geqslant 0$ such that either $a' = S^n a$ or $a = S^n a'$. Using (3), we see easily that $\equiv$ is an equivalence relation.

Consider the equivalence class of a_0. It is easy to see that it consists of the infinitely many distinct elements a_n for n in **N**, where a_n is $S^n a_0$. So this class is a copy of **N**.

Now look at any other equivalence class. Take any element b in this class, and define b_n for $n \geqslant 0$ by $b_n = S^n b$. Because b is not equivalent to a_0, axioms (1) and (2) guarantee that for every positive n there is exactly one element which is sent by S^n to b. We define b_{-n} to be this element. Since $S^{n+1}b_{-(n+1)} = S^n b_{-n}$ we see that $Sb_{-(n+1)} = b_{-n}$. By the definition of $\equiv$, our equivalence class must consist of the elements b_n for all n in **Z**. Hence this class is a copy of **Z**.■

13.2 EXPRESSIBLE SETS AND REPRESENTABLE FUNCTIONS

13.2.1 The language and axioms of number theory

The **language of number theory** is the language with equality L whose other constant, function, and predicate symbols are a constant symbol **0**, a unary function symbol **S**, two binary function symbols $+$ and $\times$, and a binary predicate symbol $\leqslant$. Plainly **N** is an L-structure, with 0 corresponding to **0**, the successor function S to $\mathbf{S}$, and so on. For convenience, when s and t are terms we shall write $s + t$ rather than $+st$, and so on.

To each natural number n we associate a closed term $\mathbf{n}$ of L, called the **numeral** of n. This is done inductively. The numeral of 0 is defined to be $\mathbf{0}$, and the numeral of $n + 1$ is defined to be $\mathbf{Sn}$. We can show by induction that $\mathbf{n}_{\mathbf{N}}$ is just n.

Let A be the following set of sentences.

(1) $\mathbf{m} + \mathbf{n} = \mathbf{p}$ for all m, n, and p with $m + n = p$
(2) $\mathbf{m} \times \mathbf{n} = \mathbf{q}$ for all m, n, and q with $mn = q$
(3) $\forall x(x \leqslant \mathbf{n} \leftrightarrow x = \mathbf{0} \vee x = \mathbf{1} \vee \ldots \vee x = \mathbf{n})$ for all n.

Plainly A is a recursive set of sentences, and $\mathbf{N}$ is a model of A. The set A is so simple that any theory which deserves to be called number theory must include A. Observe that from (3) we find that $A \vdash \mathbf{m} \leqslant \mathbf{n}$ iff $m \leqslant n$. We have not required $\neg(\mathbf{m} = \mathbf{n})$ when $m \neq n$.

Let B_0 be the finite set of sentences

β_1 is $\forall x(x + \mathbf{0} = x)$
β_2 is $\forall x \forall y(x + \mathbf{S}y = \mathbf{S}(x + y))$
β_3 is $\forall x(x \times \mathbf{0} = \mathbf{0})$
β_4 is $\forall x \forall y(x \times \mathbf{S}y = x \times y + x)$
β_5 is $\forall x(x \leqslant \mathbf{0} \leftrightarrow x = \mathbf{0})$
β_6 is $\forall x \forall y(x \leqslant \mathbf{S}y \leftrightarrow x \leqslant y \vee x = \mathbf{S}y)$.

Plainly $\mathbf{N}$ is a model of B_0 also.

Lemma 13.7 $B_0 \vdash A$.

Proof By induction. From β_1 we derive $\mathbf{m} + \mathbf{0} = \mathbf{m}$, which is the start of the induction. Let $m + n = p$, and suppose $B_0 \vdash \mathbf{m} + \mathbf{n} = \mathbf{p}$. From β_2 we derive $\mathbf{m} + \mathbf{Sn} = \mathbf{S}(\mathbf{m} + \mathbf{n})$. So $B_0 \vdash \mathbf{m} + \mathbf{Sn} = \mathbf{Sp}$. Since $\mathbf{Sn}$ and $\mathbf{Sp}$ are the numerals of $n + 1$ and $p + 1$, this proves the inductive step.

Now $\beta_3 \vdash \mathbf{m} \times \mathbf{0} = \mathbf{0}$, which starts the induction for multiplication. Let mn be q, and let $m(n + 1)$ be r. Suppose, inductively, that $B_0 \vdash \mathbf{m} \times \mathbf{n} = \mathbf{q}$. From β_4 we derive $\mathbf{m} \times \mathbf{Sn} = \mathbf{m} \times \mathbf{n} + \mathbf{m}$. Hence $B_0 \vdash \mathbf{m} \times \mathbf{Sn} = \mathbf{q} + \mathbf{m}$. We have already seen that $B_0 \vdash \mathbf{q} + \mathbf{m} = \mathbf{r}$. Hence $B_0 \vdash \mathbf{m} \times \mathbf{Sn} = \mathbf{r}$, as required.

The proof for $\leqslant$ is similar.■

13.2.2 Expressible sets and representable functions

It will be convenient to write $\phi(x_1, \ldots, x_k)$ to mean that ϕ is a formula whose free variables are among $x_1, \ldots, x_k$. We will then write $\phi(t_1, \ldots, t_k)$ to denote the result of simultaneously replacing the free occurrences of every x_i with the corresponding t_i. We shall only need this when each t_i is a very simple term; except in one case, each t_i is either a constant symbol or a variable.

Definition Let $S \subset \mathbf{N}^k$ for some k. We call S **expressible** if there is a formula $\phi(x_1, \ldots, x_k)$ such that

if $(n_1, \ldots, n_k)$ is in S then $A \vdash \phi(\mathbf{n}_1, \ldots, \mathbf{n}_k)$ and
if $\mathbf{N} \models \phi(\mathbf{n}_1, \ldots, \mathbf{n}_k)$ then $(n_1, \ldots, n_k)$ is in S.

A function $f{:}\mathbf{N}^k \to \mathbf{N}^r$ is called **representable** if its graph, which is the set $\{(n_1, \ldots, n_k, m_1, \ldots, m_r); (m_1, \ldots, m_r) = f(n_1, \ldots, n_k)\}$, is expressible.

These definitions are slightly different from those used by other authors, but I think they are more convenient in our approach. Different versions of the definition will be given in the final section of this chapter.

Our main aim is to show that the expressible sets are exactly the r.e. sets. The easiest proof uses the fact that r.e. sets are diophantine. Readers who have not covered this material will have to be content with a longer proof using modular machines and the Gödel sequencing function. For those readers who have seen the proof that r.e. sets are exponential diophantine, but have not looked at the reduction to the diophantine case, I include yet another proof, which again requires use of the Gödel sequencing function. Alternatively, we can obtain similar results by extending the language with a further binary function symbol (corresponding to exponentiation) and corresponding additional sentences in A.

Lemma 13.8 *To any polynomial P with non-negative coefficients there is a term t_P such that the formula $t_P = y$ represents P.*

Proof If $P(n_1, \ldots, n_k)$ is either n_i or a constant m, then we take t_P to be x_I or $\mathbf{m}$, respectively. Otherwise P is either $Q + R$ or QR, where Q and R have fewer additions and multiplications than P, so, inductively, we have suitable terms t_Q and t_R. We then define t_P as $t_Q + t_R$ or $t_Q \times t_R$, respectively. (Notice that Q and R are not unique, so there are several possibilities for t_P.)

Let $P = Q + R$, the other case being similar. Define a and b by $Q(n_1, \ldots, n_k) = a$ and $R(n_1, \ldots, n_k) = b$. Then, inductively, $t_Q(\mathbf{n}_1, \ldots, \mathbf{n}_k) = \mathbf{a}$ and $t_R(\mathbf{n}_1, \ldots, \mathbf{n}_k) = \mathbf{b}$ can both be derived from A, and are both true in $\mathbf{N}$.

Suppose that $P(n_1, \ldots, n_k) = m$. Then $m = a + b$, and so $A \vdash \mathbf{a} + \mathbf{b} = \mathbf{m}$. Since t_P is $t_Q + t_R$ we also have that $A \vdash t_P(\mathbf{n}_1, \ldots, \mathbf{n}_k) = \mathbf{a} + \mathbf{b}$, and so $A \vdash t_P(\mathbf{n}_1, \ldots, \mathbf{n}_k) = \mathbf{m}$.

Conversely suppose that $t_P(\mathbf{n}_1, \ldots, \mathbf{n}_k) = \mathbf{m}$ is true in $\mathbf{N}$. Then $\mathbf{m} = \mathbf{a} + \mathbf{b}$ is true in $\mathbf{N}$, and so $m = a + b$, which tells us that $P(n_1, \ldots, n_k) = m$. ■

Lemma 13.9 *Let P and Q be polynomials with non-negative coefficients. Then $\{(n_1, \ldots, n_k); P(n_1, \ldots, n_k) = Q(n_1, \ldots, n_k)\}$ is expressible.*

Proof Define m to be $P(n_1, \ldots, n_k)$. By the previous lemma,

$A \vdash t_P(\mathbf{n}_1, \ldots, \mathbf{n}_k) = \mathbf{m}$. If $P(n_1, \ldots, n_k) = Q(n_1, \ldots, n_k)$ we have $m = Q(n_1, \ldots, n_k)$, and so $A \vdash t_Q(\mathbf{n}_1, \ldots, \mathbf{n}_k) = \mathbf{m}$. Hence $A \vdash t_P(\mathbf{n}_1, \ldots, \mathbf{n}_k) = t_Q(\mathbf{n}_1, \ldots, \mathbf{n}_k)$. Conversely, if $t_P(\mathbf{n}_1, \ldots, \mathbf{n}_k) = t_Q(\mathbf{n}_1, \ldots, \mathbf{n}_k)$ is true in **N** then $t_Q(\mathbf{n}_1, \ldots, \mathbf{n}_k) = \mathbf{m}$ is true in **N**, and so $Q(n_1, \ldots, n_k) = m$, as required.■

Notice that the rule for left-introduction of $\exists$ shows that if $A \vdash \phi(\mathbf{n})$ for some n then $A \vdash \exists x\phi$, and conversely if $\exists x\phi$ is true in **N** then $\phi(\mathbf{n})$ is true in **N** for some n. Similarly, if $A \vdash \phi(\mathbf{m}, \mathbf{n})$ for some m and n then two applications of the same rule show that $A \vdash \exists x \exists y \phi$, and so on. The corollary below is now immediate.

Corollary *Diophantine sets are expressible.*

Theorem 13.10 *A set is expressible iff it is r.e.*

Proof Suppose S is expressed by ϕ. Then $(n_1, \ldots, n_k) \in S$ iff $\phi(\mathbf{n}_1, \ldots, \mathbf{n}_k) \in C(A)$. We know, by Theorem 12.9, that $C(A)$ is r.e. Since the map sending $(n_1, \ldots, n_k)$ to $\phi(\mathbf{n}_1, \ldots, \mathbf{n}_k)$ is recursive, by Church's Thesis, it follows that S, as the counter-image of a r.e. set by a recursive function, is r.e.

Conversely, let S be r.e. Then S is diophantine, by the main theorem on diophantine sets. The results follows from the previous corollary.■

Lemma 13.11 *The functions γ, J, and J^{-1} are representable. So is the function J':$\mathbf{N}^3 \to \mathbf{N}^2$ defined by $J'(m,n,r) = (J(m,n),r)$.*

Proof J and γ are both diophantine, since $J(m,n) = r$ iff $2r = (m + n)(m + n + 1) + 2m$, while $\gamma(i,t,u) = r$ iff there are q and s with $u = q(1 + (i + 1)t) + r$ and $r + s = (i + 1)t$. However, Theorem 13.10 shows that any diophantine set is expressible, and hence that any diophantine function is representable.

It is easy to see that J^{-1} is represented by $\phi(x_2,x_3,x_1)$, where ϕ is the formula representing J. Similarly J' is represented by $\phi(x_1,x_2,x_4) \wedge x_3 = x_5$.■

Lemma 13.12 *Let f:$\mathbf{N}^k \to \mathbf{N}^s$ and g:$\mathbf{N}^s \to \mathbf{N}^r$ be representable, and let S be an expressible subset of $\mathbf{N}^s$. Then the set $f^{-1}S$ is expressible and the function gf is representable.*

Proof Let S be expressed by ψ and let f and g be represented by ϕ and θ. Now $(n_1, \ldots, n_k)$ is in $f^{-1}S$ iff there is $(m_1, \ldots, m_s)$ such that $(m_1, \ldots m_s)$ is in S and $f(n_1, \ldots, n_k) = (m_1, \ldots, m_s)$. If this happens then $A \vdash \psi(\mathbf{m}_1, \ldots, \mathbf{m}_s) \wedge \phi(\mathbf{n}_1, \ldots, \mathbf{n}_k, \mathbf{m}_1, \ldots, \mathbf{m}_s)$. We see easily that $f^{-1}S$ is expressed by $\exists y_1 \ldots \exists y_s(\psi(y_1, \ldots, y_s) \wedge \phi(x_1, \ldots, x_k, y_1, \ldots, y_s))$.

Similarly gf is represented by

$\exists y_1 \ldots \exists y_s(\theta(y_1, \ldots, y_s, x_{k+1}, \ldots, x_{k+r}) \wedge \phi(x_1, \ldots, x_k, y_1, \ldots, y_s))$.■

Now suppose $A \vdash \phi(\mathbf{i})$ for all $i \leqslant n$. At the end of the discussion of equality in the previous chapter it was shown that $\mathbf{i} = x \vdash \phi(\mathbf{i}) \to \phi(x)$. Since $x = \mathbf{i} \vdash \mathbf{i} = x$, we see that $A \cup \{x = \mathbf{i}\} \vdash \phi(\mathbf{i}) \to \phi$. The rule for left-introduction of $\vee$ now tells us that $A \cup \{x = \mathbf{0} \vee \ldots \vee x = \mathbf{n}\} \vdash \phi$. By (3) of A, we have $A \cup \{x \leqslant \mathbf{n}\} \vdash \phi$. Hence $A \vdash \forall x(x \leqslant \mathbf{n} \to \phi)$. Conversely, if $\mathbf{N} \models \forall x(x \leqslant \mathbf{n} \to \phi)$ then $A \models \phi(\mathbf{i})$ for all $i \leqslant n$. Similarly, if we have $A \vdash \phi(\mathbf{i})$ for all $i < n$ then $a \vdash \forall x(x \leqslant \mathbf{n} \to x = \mathbf{n} \vee \phi)$.

Lemma 13.13 *Let f:$\mathbf{N} \to \mathbf{N}$ be representable. Then so is its iterate F:$\mathbf{N}^2 \to \mathbf{N}$.*

Proof By Lemma 9.13, $F(m,n) = r$ iff there are t and u such that $\gamma(0,t,u) = m, \gamma(n,t,u) = r$, and $\gamma(i+1,t,u) = f\gamma(i,t,u)$ for all $i < n$.

Let ϕ represent γ, and let ψ represent $f\gamma$. Suppose $F(m,n) = r$. Then there are t and u such that $A \vdash \phi(\mathbf{0},\mathbf{t},\mathbf{u},\mathbf{m}))$, $A \vdash \phi(\mathbf{n},\mathbf{t},\mathbf{u},\mathbf{r})$, and, for all $i < n$ there is some p (depending on i) such that $A \vdash \phi(\mathbf{Si},\mathbf{t},\mathbf{u},\mathbf{p}) \wedge \psi(\mathbf{i},\mathbf{t},\mathbf{u},\mathbf{p})$. Thus $\mathbf{A} \vdash \exists z(\phi(\mathbf{Si},\mathbf{t},\mathbf{u},z) \wedge \omega(\mathbf{i},\mathbf{t},\mathbf{u},z))$. By the previous discussion, we see that $A \vdash \phi(\mathbf{0},\mathbf{t},\mathbf{u},\mathbf{m}) \wedge \phi(\mathbf{n},\mathbf{t},\mathbf{u},\mathbf{r}) \wedge \alpha(\mathbf{n},\mathbf{t},\mathbf{u})$, where α is the formula

$$\forall w(w \leqslant x_2 \to w = x_2 \vee \exists z(\phi(\mathbf{S}w,y_1,y_2,z) \wedge \psi(w,y_1,y_2,z))).$$

Let θ be the formula

$$\phi(\mathbf{0},y_1,y_2,x_1) \wedge \phi(x_2,y_1,y_2,x_3) \wedge \alpha(x_2,y_1,y_2).$$

Then $A \vdash \exists y_1 \exists y_2 \theta(\mathbf{m},\mathbf{n},\mathbf{r},y_1,y_2)$. Reversing this procedure, we see that if $\mathbf{N} \models \exists y_1 \exists y_2 \theta(\mathbf{m},\mathbf{n},\mathbf{r},y_1,y_2)$ then $F(m,n) = r$. Hence F is represented by $\exists y_1 \exists y_2 \theta(x_1,x_2,x_3,y_1,y_2)$.■

Corollary 1 *Let g:$\mathbf{N}^2 \to \mathbf{N}^2$ be representable. Then so is its iterate G:$\mathbf{N}^3 \to \mathbf{N}^2$.*

Proof Let f be JgJ^{-1}. Then f is representable. Hence so is the iterate F of f. Since G is $J^{-1}FJ'$, G is also representable.■

Corollary 2 *The functions sending n to m^n and to m^{n+1} are representable.*

Proof The result is in fact true when we regard these functions as functions of the two variables m and n. However, for the application it is enough to know it for m fixed.

Now the function sending r to mr is obviously representable. Hence its iterate, which sends (n,r) to $m^n r$, is also representable, by ϕ, say. The functions we want are obtained by taking r to be 1 or m, so are represented by $\phi(x_1,\mathbf{1},x_2)$ and $\phi(x_1,\mathbf{m},x_2)$.■

If we have shown that r.e. sets are base-2 exponential diophantine, but have not shown they are diophantine, we can now give another proof of Theorem 13.10. A third proof will follow, using modular machines, and not requiring any knowledge of diophantine properties.

Proof of Theorem 13.10 (second proof) Let S be r.e. Then there are polynomials P and Q with non-negative coefficients such that $n \in S$ iff there exist $n(1), \ldots, n(k), m(1), \ldots, m(k)$ with
$P(n, n(1), \ldots, n(k), m(1), \ldots, m(k)) =$
$Q(n, n(1), \ldots, n(k), m(1), \ldots, m(k))$ and
$m(i) = 2^{n(i)}$ for $i = 1, \ldots, k$. This condition is expressed by
$\exists x_1 \ldots \exists x_k \exists y \ldots \exists y_k (t_P(x, x_1, \ldots, y_k)) = t_Q(x, x_1, \ldots y_k \wedge \phi(x_1, y_1) \wedge \ldots \wedge \phi(x_k, y_k))$, where ϕ represents the function 2^n.■

Proposition 13.14 *Let M be a modular machine. Then the halting set $H(M)$ of M is expressible.*

Proof Let the quadruples of M be (a_i, b_i, c_i, R) for i in some index set I and (a_j, b_j, c_j, L) for j in some index set J. Let the pairs beginning no quadruple be (a_k, b_k) for k in some index set K. If we are given formulae ϕ_i for all $i \in I$ we shall denote their disjunction by $\vee_I \phi_i$, and so on.

We know that (α, β) is terminal iff there are u and v such that $\alpha = mu + a_k$ and $\beta = mv + b_k$ for some k in K. It is then easy to see that Term_M is represented by $\exists y_1 \exists y_2 \vee_K (x_1 = \mathbf{m} \times y_1 + \mathbf{a}_k \wedge x_2 = \mathbf{m} \times y_2 + \mathbf{b}_k)$.

The function Next_M is defined by $\text{Next}_M(\alpha, \beta) = (\alpha', \beta')$ iff, for some u and v, either

$\alpha = mu + a_i, \beta = mv + b_i, \alpha' = m^2u + c_i$, and $\beta' = v$, for some i in I, or
$\alpha = mu + a_j, \beta = mv + b_j, \alpha' = u$, and $\beta' = m^2v + c_j$, for some j in J, or
$\alpha = mu + a_k, \beta = mu + b_k, \alpha' = \alpha$, and $\beta' = \beta$, for some k in K.

Hence Next_M is represented by $\exists y_1 \exists y_2 (\theta \vee \phi \vee \psi)$, where θ is $\vee_I \theta_i$, ϕ is $\vee_J \phi_j$, and ψ is $\vee_K \psi_k$, and

θ_i is $x_1 = \mathbf{m} \times y_1 + \mathbf{a}_i \wedge x_2 = \mathbf{m} \times y_2 + \mathbf{b}_i \wedge x_3 = \mathbf{m} \times \mathbf{m} \times y_1 + \mathbf{c}_i \wedge x_4 = y_2$,
ϕ_j is $x_1 = \mathbf{m} \times y_1 + \mathbf{a}_j \wedge x_2 = \mathbf{m} \times y_2 + \mathbf{b}_j \wedge x_3 = y_1 \wedge x_4 = \mathbf{m} \times \mathbf{m} \times y_2 + \mathbf{c}_j$,
ψ_k is $x_1 = \mathbf{m} \times y_1 + \mathbf{a}_k \wedge x_2 = \mathbf{m} \times y_2 + \mathbf{b}_k \wedge x_3 = x_1 \wedge x_4 = x_2$.

It follows, from Corollary 1 to Lemma 13.13, that Comp_M, which is the iterate of Next_M, is also representable. Hence the set $\{(\alpha, \beta, t);\ \text{Comp}_M(\alpha, \beta t) \in \text{Term}_M\}$ is expressible; let ϕ express it.

Now $(\alpha, \beta) \in H(M)$ iff there is t with (α, β, t) in the latter set. Hence $H(M)$ is expressed by $\exists y \phi(x_1, x_2, y)$.■

Proof of Theorem 13.10 (third proof) Let S be r.e. Then its partial characteristic function is partial recursive, and so can be computed by a special modular machine M. We have seen that $H(M)$ is expressible. As $n \in S$

iff $\text{In}_M n \in H(M)$, it is enough, by Lemma 13.12, to show that In_M is representable.

Now $\text{In}_M n = (m^{n+1} - 1)/(m-1)$. Thus $\text{In}_M n = r$ iff $(m-1)r + 1 = m^{n+1}$. We know that the function m^{n+1} is representable, by θ, say. Let p be $m-1$. Then In_M is represented by $\theta(x_1, \mathbf{S}(\mathbf{p} \times x_2))$.■

13.3 THE MAIN THEOREMS

The main Undecidability and Incompleteness Theorems are now fairly easy, as the hard work was done in the last section.

Two subsets A and B of a countable set X are called **recursively separable** if there is a recursive set R with $A \subset R$ and $R \cap B$ empty. If A and B are not recursively separable we say that they are **recursively inseparable**.

We shall denote the sets $\{\sigma; \mathbf{N} \models \sigma\}$ and $\{\sigma; \mathbf{N} \not\models \sigma\}$ by **T** and **F** in this section.

Theorem 13.15 (General Undecidability Theorem) *$C(A)$ and* **F** *are recursively inseparable.*

Proof Let Y be a set containing $C(A)$ and not meeting **F**. Let S be a subset of **N** (or $\mathbf{N}^2$) which is expressible, and let ϕ express it. The map sending n to $\phi(\mathbf{n})$ is recursive, by Church's Thesis, and S is the counter-image of Y under this map. It follows that if $C(A)$ and **F** were recursively separable, then any expressible subset of **N** (or $\mathbf{N}^2$, similarly) would be recursive.

There are now two easy ways to complete the proof. Theorem 4.7 gives an r.e. set which is not recursive, and this set is expressible by Theorem 13.10. Alternatively, Theorem 7.11 gives a modular machine whose halting set is not recursive, and Proposition 13.14 shows this set is expressible.■

Theorem 13.16 (Undecidability of Number Theory) *Let Γ be a set of sentences such that $\Gamma \vdash A$ and* **N** *is a model of Γ. Then $C(\Gamma)$ is undecidable.*

Proof Since $\Gamma \vdash A$ and $\mathbf{N} \models \Gamma$ we know that $C(A) \subset C(\Gamma) \subset \mathbf{T}$. The theorem is immediate from the previous theorem.■

Theorem 13.17 (Undecidability of Logic) *Let L be the language of number theory. Then* ThL *is undecidable, whether we regard L as a language with equality or as a pure language.*

Proof The theory $C(B_0)$ is undecidable, by the previous theorem, where B_0 is the set of six sentences discussed in the previous section. Let β be the conjunction of these six sentences.

Then $C(B_0)$ is $\{\sigma; \beta \vdash \sigma\}$, and so is also $\{\sigma; \vdash \beta \to \sigma\}$. Since $C(B_0)$ is undecidable, and the map sending σ to $\beta \to \sigma$ is recursive, the set $\{\sigma; \vdash \sigma\}$, which is Th$L$, must also be undecidable.

The above holds when L is regarded as a language with equality. If we regard L as a pure language we have just shown that $\{\sigma; E \vdash \sigma\}$ is undecidable, where E is the set of axioms of equality. The set E, by its definition, is finite. Then the same argument as in the previous paragraph shows that ThL (regarding L as a pure language) is also undecidable.■

Theorem 13.18 (Undecidability of Truth) **T** *is not r.e.*

Proof We already know that **T** is not recursive. If **T** were r.e. then $\{\sigma; \mathrm{N} \models \neg\sigma\}$ would also be r.e. But this set is just **F**, which is the complement of **T**. As **T** is not recursive, it is impossible for both **T** and **F** to be r.e.■

Theorem 13.19 (Weak Form of Gödel's Incompleteness Theorem) *Let* Γ *be an r.e. set of sentences of* L *such that* $\Gamma \vdash A$ *and* $\mathbf{N} \models \Gamma$. *Then there is a sentence in* T *which is not in* $C(\Gamma)$.

Proof We know that $C(\Gamma)$ is r.e., and that **T** is not r.e. Since $C(\Gamma) \subset \mathbf{T}$, the result is obvious.■

The Completeness Theorem tells us that $\Gamma \vdash \sigma$ iff σ is true in all models of Γ. The Incompleteness Theorem simply says that there is a sentence σ true in the model **N** of Γ such that $\Gamma \not\models \sigma$. It follows (the same result is immediate from the Lowenheim–Skolem Theorems) that Γ has a model different from **N**.

Notice that neither σ nor $\neg\sigma$ are in $C(\Gamma)$. By definition σ is not in $C(\Gamma)$, while $\neg\sigma$ is not in $C(\Gamma)$ because $\neg\sigma$ is not true in **N**. Thus $C(\Gamma)$ is not complete according to the earlier definition of completeness. Conversely, if $C(\Gamma)$ is incomplete then there is a sentence σ such that neither σ nor $\neg\sigma$ is in $C(\Gamma)$. Since one of σ and $\neg\sigma$ is true in **N**, one of the two (but we do not know which) is true in **N** but is not in $C(\Gamma)$. This explains the name 'Incompleteness Theorem'. This version of the theorem is called the Weak Form because it gives no indication of how to find such a sentence, but simply asserts that there must be a suitable sentence. If we look at the proof we see that it uses the following results. First, $C(\Gamma)$ is r.e. if Γ is r.e. If we follow the proof using modular machines, the second fact needed is that $H(M)$ is expressible for any modular machine M, and the third is that there is a modular machine with undecidable halting problem. This last comes from the existence of a non-recursive r.e. set, which uses the diagonal argument. In the other proof we also need the facts that there is a non-recursive r.e. set and that all r.e. sets are expressible. If we put these pieces together in a different way we can obtain the Strong Form of the theorem.

Theorem 13.20 (Strong Form of Gödel's Incompleteness Theorem) *Suppose we are given* Γ *as before. Then we can find (explicitly and uniformly) a sentence* σ *which is in* T *but not in* $C(\Gamma)$.

Remark We first have to see what is meant by saying Γ is given. If Γ were finite, we could simply give all the elements of Γ, but we need to consider the infinite case. Since we are referring to Γ as r.e., we must have in mind some

Gödel numbering of the set of all sentences. Then the set of Gödel numbers of members of Γ is r.e., and hence is the image of a partial recursive function f. To be given Γ can be regarded as the same as being given f. So how are we given a partial recursive function? This could be done by giving an abacus machine which computes f; that is, by being given a string on a certain alphabet. Better would be to specify some universal partial recursive function, and then f can be given by giving an index of f. Hence we can regard giving Γ as being the same as giving some integer k.

Now what is meant by saying that σ can be obtained explicitly and uniformly? To say that σ is given uniformly means that there is a recursive function g such that gk is the Gödel number of a suitable sentence σ if k is an integer defining Γ as above. To say that σ is given explicitly means that we are explicitly able to define (by an abacus machine, or whatever method seems convenient) a suitable function g, and that we are not simply asserting that such a g actually exists.

The detailed construction of g would be difficult. Among other things, it would require us to construct an integer defining $C(\Gamma)$ from an integer defining Γ. Nonetheless, it is clear that we know $C(\Gamma)$ when Γ is known, and that in fact Theorem 12.9 and Lemma 12.10 can be used to construct $C(\Gamma)$ from Γ; in principle, then, we could explicitly find a computable function which send the integer defining Γ to the integer defining $C(\Gamma)$.

Proof of Theorem Let the sentences δ_n be defined as follows. If n is the Gödel number of a formula ϕ with at most one free variable then δ_n is $\phi(\mathbf{n})$. Otherwise δ_n is to be $\mathbf{0} = \mathbf{0}$. Now we can tell of any n whether or not it is the Gödel number of some formula ϕ, and, if so, find the relevant ϕ and check whether it has at most one free variable. Hence, by Church's Thesis, the function sending n to δ_n is recursive. Notice that the function δ is constructed to be a kind of diagonal function.

As $C(\Gamma\}$ is r.e., because Γ is r.e., it follows that $\{n; \Gamma \vdash \delta_n\}$ is r.e. Hence this set is expressible. Let ψ express it, and let θ be $\neg\psi$.

By definition, if $\Gamma \vdash \delta_n$ then $A \vdash \psi(\mathbf{n})$. Hence $\psi(\mathbf{n})$ is in $\mathbf{T}$, and $\theta(\mathbf{n})$ is in $\mathbf{F}$ in this case. Also, if $\Gamma \nvdash \delta_n$, then $\psi(\mathbf{n})$ is in $\mathbf{F}$, and so $\theta(\mathbf{n})$ is in $\mathbf{T}$. Notice also that if $\Gamma \vdash \delta_n$ then δ_n is in $\mathbf{T}$.

Now θ is a formula with one free variable. Let its Gödel number be k. Using the diagonal argument, we consider δ_k. Then, by definition, δ_k is just $\theta(\mathbf{k})$. So the above tells us that if $\Gamma \vdash \delta_k$ then δ_k is in both $\mathbf{T}$ and $\mathbf{F}$, which is impossible.

Hence we have both $\Gamma \nvdash \delta_k$ and δ_k in $\mathbf{T}$, as required. ■

Notice that the truth of δ_k amounts to saying that $\theta(\mathbf{k})$ is not derivable from Γ. Since δ_k is just $\theta(\mathbf{k})$, this amounts to saying that the meaning of δ_k is, 'This statement is unprovable from Γ', which is the very informal version which we mentioned in the first chapter.

The sentence σ constructed by this is highly artificial, even when Γ is chosen to be very nice. It was an open question for a long time whether there was a sentence which was of genuine mathematical interest and which was true but not derivable from reasonable axioms. Such sentences have been found in recent years, but they are too complicated to give here. Though subtle, their truth is accessible to an advanced undergraduate. The fact that they are not provable is much deeper, however. The results usually state that to each n there is an m with a certain property, and the unprovability is connected with the fact that m increases very rapidly with n; in fact m, as function of n, increases much faster than any function which can be proved total.

We now look at a concept somewhat weaker than expressibility. A set S is called **definable in N** if there is a formula ϕ such that $(n_1, \ldots, n_k) \in S$ iff $\phi(\mathbf{n}_1, \ldots, \mathbf{n}_k)$ is true in $\mathbf{N}$, and a function is called **definable in N** iff its graph is definable in $\mathbf{N}$. Plainly any expressible set and any representable function, that is, any r.e. set and any partial recursive function, are definable in $\mathbf{N}$. If S is definable in $\mathbf{N}$ by the formula ϕ, then the complement of S is plainly definable in $\mathbf{N}$ by the formula $\neg\phi$. In particular, there are definable sets which are not r.e. (for instance, the complement of a set which is r.e. but not recursive). It is also easy to see, as in Lemma 13.12, that the counter-image of a definable set by a definable function is itself definable.

Is the set of Gödel numbers of sentences true in $\mathbf{N}$ a set definable in $\mathbf{N}$? At first glance this question might appear pointless, since it only replaces the question of the truth of one sentence with the question of the truth of another sentence. But if this set were definable by a formula ϕ (we shall soon see that it is not) then questions about the truth of arbitrary sentences could be reduced to questions about the truth of the sentence $\phi(\mathbf{n})$ for varying n; thus ϕ would have in some sense a universal property.

Theorem 13.21 (Undefinability of Truth) *The set of Gödel numbers of true sentences is not definable in* $\mathbf{N}$.

Proof Suppose this set were definable. By the preceding remarks, the set $\{n; \delta_n \text{ is true in } \mathbf{N}\}$ would also be definable, where δ_n is defined in the proof of Theorem 13.20. Let ϕ define this set, and let θ be $\neg\phi$. Then δ_n is true in $\mathbf{N}$ iff $\theta(\mathbf{n})$ is false in $\mathbf{N}$. Let k be the Gödel number of θ. Then δ_k is just $\theta(\mathbf{k})$, and we get the usual contradiction.■

13.4 FURTHER RESULTS

Gödel's original proof did not refer to truth in $\mathbf{N}$, but instead showed that $C(\Gamma)$ is incomplete. One advantage of this is that no mention is made of truth, which is a concept requiring an infinite process to check, but only of derivability, and we can check whether or not a tree is a derivation by a finite process. A further advantage is that the theorem can be proved in cases where Γ does not have $\mathbf{N}$ as a model. In order to prove these versions we need stronger forms of expressibility and representability.

The set B of sentences consists of the six sentences of B_0 together with the three further sentences

$$\beta_7: \forall x(\neg(\mathbf{S}x = \mathbf{0}))$$

β_8: $\forall x \forall y(\mathbf{S}x = \mathbf{S}y \rightarrow x = y)$
β_9: $\forall x \forall y(x \leqslant y \vee y \leqslant x)$.

We shall write $x < y$ instead of $x \geqslant y \wedge -(x = y)$.

Lemma 13.22 *If $m \neq n$ then $B \vdash \neg(\mathbf{m} = \mathbf{n})$.*

Proof We use induction on $\min(m,n)$. We may assume $m > n$. Then $m \neq 0$, and we may write m as $p+1$ for some p. Then **m** is **Sp**, so $\beta_7 \vdash \neg(\mathbf{m} = \mathbf{0})$, and the result follows if $n = 0$. If $n \neq 0$ we write n as $q+1$, and so **n** is **Sq**. Since $\beta_8 \vdash (\mathbf{m} = \mathbf{n} \rightarrow \mathbf{p} = \mathbf{q})$, and, inductively, $B \vdash \neg(\mathbf{p} = \mathbf{q})$, we have $B \vdash \neg(\mathbf{m} = \mathbf{n})$, as needed.■

Definition The subset S of $\mathbf{N}^k$ is **strongly expressible** if there is a formula ϕ such that $B \vdash \phi(\mathbf{n}_1, \ldots, \mathbf{n}_k)$ if $(n_1, \ldots, n_k) \in S$ and $B \vdash \neg\phi(\mathbf{n}_1, \ldots, \mathbf{n}_k)$ if $(n_1, \ldots, n_k) \notin S$.

The function $f: \mathbf{N}^k \rightarrow \mathbf{N}$ is **strongly representable** if there is a formula ϕ such that $B \vdash \phi(\mathbf{n}_1, \ldots, \mathbf{n}_k, \mathbf{m})$ if $f(n_1, \ldots, n_k) = m$ and $f(n_1, \ldots, n_k) = m$ if $\psi(\mathbf{n}_1, \ldots, \mathbf{n}_k, \mathbf{m})$ is true in **N** and such that

$$B \vdash \forall x_1 \ldots \forall x_k \forall y \forall z(\phi(x_1, \ldots, x_k, y) \wedge \phi(x_1, \ldots, x_k, z) \rightarrow y = z).$$

We shall see that all partial recursive functions are strongly representable, while strong expressibility is equivalent to recursiveness.

Lemma 13.23 *Let $f_1, \ldots, f_r: \mathbf{N}^k \rightarrow \mathbf{N}$ and $g: \mathbf{N}^r \rightarrow \mathbf{N}$ be strongly representable. Then so is their composite $h: \mathbf{N}^k \rightarrow \mathbf{N}$.*

Proof Let the functions be strongly represented by $\phi_1, \ldots, \phi_r$, and ψ, respectively. Let θ be the formula

$$\exists y_1 \ldots \exists y_r(\phi_1(x_1, \ldots x_k, y_1) \wedge \ldots \wedge \phi_r(x_1, \ldots, x_k, y_r) \wedge \wedge \psi(y_1, \ldots, y_1, \ldots, y_r, y)).$$

As in Lemma 13.12, θ satisfies the first two conditions for it to strongly represent h.

We need only show that

$$B \cup \{\phi_1(x_1, \ldots, x_k, y_1) \wedge \ldots \wedge \phi_r(x_1, \ldots, x_k, y_r) \wedge \psi(y_1, \ldots, y_r, y), \phi_1(x_1, \ldots, x_k, z_1) \wedge \ldots \wedge \phi_r(x_1, \ldots, x_k, z_r) \wedge \psi(z_1, \ldots, z_r, z)\} \vdash y = z.$$

For then the law of introduction of $\exists$ (together with changing the names of bound variables) tells us that $B \cup \{\theta(x_1, \ldots, x_k, y), \theta(x_1, \ldots, x_r, z)\} \vdash y = z$, from which we can obtain the result we want. Now, from the given set of formulae we can derive $y_i = z_i$ for $i = 1, \ldots, r$, since each ϕ_i strongly represents f_i. From the axioms of equality we have immediately that

$$y_1 = z_1 \wedge \ldots \wedge y_r = z_r \vdash \psi(y_1, \ldots, y_r, y) \to \psi(z_1, \ldots, z_r, y).$$

Since ψ strongly represents g we can then derive $y = z$, as needed.■

For the proof of the next lemma, a few results on derivations are helpful. We exhibit a number of formulae, such that each formula in a pair can be derived from the other. We know, as examples and exercises, that this holds for the following pairs. First $\theta \wedge (\phi \vee \psi)$ and $(\theta \wedge \phi) \vee (\theta \wedge \psi)$. Second, $\exists x(\phi \vee \psi)$ and $\exists x\phi \vee \exists x\psi$. Third, when ψ does not have x as a free variable, $\exists x(\phi \wedge \psi)$ and $\exists x\phi \wedge \psi$. Also, easily, we have the pair $\beta \leftrightarrow \gamma$ and $(\beta \wedge \gamma) \vee (\neg\beta \wedge \neg\gamma)$. Combining these, we see that, if x does not occur free in γ then the same result holds for $\exists x(\alpha \wedge (\beta \leftrightarrow \gamma))$ and $(\exists x(\alpha \wedge \beta) \wedge \gamma) \vee (\exists x(\alpha \wedge \neg\beta) \wedge \neg\gamma)$. Further, we note that $\phi(x) \wedge x = \mathbf{0} \vdash \phi(\mathbf{0})$, and so $\exists x(\phi(x) \wedge x = \mathbf{0}) \vdash \phi(\mathbf{0})$. Finally, note that $\{\gamma, \theta \wedge \gamma\} \vdash \theta$, obviously, while $\{\gamma, \phi \wedge \neg\gamma\} \vdash \theta$, because the set is inconsistent. Hence $\{\gamma, (\theta \wedge \gamma) \vee (\phi \wedge \neg\gamma)\} \vdash \theta$.

Lemma 13.24 *Let $g:\mathbf{N}^k \to \mathbf{N}$ come from $f:\mathbf{N}^{k+1} \to \mathbf{N}$ by minimisation. Then g is strongly representable if f is.*

Proof Let ϕ strongly represent f. For convenience of notation we take k to be 1. Let ψ be the formula

$$\forall w(w \leqslant z \to \exists y(\phi(x, w, y) \wedge (y = \mathbf{0} \leftrightarrow w = z))).$$

Then $\psi(x, z) \vdash \phi(x, z, \mathbf{0})$, by the above comments. Also, with u being a new variable, $B \vdash \phi(x, u, y) \wedge \phi(x, u, \mathbf{0}) \to y = \mathbf{0}$. Hence $B \cup \{\phi(x, u, \mathbf{0})\} \vdash \neg(\phi(x, u, y) \wedge \neg(y = \mathbf{0}))$, from which it follows that $B \cup \{\phi(x, u, \mathbf{0})\} \vdash \forall y \neg(\phi(x, u, y) \wedge \neg(y = \mathbf{0}))$. Since $\psi(x, z) \vdash u < z \to \exists y(\phi(x, u, y) \wedge \neg(y = \mathbf{0}))$, we find that $B \cup \{u < z \wedge \psi(x, z)\} \vdash \neg\phi(x, u, \mathbf{0})$. Hence $B \cup \{u < z \wedge \psi(x, z)\} \vdash \neg\psi(x, u)$. Using the last axiom in B, we find that $B \vdash \psi(x, z) \wedge \psi(x, u) \to u = z$, giving us the third condition for strong representability.

Suppose $\psi(\mathbf{m}, \mathbf{n})$ is true in $\mathbf{N}$. Then $\phi(\mathbf{m}, \mathbf{n}, \mathbf{0})$ is true in $\mathbf{N}$, so $f(m, n) = 0$. Also, if $r < n$ then $\exists y\phi(\mathbf{m}, \mathbf{r}, y)$ is true in $\mathbf{N}$. Hence there is some s with $\phi(\mathbf{m}, \mathbf{r}, \mathbf{s})$ true in $\mathbf{N}$. Thus $f(m, r) = s$. Also, as $\phi(\mathbf{m}, \mathbf{n})$ is true in $\mathbf{N}$, we must have $s \neq 0$, so that $gm = n$.

Finally, let $gm = n$. Then $f(m, n) = 0$, while $f(m, r) = s$ for some non-zero s if $r < n$. Hence $B \vdash \phi(\mathbf{m}, \mathbf{n}, \mathbf{0})$, and, if $r < n$, we have $B \vdash \phi(\mathbf{m}, \mathbf{r}, \mathbf{s})$ where $s \neq 0$. We then know that $B \vdash \neg(\mathbf{s} = \mathbf{0})$. Hence $B \vdash \phi(\mathbf{m}, \mathbf{r}, \mathbf{s}) \wedge (\mathbf{s} = \mathbf{0} \leftrightarrow \mathbf{r} = \mathbf{n})$ for all $r < n$. It follows that $B \vdash \exists y(\phi(\mathbf{m}, \mathbf{r}, y) \wedge (y = \mathbf{0} \leftrightarrow \mathbf{r} = \mathbf{n}))$ for all $r \leqslant n$. But we know that $B \vdash \forall w(w \leqslant \mathbf{n} \leftrightarrow w = \mathbf{0} \vee \ldots \vee w = \mathbf{n})$, As remarked before Lemma 13.19, this ensures that $B \vdash \psi(\mathbf{m}, \mathbf{n})$.■

Theorem 13.25 *A function is strongly representable iff it is partial recursive.*

Proof Let $f:\mathbf{N}^k \to \mathbf{N}$ be strongly represented by ϕ. Then $f(n_1, \ldots, n_k) = m$ iff $\phi(\mathbf{n}_1, \ldots, \mathbf{n}_k, \mathbf{m})$ is in $C(B)$. Since $C(B)$ is r.e. and the function sending $(n_1, \ldots, m)$ to $\phi(\mathbf{n}_1, \ldots, \mathbf{m})$ is recursive, the graph of f will be r.e., being the counter-image of $C(B)$ by that function. Hence f is partial recursive.

We have seen that the set of strongly representable functions is closed under composition and minimisation. We will show this set contains addition, multiplication, the projection functions, and the function $c:\mathbf{N}^2 \to \mathbf{N}$ given by $c(m,n) = 1$ if $m = n$ and 0 otherwise. It then follows, by Theorem 9.15 about min-computable functions, that the set of strongly representable functions contains all partial recursive functions.

Plainly addition and multiplication are strongly represented by $x + y = z$ and $x \times y = z$ respectively, while the projection function π_{ni} is strongly represented by $x_i = y$.

Let ϕ be the formula $(x = y \wedge z = \mathbf{0}) \vee (\neg(x = y) \wedge z = \mathbf{1})$. If $\phi(\mathbf{m}, \mathbf{n}, \mathbf{r})$ is true in $\mathbf{N}$ then either $m = n$ and $r = 0$ or $m \neq n$ and $r = 1$, and so $r = c(m,n)$. Conversely, let $c(m,n) = r$. If $m = n$ then $r = 0$ and $B \vdash \phi(\mathbf{m}, \mathbf{n}, \mathbf{r})$ immediately. If $m \neq n$ then we know that $B \vdash \neg(\mathbf{m} = \mathbf{n})$, and, because $r = 1$, again $B \vdash \phi(\mathbf{m}, \mathbf{n}, \mathbf{r})$.

We still have to prove the final condition on ϕ. It is enough to prove that $B \cup \{\phi(x,y,z) \wedge \phi(x,y,w)\} \vdash z = w$. Now $\vdash (x = y) \vee \neg(x = y)$. So the rule for left-introduction of $\vee$ tell us that it is enough to show that $B \cup \{\phi(x,y,z) \wedge \phi(x,y,\mathrm{w}), x = y\} \vdash z = w$ and also that $B \cup \{\phi(x,y,z) \wedge \phi(x,y,\mathrm{w}), \neg(x = y)\} \vdash z = w$. Now from the first of these we can derive both $z = 0$ and $w = 0$, and so we can derive $z = w$, while from the second we can derive both $z = 1$ and $w = 1$, and so we can again derive $z = w$.■

Proposition 13.26 *A set is strongly expressible iff it is recursive.*

Proof For convenience of notation, we will consider only subsets of $\mathbf{N}$. Suppose S is strongly expressible by the formula ϕ. Then $n \in S$ iff $\phi(\mathbf{n}) \in C(B)$, so, as usual, S is r.e. But also $n \in \mathbf{N} - S$ iff $\neg\phi(\mathbf{n}) \in C(B)$, so, for the same reason, $\mathbf{N} - S$ is also r.e., and hence S is recursive.

Now let S be recursive, and let ψ strongly represent the characteristic function of S. Then $B \vdash \psi(\mathbf{n}, \mathbf{0})$ if $n \in S$, while $B \vdash \psi(\mathbf{n}, \mathbf{1})$ if $n \notin S$. Now $B \vdash \neg(\mathbf{1} = \mathbf{0})$, and, by strong representability, $B \cup \{\psi(\mathbf{n}, \mathbf{1}) \wedge \psi(\mathbf{n}, \mathbf{0})\} \vdash 1 = \mathbf{0}$. Hence if $n \notin S$ we have $B \vdash \neg\psi(\mathbf{n}, \mathbf{0})$. Hence $\psi(x, \mathbf{0})$ strongly expresses S.■

It is sometimes useful to add a further condition to strong representability. This is done in the next proposition.

Proposition 13.27 *The function $f:\mathbf{N}^k \to \mathbf{N}$ can be strongly represented by a formula ϕ such that $B \vdash \forall x_1 \ldots \forall x_k \exists y \phi(x_1, \ldots x_k, y)$ iff f is recursive.*

Proof Take k to be 1 for convenience of notation. Suppose there is such a formula ϕ. Since ϕ strongly represents f, we know that f is partial recursive. We also know that $\forall x \exists y \phi$ is true in $\mathbf{N}$. Hence for every m there is n such that $\phi(\mathbf{m}, \mathbf{n})$ is true in $\mathbf{N}$. It follows that $fm = n$, and so f is total, as needed.

Conversely, let f be recursive and let ψ strongly represent f. Let ϕ be the formula $(\psi \wedge \exists y \psi) \vee (y = \mathbf{0} \wedge \neg \exists y \psi)$ (here y is used both as a bound and a free variable). We know that for any formulae α and β we have $\vdash \exists y(\alpha \vee \beta) \leftrightarrow \exists y \alpha \vee \exists y \beta$, and that $\vdash \exists y(\alpha \wedge \beta) \leftrightarrow \exists y \alpha \wedge \beta$ provided y does not occur free in β. Using these we see easily that $\vdash \exists y \phi \leftrightarrow \exists y \psi \vee \neg \exists y \psi$, and so $\vdash \exists y \phi$.

If $fm = n$ then $B \vdash \psi(\mathbf{m}, \mathbf{n})$ and so $B \vdash \exists y \phi(\mathbf{m}, y)$. Hence $B \vdash \phi(\mathbf{m}, \mathbf{n})$. As f is total we know that $B \vdash \exists y \phi(\mathbf{m}, y)$ for every m. Hence if $\phi(\mathbf{m}, \mathbf{n})$ is true in $\mathbf{N}$, then so is $\psi(\mathbf{m}, \mathbf{n})$, and so $fm = n$.

Finally we have to show that $B \cup \{\phi(x,z) \wedge \phi(x,w)\} \vdash z = w$. As in the previous proposition, it is enough to show both that $B \cup \{\phi(x,z) \wedge \phi(x,w), \exists y \psi\} \vdash z = w$ and that $B \subset \{\phi(x,z) \wedge \phi(x,w), \neg \exists y \psi\} \vdash z = w$. From the former we can derive $\psi(x,z)$ and $\psi(x,w)$, from which we can derive $z = w$ by the fact that ψ strongly represents f. From the latter we can derive both $z = \mathbf{0}$ and $w = \mathbf{0}$, so again we can derive $z = w$. ■

We can now obtain the Undecidability and Incompleteness Theorems in a somewhat more general setting than previously.

Proposition 13.28 *Let $F: \mathbf{N}^2 \to \mathbf{N}$ be a universal partial recursive function, and let K_0 and K_1 be $\{n; F(n,n) = 0\}$ and $\{n; F(n,n) = 1\}$ respectively. Then K_0 and K_1 are recursively inseparable.*

Proof Suppose the recursive set R separates them, and let χ be the chracteristic function of $\mathbf{N} - R$. Then $\chi n = 1$ if $F(n,n) = 0$ and $\chi n = 0$ if $F(n,n) = 1$. There is an integer k such that $\chi n = F(k,n)$ for all n. In particular, the function $F(k,n)$ is a total function of n which takes on only the values 0 and 1. Replacing n by k we get the usual contradiction. ■

Theorem 13.29 *The sets $C(B)$ and $\{\sigma; \vdash \neg \sigma\}$ are recursively inseparable.*

Proof Let F, K_0, and K_1 be as above. It is enough to find a formula ϕ such that $B \vdash \phi(\mathbf{n})$ for $n \in K_0$ and $\vdash \neg \phi(\mathbf{n})$ for $n \in K_1$. For if R separates $C(B)$ and $\{\sigma; \vdash \neg \sigma\}$ then its counter-image by the recursive function which sends n to $\phi(\mathbf{n})$ would separate K_0 and K_1. Since no recursive set separates these two, R could not be recursive.

Let ψ strongly represent the partial recursive function $F(n,n)$. Let β be the conjunction of the nine sentences in B. Let ϕ be $\beta \wedge \psi(x, \mathbf{0})$. If $n \in K_0$ then $B \vdash \psi(\mathbf{n}, \mathbf{0})$, and so $B \vdash \phi(\mathbf{n})$. If $n \in K_1$ we have $B \vdash \psi(\mathbf{n}, \mathbf{1})$. Since we also know that $B \cup \{\psi(\mathbf{n}, \mathbf{1}) \wedge \psi(\mathbf{n}, \mathbf{0})\} \vdash \mathbf{1} = \mathbf{0}$ and $B \vdash \neg(\mathbf{1} = \mathbf{0})$, we see that if $n \in K_1$ then $B \cup \{\psi(\mathbf{n}, \mathbf{0})\} \vdash \perp$. In this case it follows that $\vdash \neg(\beta \wedge \psi(\mathbf{n}, \mathbf{0}))$, as needed. ■

Theorem 13.30 *Let Γ be a set of sentences in a language containing L. If $\Gamma \cup B$ is consistent, then $C(\Gamma)$ is not recursive.*

Proof Let β be as above. Then we know that $\Gamma \cup B \vdash \sigma$ iff $\Gamma \vdash \beta \rightarrow \sigma$. It follows that $C(\Gamma \cup B)$ would be recursive if $C(\Gamma)$ were recursive. So we may assume that $B \subset \Gamma$.

The set of sentences of L is a recursive subset of the set of sentences of our larger language. Hence, if $C(\Gamma)$ were recursive then $\{\sigma; \sigma$ is a sentence of L such that $\Gamma \vdash \sigma\}$ would also be recursive. Since $B \subset \Gamma$, this set contains $C(B)$. Since Γ is consistent, this set cannot meet $\{\sigma; \vdash \neg\sigma\}$, so by the previous theorem this set cannot be recursive.■

Gödel's original proof of his Incompleteness Theorem required a special condition on the set Γ. We first obtain a proof under this additional condition, and then show how this condition can be got rid of by using a rather more complicated formula. Where previously we looked at the r.e. set of formulae which could be derived from Γ, we now look at the recursive set of pairs (D, σ), where D is a derivation of σ from a recursive set Γ.

Suppose we have a set Γ of sentences in a language containing L and a formula ϕ such that $\Gamma \vdash \phi(\mathbf{n})$ for all n. It is not necessarily true that $\Gamma \vdash \forall x\phi$. (Since we are not even requiring that **N** is a model of Γ, this is clear. For a more interesting example, observe that if ϕ is $\mathbf{0} \leqslant x$ then $A \vdash \phi(\mathbf{n})$ for all n. But there is a model of A in which $\forall x\phi$ is not true, so that $A \nvdash \forall x\phi$. This model can be obtained by adding one extra element to **N**, with suitable extended definitions for the functions.) In fact, if $\Gamma \vdash \forall x\phi$ using a derivation which has k steps then for every n there is a derivation of $\phi(\mathbf{n})$ with $k + 1$ steps. (For many formulations of logic, the requirement that $\Gamma \vdash \phi(\mathbf{n})$ for all n with the derivation for each n requiring at most m steps for some fixed m is enough to ensure that $\Gamma \vdash \forall x\phi$). But we can reasonably hope that at least we do not have $\Gamma \vdash \neg\forall x\phi$. A set Γ such that for no formula ϕ do we have $\Gamma \vdash \phi(\mathbf{n})$ for all n and also $\Gamma \vdash \neg\forall x\phi$ is called **ω-consistent**. Plainly an ω-consistent set must be consistent (since every formula can be derived from an inconsistent set). Also Γ must be ω-consistent if it has **N** as a model. Further, if $\Gamma \vdash B$ then $\Gamma \vdash \sigma$ iff $\Gamma \cup B \vdash \sigma$, and so $\Gamma \cup B$ will be ω-consistent iff Γ is ω-consistent.

Theorem 13.31 (Original Form of Gödel's Incompleteness Theorem) *Let Γ be a r.e. set of sentences in a language containing L. Then we can find a sentence σ such that $\Gamma \nvdash \sigma$ if $\Gamma \cup B$ is consistent and such that $\Gamma \nvdash \neg\sigma$ if $\Gamma \cup B$ is ω-consistent.*

Proof We may assume Γ is recursive and not just r.e., by Lemma 12.10. We may also assume that $B \subset \Gamma$, replacing Γ by $\Gamma \cup B$.

Since Γ is recursive, the set of derivations from Γ is recursive, and, in particular, $\{(s,t); t$ is the Gödel number of a derivation from Γ of the sentence with Gödel number $s\}$ is recursive.

Let δ_n be, as before, $\phi(\mathbf{n})$ if n is the Gödel number of a formula ϕ with one free variable, and let δ_n be $\mathbf{0} = \mathbf{0}$ for all other n. Let P be the set $\{(n,m);$ m is the Gödel number of a derivation of δ_n from $\Gamma\}$. Then P is recursive, and hence P is strongly expressible. Let α strongly express P. (Then $\exists y\alpha$ expresses $\{n; \Gamma \vdash \delta_n\}$, which connects the current proof with the previous one.) Let ϕ be $\neg \exists y\alpha$, and let k be the Gödel number of ϕ.

If $\Gamma \vdash \delta_n$ there is some m with $(n,m) \in P$, and so $B \vdash \alpha(\mathbf{n}, \mathbf{m})$, and then (since we are assuming $B \subset \Gamma$) $\Gamma \vdash \neg \phi(\mathbf{n})$. In particular, since $\phi(\mathbf{k})$ is δ_k, if $\Gamma \vdash \phi(\mathbf{k})$ then also $\Gamma \vdash \neg \phi(\mathbf{k})$, so $\Gamma \nvdash \phi(\mathbf{k})$ if Γ is consistent.

Since δ_k *is* $\phi(\mathbf{k})$, we have just shown that $\Gamma \nvdash \delta_k$. We will show that if Γ is ω-consistent and n is such that $\Gamma \nvdash \delta_n$ then $\Gamma \nvdash \neg \phi(\mathbf{n})$. In particular, if Γ is ω-consistent then $\Gamma \nvdash \neg \phi(\mathbf{k})$.

Suppose that $\Gamma \nvdash \delta_n$. Then $(n,m) \notin P$ for every m. By strong expressibility, it follows that $\Gamma \vdash \neg \alpha(\mathbf{n}, \mathbf{m})$ for every m. If Γ is ω-consistent then $\Gamma \nvdash \forall y \neg \alpha(\mathbf{n}, y)$. This is the same as saying that $\Gamma \nvdash \neg \phi(\mathbf{n})$, as required.■

Theorem 13.32 (Rosser's Extension of the Original Form) *Let Γ be an r.e. set of sentences in a language containing L such that $\Gamma \cup B$ is consistent. Then we can find a sentence σ such that $\Gamma \nvdash \sigma$ and $\Gamma \nvdash \neg \sigma$.*

Proof As before, we may assume that Γ is recursive and contains B. Let δ_n and α be as before. Then the set $Q = \{(n,m);$ m is the Gödel number of a derivation of $\neg \delta_n$ from $\Gamma\}$ is recursive, and we choose a formula β which strongly expresses it. We take ϕ to be the formula $\forall y(\alpha \rightarrow \exists z(z \leq y \wedge \beta(x,z)))$ and let the Gödel number of ϕ be r. Intuitively the meaning of ϕ is that if there is a derivation of δ_n then there is an earlier derivation of $\neg \delta_n$.

We shall show that both $\Gamma \vdash \delta_r$ and $\Gamma \vdash \neg \delta_r$ lead to contradictions. Recall that δ_r is just $\phi(\mathbf{r})$. Suppose first that $\Gamma \vdash \delta_r$. Then there is some m with $\Gamma \vdash \alpha(\mathbf{r}, \mathbf{m})$ (since $B \subset \Gamma$). Since Γ is assumed consistent, we cannot have $\Gamma \vdash \neg \delta_r$. Hence $B \vdash \neg \beta(\mathbf{r}, \mathbf{n})$ for all $n \leq m$ (in fact, for all n). It follows, as in the remark before Lemma 13.13, that $B \vdash \forall z(z \leq \mathbf{m} \rightarrow \neg \beta(\mathbf{r}, z))$. However, $\Gamma \vdash \phi(\mathbf{r})$ and $\phi(\mathbf{r})$ is $\forall y(\alpha(\mathbf{r}, y) \rightarrow \exists z(z \leq y \wedge \beta(\mathbf{r}, z)))$. Since we already know that $\Gamma \vdash \alpha(\mathbf{r}, \mathbf{m})$ we deduce that $\Gamma \vdash \exists z\, (z \leq \mathbf{m} \wedge \beta(\mathbf{r}, z))$. This is a contradiction, since Γ is consistent and contains B.

Now suppose that $\Gamma \vdash \neg \delta_r$. As Γ is consistent we know that $\Gamma \nvdash \delta_r$. This time there will be some m such that $B \vdash \beta(\mathbf{r}, \mathbf{m})$, while $B \vdash \neg \alpha(\mathbf{r}, \mathbf{n})$ for all $n \leq m$ (in fact, for all n). As before, we have that $B \vdash y \leq \mathbf{m} \rightarrow \neg \alpha(\mathbf{r}, y)$. Hence $B \cup \{\alpha(\mathbf{r}, y)\} \vdash \neg (y \leq \mathbf{m})$. By the last sentence in B it follows that $B \cup \{\alpha(\mathbf{r}, y)\} \vdash \mathbf{m} \leq y$. Since $B \vdash \beta(\mathbf{r}, \mathbf{m})$ we see that $B \cup \{\alpha(\mathbf{r}, y) \vdash \exists z(z \leq y \wedge \beta(\mathbf{r}, z))$. Hence $B \vdash \forall y(\alpha(\mathbf{r}, y) \rightarrow \exists z(z \leq y \wedge \beta(\mathbf{r}, z)))$; that is, $B \vdash \phi(\mathbf{r})$, giving the required contradiction.■

These results can be extended further, to certain sets of sentences in some languages that do not contain L. It is necessary to show how we can interpret the symbols of L in the new language in such a way that the translations of the sentences in B can be derived from the given set. This is not particularly difficult, but is rather technical. The most useful application is to extend Gödel's Theorems to the language of formal set theory. But in order to prove that we can make the necessary translations we would need to go into considerable detail about axiomatic set theory, which would take too much space.

There is a second Incompleteness Theorem of Gödel. This states that, under reasonable circumstances, if a set Γ of sentences is consistent then we cannot derive from Γ the consistency of Γ (more precisely, there is a sentence σ whose intuitive meaning is that Γ is consistent but which cannot be derived from Γ). The general idea is to formalise precisely within number theory all the informal arguments we have used to prove the previous theorems. Though the idea is reasonably understandable, the details are exceptionally long and technical, and are beyond the scope of texts much more advanced than this one.

14

The natural numbers under addition

We have seen that the theory of **N** *with the operations of addition and multiplication is undecidable. In surprising contrast, the theory of* **N** *under addition (but not including multiplication) is decidable. This will be proved in the second section. The technique is quite difficult, and a simpler example of the same technique is given in the first section.*

14.1 THE ORDER RELATION ON Q

In this section we look at the theory of **Q** with the order relation $<$ and the relation $=$. We have already shown, in section 13.1, that this theory is decidable. We give a new proof of this result. The current proof is harder than the earlier one, and proves a slightly weaker result. Nonetheless this proof is worth giving, as it shows the main steps needed in the next section in a simpler form; other decidability results may also be proved in a similar framework.

We now state a crucial lemma. The proof of this lemma will be left until later in the section. Before proving the lemma, we will show how it leads to a solution of the decision problem. In other situations, it is the proof of a similar lemma that takes the hard work; most of the other steps follow as in this section.

We have already defined equivalence of formulae. More generally, we say ϕ is equivalent to ψ in the structure **A** if $v_{\mathbf{A}}i\phi = v_{\mathbf{A}}i\psi$ for any interpretation i in **A**.

Lemma 14.1 (Elimination of Quantifiers) *Let ϕ be a formula with no quantifiers, and let x be a variable. Then we can find explicitly a formula ϕ' such that $\exists x\phi$ is equivalent in* **Q** *to ϕ', and such that ϕ' does not contain x and does not contain any variable which is not in ϕ.*

Once this is proved, we can also find ϕ'' such that $\exists x \neg\phi$ is equivalent in **Q** to ϕ'', with ϕ'' not containing x or any variable not in ϕ. Then $\forall x\phi$ is equivalent in **Q** to $\neg\phi''$.

It follows immediately, by induction, that any sentence in prenex normal form (see section 12.8) is equivalent in **Q** to a sentence containing no variables, and that the latter can be found explicitly from the original sentence.

We have shown in Proposition 12.33 that to any sentence we can find an equivalent sentence in prenex normal form. The procedure in Proposition 12.33 is constructive, so the required sentence can be found explicitly using that procedure. Alternatively, knowing that such a sentence exists we could, given a sentence σ, run through all derivations until we found one which showed us that $\vdash \sigma \leftrightarrow \sigma'$ with σ' a sentence in prenex normal form. Similar remarks apply at other stages of our results. It follows that given any sentence we can find a sentence containing no variables which is equivalent in **Q** to the original sentence. Hence to solve the decision problem for the theory of **Q** it is enough to solve it for sentences containing no variables.

Since there are no constant symbols, the only atomic formula containing no variable is $\perp$. Any sentence containing no variables is built up from $\perp$ using only $\neg$, $\vee$, $\wedge$ and $\rightarrow$. It is then easy to determine whether or not such a sentence is true in **Q.** One method is to obtain (which can be done explicitly) an equivalent sentence in disjunctive normal form. This sentence will be true in **Q** iff at least one of its disjuncts is true in **Q**. But each disjunct is a conjunction of atomic formulae and negations of atomic formulae, and it is true in **Q** iff the formulae of which it is a conjunction are all true in **Q.** But these formulae must be either $\perp$ or $\neg\perp$, and only the latter is true in **Q.**

Proof of Lemma 14.1 We know (see Proposition 12.34) that to any formula which has no quantifiers we can find an equivalent formula in disjunctive normal form, and the two formulae will involve the same variables (in fact, they will involve the same atomic subformulae other than $\perp$). Thus it is enough to prove the lemma when ϕ is in disjunctive normal form. Further, the discussion preceding Proposition 11.20 shows that unless ϕ is $\perp$ or $\neg\perp$ we may assume that $\perp$ does not occur in ϕ. Note that the only atomic formulae other than $\perp$ are $y{=}z$ and $y{<}z$ for some variables y and z.

Further, the formulae $\neg(y{=}z)$ and $\neg(y{<}z)$ are equivalent in **Q** to $y{<}z \vee z{<}y$ and $y{=}z \vee z{<}y$ respectively. Using the distributive laws relating $\vee$ and $\wedge$, we may assume that ϕ, which is assumed in disjunctive normal form, has its atomic subformulae appearing without negation. Also, altering the order of the subformulae produces an equivalent formula, so this will be done when convenient without further comment.

Now $\exists x(\phi' \vee \phi'')$ is equivalent to $\exists x\phi' \vee \exists x\phi''$, so we may assume ϕ consists of a conjunction of atomic formulae. Further, if ϕ' does not involve x the formula $\exists x(\phi' \wedge \phi'')$ is equivalent to $\phi' \wedge \exists x\phi''$, so we may assume each atomic formula in ϕ involves x.

Plainly $\exists x(x{<}x \wedge \phi')$ is equivalent in **Q** to $\perp$. Also $\exists x(x{=}x \wedge \phi')$ is equivalent to $\exists x\phi'$, while $\exists x(x=x)$ is equivalent to $\neg\bot$. Thus we may assume each atomic formula in ϕ involves x on only one side.

Now $\exists x(x=y\wedge\phi')$ and $\exists x(y=x\wedge\phi')$ are both equivalent to $\phi'(y/x)$ (since ϕ' has no quantifiers), so we have what we want if ϕ has either $x=y$ or $y=x$ as one of its atomic subformulae.

We are left with three cases. The first is that ϕ is a conjunction of formulae $y_i<x$ for some set of variables y_i, and the second is that ϕ is a conjunction of formulae $x<y_i$. In both these cases ϕ is true in **Q**, and so is equivalent in **Q** to $\neg\bot$.

The third case is that ϕ is $\phi'\wedge\phi''$, where ϕ' is a conjunction of formulae $y_i<x$ for variables y_i, where i runs over an index set I, and ϕ'' is a conjunction of formulae $x<z_j$ for variables z_j, where j runs over an index set J. It is then easy to see that $\exists x\phi$ is equivalent to **Q** to the conjunction, for all i in I and all j in J, of the formulae $y_i<z_j$. This completes the proof of the lemma, and so shows that the theory of **Q** is decidable.■

In section 13.1 we showed that the theory DLO was complete, and hence that DLO was decidable and was the theory of **Q**. The proof above can be modified to give this result. Any time we have formulae θ and ψ which are equivalent, we will need to show that $\vdash\theta\leftrightarrow\psi$. In each relevant case this is easy to prove directly; alternatively, it follows from the Completeness Theorem. Also, whenever we showed that formulae θ and ψ were equivalent in **Q**, we will need to show that DLO $\vdash\theta\leftrightarrow\psi$. For most of the times this is needed it is easy to show. The biggest difficulty is to obtain the result when θ is the formula $\exists x\phi$ of the previous paragraph and ψ is the conjunction of the formulae $y_i<z_j$ for all i in I and j in J, in the notation of the previous paragraph. It is still easy to show that DLO $\vdash\exists x\phi\rightarrow\psi$. We also need to show that DLO $\vdash\psi\rightarrow\exists x\phi$. This is most easily done by showing that DLO $\vdash\alpha_r\wedge\psi\rightarrow\exists x\phi$, where α_r runs over a set of formulae whose disjunction is in DLO. The relevant formulae α_r give the possible orderings of the y_i and of the z_j. The details are left to the reader.

14.2 THE NATURAL NUMBERS UNDER ADDITION

We have seen that the theory of the natural numbers under addition and multiplication is undecidable. Surprisingly, if we forget about multiplication, and consider the theory under addition only we get a decidable theory.

We shall always be looking at truth of sentences in **N**, and will not be considering derivations of sentences. Because of this, the distinction between symbols of the language and the corresponding functions and sets is unimportant. To make easier reading, we will no longer bother to use bold type to make this distinction. However, it is still convenient to distinguish the natural number n and the corresponding numeral **n**.

The symbols of the language will be **0**, S, +, <, =, and a further set of unary predicate symbols D_p for all $p>1$. The subset $D_{p\mathbf{N}}$ is $\{n; p \text{ divides } n\}$. We shall see in the exercises that the method of elimination of quantifiers cannot be used without the extra symbols (or some variation of them).

Let k be a positive integer and t any term. We write kt to denote the term $t+(t+(t+\ldots))$, where t occurs k times. Notice that this is simply an abbreviation, and is not a disguised way of re-introducing multiplication.

It is easy to see by induction that given any term t whose only variables are $x_{i(1)},\ldots,x_{i(n)}$, we can find $k_1,\ldots,k_n$, and m such that $t=\mathbf{m}+k_1x_{i(1)}+\ldots+k_nx_{i(n)}$ is true in **N.**

Lemma 14.2 *The set of quantifier-free sentences which are true in* **N** *is decidable.*

Proof Such a sentence involves no variables. Consequently, each term which occurs is closed, and we can find explicitly (for instance, by induction) to each such term t a number n such that $t=\mathbf{n}$ is true in **N.** If we replace each t by the corresponding **n**, we get a formula, equivalent in **N** to the original one, in which all terms that occur are numerals. We can replace this formula by an equivalent one in disjunctive normal form (and this can be explicitly found). To determine whether or not such a formula is true in **N,** we only need to consider the truth in **N** of the atomic subformulae. So we only need to find out whether or not certain formulae of form $\mathbf{m}<\mathbf{n}$, $\mathbf{m}=\mathbf{n}$, and $D_p(\mathbf{n})$ are true in **N,** and this is immediate.■

The arguments of the previous section can now be used to tell us that in order to show the theory of **N** is decidable it is enough to prove the following lemma.

Lemma 14.3 (Elimination of Quantifiers) *Let x be a variable and ϕ a quantifier-free formula. Then we can find a quantifier-free formula θ, such that θ does not involve x and only involves those variable which are involved in ϕ, for which $\exists x\phi$ is equivalent in* **N** *to θ.*

Proof As in the previous section we can assume ϕ is in disjunctive normal form. We can replace $\neg(t<t')$ by $t'<t+\mathbf{1}$, $\neg(t=t')$ by $(t<t')\vee(t'<t)$, and $t=t'$ by $(t<t'+\mathbf{1})\wedge(t'<t+\mathbf{1})$ and get a formula equivalent to ϕ in **N.** Further, we can replace $\neg D_pt$ by the disjunction of the formulae $D_p(t+\mathbf{r})$ for $0<r<p$, and the result will still be equivalent in **N** to ϕ. This is because p divides exactly one of $n, n+1,\ldots, n+p\neg 1$, for any n in **N.**

If we make the above changes and then use the distributive laws, we may assume that ϕ is in disjunctive normal form, and that the atomic formulae occurring do not involve $=$ and do not occur negated.

Now any term is equal in **N** to a term $t+kx$, where t is a term not involving x and $k\in\mathbf{N}$; here we allow $k=0$ to include the case where x does not occur. The formula $t+kx<t'+k'x$ is equivalent in **N** to one of three formulae; if $k<k'$ it is equivalent to $t<t'+(k'-k)x$, if $k>k'$ it is equivalent to $t+(k-k')x<t'$, while if $k=k'$ it is equivalent to $t<t'$. Thus we may assume our formula does not have x occurring on both sides of an inequality.

As in the previous section we can further assume our formula consists only of one disjunct, and that x occurs in every atomic subformula. Thus we are assuming that ϕ is the conjunction of various formulae of forms $t<t'+kx$, $t+kx<t'$ and $D_p(t+kx)$, where t and t' are terms not involving x, and not involving any new variables.

Now, for any positive integer n and any terms t and t' (whether or not they involve x) we know that in **N** $t<t'$ is equivalent to $nt<nt'$ and D_pt is equivalent to $D_{np}(nt)$, and that $n(t+t')=nt+nt'$ is true in **N.** It follows that the various integers k which occur in the atomic formulae may be replaced in this manner by their least common multiple. That is, we may assume that there is a fixed integer k such that ϕ is the conjunction of various formulae $t<t'+kx, t+kx<t'$, and $D_p(t+kx)$.

Let ϕ' be obtained from ϕ by replacing each kx by x. It is easy to check that $\exists x\phi$ is equivalent in **N** to $\exists x(\phi' \wedge D_kx)$. It is because of this step that we need the predicate symbols D_p, as one such symbol has now been introduced even if the original formula did not involve any such symbol.

So we can now assume that ϕ is the conjunction of formulae of the forms $t<t'+x$, $t+x<t'$, and $D_p(t+x)$, where t and t' are terms not involving x. As $\exists x\phi$ is equivalent in **N** to $\exists x(\phi \wedge \mathbf{0}<\mathbf{1}+x)$, we can assume that $\mathbf{0}<\mathbf{1}+x$ is one of the conjuncts occurring.

We now turn our attention from **N** to **Z**, the set of all integers. Because of the presence of the conjunct $\mathbf{0}<\mathbf{1}+x$, any interpretation in **Z** under which ϕ is true must send x to an element of **N**. Now any interpretation i in **N** can also be regarded as an interpretation in **Z**. By the previous remark, $\exists x\phi$ is true under i when i is regarded as an interpretation in **N** iff it is true when i is regarded as an interpretation in **Z**. Also, any quantifier-free formula is true for i in **N** iff it is true for i in **Z**. Thus we need only find a formula θ of the required form such that $\exists x\phi$ is equivalent to θ in **Z**.

We can now extend our language by including a symbol for subtraction. Once this is done, ϕ can be regarded as the conjunction of formulae of the forms $u<x$ (where u runs over some set U), $x<v$, and $D_p(t+x)$, where u and v are terms of form $t-t'$. There will be at least one formula $u<x$, because of the conjunct $\mathbf{0}<\mathbf{1}+x$.

Let m be the least common multiple (or, indeed, any common multiple) of all the integers p occurring in the various D_p. Choose some u occurring in some inequality $u<x$ and some r with $1\leqslant r\leqslant m$. Various formulae $x<v$ and $D_p(t+x)$ will occur in ϕ. For each chosen u and r, let ψ_{ur} be the conjunction of corresponding formulae $\mathbf{r}+u<v$ and $D_p((t+\mathbf{r})+u)$. Let ψ_u be the disjunction of all ψ_{ur}, and let ψ be the conjunction of all ψ_u. We shall show that $\exists x\phi$ is equivalent in **Z** to ψ, and will then find a formula θ not involving subtraction which is equivalent in **Z** to ψ.

Let i be any interpretation in **Z.** First suppose that ψ is true under i. Then ψ_u is true under i for each u, and so for each u there is r with ψ_{ur} true under i. We may choose u_0 in U such that $iu_0\geqslant iu$ for every u in U. Let r_0 be the corresponding value of r. Then, by definition of r_0, all the formulae $\mathbf{r}_0+\mathrm{u}_0<\mathrm{v}$ and $D_p(t+(\mathbf{r}_0+\mathrm{u}_0))$ are true under i. By the choice of u_0, all the formulae $u<\mathbf{r}_0+\mathrm{u}_0$ are also true under i. Thus ϕ is true under i', where $i'x=r_0+iu_0$ and $i'y=iy$ for all other variables y. Hence $\exists x\phi$ is true under i.

Conversely, suppose $\exists x\phi$ is true under i. Then there is n in **Z** such that n

satisfies, for all u, v, t, and p occurring, the conditions $iu<n$, $n<iv$, and p divides $it+n$. Choose any u. Then, as $n-iu>0$, there will be some $q\geqslant 0$ and some r with $1\leqslant r\leqslant m$ such that $r+iu=n-qm$. Since m is a multiple of every p occurring, we know that p also divides $it+r+iu$. Since $q\geqslant 0$, we still have $r+iu<iv$. Thus ψ_{ur} is true under i with this choice of r. Hence ψ_u is true under i. Since this holds for every u, ψ is true under i.

Finally, we have to replace ψ by a formula not using the subtraction sign. We begin by replacing the terms of the extended language which occur in ψ by terms which are equal to them in **Z** and which are of the form $t-t'$ for some terms t and t' of the original language. This is easily done, using the rules for addition and subtraction in **Z**. Next, any formula $t_1-t_2<t_3-t_4$ is equivalent in **Z** to $t_1+t_4<t_2+t_3$. Also, if n and n' are in **N**, p divides $n-n'$ iff there is some r with $1\leqslant r\leqslant p$ such that p divides both $n+r$ and $n'+r$. Hence $D_p(t-t')$ is equivalent in **Z** to the disjunction, for all r with $1\leqslant r\leqslant p$, of the formulae $D_p(t+\mathbf{r})\wedge D_p(t'+\mathbf{r})$. Making these changes gives the formula θ we need.■

A topic of great interest which we are not considering in this book is that of complexity of computation. In that context it can be shown that the theory of **N** under addition has a difficult decision problem (in terms, for instance, of the number of steps necessary to decide whether or not a sentence of given length is true in **N**).

We should also consider the theory of **N** under multiplication, not permitting the addition symbol. This theory is undecidable. In fact, we can define addition in terms of multiplication. Hence any sentence involving addition can be replaced by an equivalent one without addition; this means that if we could find a decision procedure for the theory of multiplication we could obtain a decision procedure for the theory with both multiplication and addition, which we know is impossible. To define addition in terms of multiplication observe that in **N** we have $a+b=c$ iff either $a=b=c=0$ or $c\neq 0$ and $(ab+1)c^2+1=(ac+1)(bc+1)$. Notice that addition of 1 is permitted, since our language contains the symbol S, whose corresponding function is just the successor function.

Exercise 14.1 Let θ be a formula in the language of addition (without the additional symbols D_p) which is quantifier-free and whose only variable is x. Show that the set $\{n;\mathbf{N}\models\theta(n)\}$ is either finite or has finite complement. Deduce that there is no quantifier-free formula involving only the variable x which is equivalent in **N** to $\exists y(x=y+y)$.

Exercise 14.2 Use elimination of quantifiers to give another proof that the theory of **N** under the successor function is decidable.

Exercise 14.3 Extend the previous exercise to give an elimination of quantifiers argument that the theory SUC is complete (and hence is decidable and also equals the theory of **N** under the successor function).

Exercise 14.4 Complete the elimination of quantifiers argument indicated in the first section to show that DLO is complete (and hence is decidable and equal to the theory of **Q** under the order relation).

Notes

CHAPTER 1

Gödel proved his incompleteness theorem in 1931 C.E. (The abbreviation AD for the date is offensive to those of us who are not Christian, and scholars now use C.E. and B.C.E. These are usually described as meaning (Before) the Common Era. But it seems to me that it is also offensive to describe this system of dating as common — though of course it is widely used — and it is best referred to as the Christian era, meaning the way Christians reckon dates.)

Russell published his paradox in 1903 C.E.

Cantor founded the theory of infinite sets in a series of papers during the last quarter of the nineteenth century C.E.

Epimenides was not aware of the paradoxical nature of his remark. The original Epimenides was a semi-legendary Cretan of about 630 B.C.E. (my classical knowledge is not good enough to use the Greek system of dating). An account of the origins of the gods written about 430 B.C.E. had its authorship ascribed to Epimenides (it was a common practice to claim ancient authorship for recently produced work). It contains a phrase translated as 'Cretans, ever liars, wretched creatures, idle bellies'. For more information about Epimenides, see West, M. L. (1983), *The Orphic poems,* Oxford, Clarendon Press, pp. 48–53.

CHAPTER 2

The word 'algorithm' looks as if it is related to the word 'logarithm'. In fact the latter word comes from Greek and the former from Arabic. But the spelling 'algorithm' has come from association with 'logarithm'; strictly the word should be 'algorism', which was the older form.

The word comes from the name of the great Islamic mathematician Abu Ja'far Muhammad ibn Musa al Khowarismi (spellings differ slightly in different books). The last part of his name, al Khowarismi, which means a man from the town of Khowarism, gave rise to the word 'algorism'. His book, written about 210 A.H. (after the hejira — the Islamic method of dating seems appropriate for an Islamic mathematician; the date corresponds to approximately 825 C.E.) called *Al-jabr wa'l muqabalah,* has given us the word 'algebra' from its first words (words beginning with 'al' are often derived from the Arabic). Another of his books introduced to Europe the Hindu system of numbers; because they first reached us from Arabic mathematicians this system is known to us as the Arabic numerals.

For more information about al-Khowarismi see Boyer, C. B. (1968) *A history of mathematics,* New York, John Wiley, and also McNaughton, R. (1982) *Elementary computability, formal languages, and automata,* Englewood Cliffs, New Jersey, Prentice-Hill, p. 12.

CHAPTER 3

While nested recursion leads, as we have seen, out of the class of primitive recursive functions, if a function so defined happens to be bounded by a primitive recursive function then it is primitive recursive. For this and many other results on multiple and nested recursion, and other properties, see Peter, R. (1967) *Recursive functions,* New York, Academic Press (translated from the 1957 German text published in Budapest).

CHAPTER 5

The name 'abacus machine' comes from the book by Boolos and Jeffrey listed under 'Further reading', but my definition is not theirs. I learned about these machines from Egon Börger, but they are due originally to Rödding.

Unlimited register machines were introduced in Shepherdson, J. C. and Sturgis, H. E. (1963) Computability of recursive functions, *J. Ass. Computing Machinery* **10,** 217–255. As a result these machines are often called Shepherdson–Sturgis machines. In their paper they also define directly partial recursive functions with domain the strings over any finite alphabet, and show that this definition is equivalent to the usual one. They are then able to prove that partial recursive functions are Turing computable using any finite alphabet.

It is known that, with a suitable coding of n-tuples, any partial recursive function can be computed by an abacus machine with only three registers. Further, any partial recursive function can be computed by a register program with only two registers (but not by an abacus machine with only two registers). These results are proved in Börger, E. (1975) Recursively unsolvable algorithmic problems and related questions re-examined, in: Müller, G. H., Oberschelp, A., and Potthoff, K. (eds.), *ISILC logic conference: proceedings of the international summer institute and logic colloquium, Kiel,* 1974, Berlin, Springer (Lecture notes in mathematics, 499). These results are also in Börger's book given under 'Further reading'.

CHAPTER 6

The paper from which I have quoted is Turing, A. M. (1936) On computable numbers, with an application to the Entscheidungsproblem, *Proc. London Math. Soc.* (series 2) **42,** 230–265. I am grateful to the London Mathematical Society for permission to quote from this paper.

CHAPTER 7

Modular machines were originally introduced by Stal Aanderaa in a talk given to the conference on 'Decision problems in algebra' at Oxford in 1976, where he showed they provide a particularly simple approach to the unsolvability of the word problem for groups. This work was expanded and written up jointly with myself as Aanderaa, S. and Cohen, D. E. (1980) Modular machines, the word problem for finitely presented groups and Collins' theorem, in: Adjan, S. I., Boone, W. W., and Higman, G. (eds.), *Word problems* II: *The Oxford book,* Amsterdam, North-Holland, pp. 1–16, and the following paper in the same book. I later carried out further development of the theory of modular machines. The material in the present chapter and Chapter 13 on modular machines was originally presented in Cohen, D. E. (1984) Modular machines, undecidability and incompleteness, in: Börger, E., Hasenjager, G., and Rödding, D. (eds.), *Logic and machines: decision problems and complexity,* Berlin, Springer (Lecture notes in computer science, 171).

CHAPTER 9

The first proof that every r.e. set is exponential diophantine was given in Davis, M., Putnam, H., and Robinson, J. (1961) The decision problem for exponential diophantine equations, *Annals of Mathematics* (series 2) **74,** 425–436. The main reason I included in the current book an exercise stating that r.e. sets are closed under bounded universal quantification was that their proof needed to look at bounded universal quantification of diophantine sets. The proof given here that r.e. sets are exponential diophantine comes from Jones, J. P. and Matiyasevič, Y. V. (1984) Register machine proof of the theorem on exponential diophantine representation of enumerable sets, *Journal of symbolic logic* **49,** 818–829.

It was known from the time of the first cited paper that if there is any function of roughly exponential growth which is diophantine then all r.e. sets are diophantine. The first detailed construction of such a function occurs in Matijasevič, Y. V. (1972) Diophantine sets. *Uspehi mat. nauk* **27,** 185–222, translated as *Russian mathematical surveys* **27,** 124–164. The proof given here follows (until the last step) Davis, M. (1973) Hilbert's tenth problem is unsolvable, *American mathematical monthly* **80,** 233–269, which is a very readable paper.

Some of the uses of diophantine properties (for instance, to obtain Kleene's Normal Form Theorem) are folklore; I have not encountered detailed proofs elsewhere.

There are many other interesting questions about diophantine properties. We can ask for an explicit polynomial whose positive values are the primes. We can ask what bounds can be found on the degree and the number of variables in the universal polynomial. More information about such questions can be found in Davis, M., Matijasevič, Y. V., and Robinson, J. (1974) Hilbert's tenth problem. Diophantine equations: positive aspects of a negative solution, in: *Mathematical developments arising from Hilbert problems,* Providence, American Mathematical Society (Proceedings of symposia in pure mathematics, 28).

CHAPTER 11

We tend in our everyday use of expressions like 'If A then B' to mean something closer to 'A. Therefore B' than to the formal meaning of $\rightarrow$. About the only time our use is identical to the formal one is when we express our disbelief in some statement by saying, for instance, 'If *Rambo* is a good film then I'm a Dutchman'.

Russell is supposed to have been asked once, 'Given that $1=2$, show that you are the Pope'. Strictly speaking, this is a misunderstanding of the use of 'If... then'. However, Russell replied, 'The Pope and I are two. Therefore the Pope and I are one'.

Lewis Carroll, author of the 'Alice' books, was a logician. In a poem in his lesser-known book 'Sylvie and Bruno' he wrote, well before Russell,

> He thought he saw an argument
> That proved he was the Pope.

I have often wondered if he had discovered Russell's argument.

CHAPTER 13

A readable account of interesting true but unprovable theorems is given in C. Smorynski's papers in *Harvey Friedman's research on the foundations of mathematics,* Harrington, I. A. *et al.*, Amsterdam, North-Holland (1985) (Studies in logic and the foundations of mathematics, 117).

Further reading

COMPUTABILITY

Cutland, N.J. (1980) *Computability*. Cambridge University Press.

This text is intended for the same audience as my book. It bases the major developments on (a variant of) register programs. The similarities to and differences from the current book make it excellent supplementary reading.

Boolos, G.S. and Jeffrey, R.C. (1974) *Computability and logic*. Cambridge University Press.

An interesting approach to computability, and some extensive results on logic. The flow diagram approach permits easy construction of various Turing machines which are messy to define by quintuples; unfortunately, they never give a precise definition of a flow diagram, but leave the concept to be understood by examples.

Machtey, M. and Young, P. (1978) *An introduction to the general theory of algorithms*. New York, North-Holland.

This is my favourite among the many books using a programming approach. It has a good treatment of questions about complexity of computation.

Hopcroft, J.E. and Ullman, J.D. (1979) *Introduction to automata theory, languages, and computation*. Reading, Mass, Addison–Wesley.

I was probably exaggerating in calling the theory of computable functions the pure mathematics of computer science. It is undoubtedly an important part of such a concept. But the theory of formal languages and automata should probably also be regarded as part of the pure mathematics of computer science. This book is a classic in its treatment of all these topics.

Börger, E. (1987) *Computability, complexity, logic*. Amsterdam, North-Holland.

This is a translation from a German original (*Berechenbarkeit, Komplexität, Logik*. Wiesbaden, Vieweg) of which I have only seen the contents list and the publicity material. It covers a great deal of material, both in computability theory and in logic. It is intended both as a reference book and as a text. The author, whose papers I have always found very clear, uses very similar techniques to the ones I use where we cover similar material. I expect this book to be an excellent one for a further course on computability and logic.

LOGIC

van Dalen, D. (1980) *Logic and structure*. Berlin, Springer.

A more detailed account of logic than mine, also using natural deduction.

Enderton, H.D. (1972) *A mathematical introduction to logic*. New York, Academic Press.

This is the clearest account I know using the axiomatic method to define derivations.

Bell, J. L. and Machover, M. (1977) *A course in mathematical logic*. Amsterdam, North-Holland.

This book is intended for a one-year M.Sc. course in logic and computability. It is a very large book, useful as a reference text. It uses the tableau method for derivations.

Hodges, W. (1977) *Logic*. Harmondsworth (Middlesex, England), Penguin Books.

A very good introduction to logic. It covers philosophical and formal aspects, and is particularly careful to show the problems involved in looking formally at ordinary sentences. His method of derivation is the tableau method.

GENERAL READING

Newman, J.R. (ed.) (1956) *The world of mathematics* (first published in the USA; first publication in England 1960, London, George Allen & Unwin).

A fascinating collection of essays on many mathematical topics, by many authors. Especially relevant are the essays in Part XIX, Mathematical machines: can a machine think? and the article by E. Nagel and J.R. Newman on Gödel's proof (also published as a book entitled *Gödel's proof*).

Anderson, A.R. (ed.) (1964) *Minds and machines*. Englewood Cliffs, New Jersey, Prentice-Hall.

A collection of articles about computers and thought, with arguments for and against the suggestions that computers can (in principle) think and that human brains are necessarily different from computers.

Hofstadter, D. R. (1979) *Gödel, Escher, Bach: an eternal golden braid.* New York, Basic Books (first UK publication by Harvester Press, currently available in Penguin).

A long and rambling book, which you will either love or hate. It is a series of variations on the theme of self-reference in its many aspects, including Gödel's Theorem. It is intended for the interested reader with no special knowledge of mathematics.

Smullyan, R. (1978) *What is the name of this book?* Englewood Cliffs, New Jersey, Prentice-Hall.

Smullyan, R. (1982) *The lady or the tiger?* New York, Knopf.

Ray Smullyan is well known both as a logician (whose two books on formal logic take an unusual and interesting viewpoint) and as a magician. I have once seen him give a proof, using magic and logic, that mathematics is inconsistent. He said, 'Either mathematics is inconsistent or the coin in my hand is a quarter'. I looked carefully and saw that the coin in his hand was a quarter, so his statement was true. I didn't stop looking at the coin in his hand, but he showed me that the coin was in fact a dime!

In the books mentioned (and other popular books of his, published in England by Penguin Books; he also has some books of unusual chess problems) he talks about various aspects of self-reference. He then goes on to provide a collection of logical puzzles. They are mostly complicated extensions of the old puzzle about the island where there are two tribes, one of which always tells the truth, the other always lies, and one wants to find out what tribes certain people belong to. Probably only Smullyan could finish up with versions of such puzzles which essentially amount to the Incompleteness or Undecidability Theorems.

Index of special symbols

A (Ackermann's function), 53
A (axioms for number theory), 207
a_k, 67, 70, 76, 89
Add_k, 89

B (axioms for number theory), 215
B_0 (axioms for number theory), 207

c (index for composition), 130
Cat, 130
Clear_k, 70
co, 38
Comp, 105, 106
COMP, 100
Comp_M, 98
$\mathrm{Copy}_{p,q,r}$, 70
$C(\Delta)$ (consequences of Δ), 203

D_k, 125
$\mathrm{Descopy}_{p,q}$, 70
DLO, 204
dom, 20

exp_n, 44
exq, 47

F (false), 155
F (sentences false in **N**), 212

In, 105, 106
IN, 100
In_M, 97
$\mathrm{In}_{M,r}$, 97
$\mathrm{In}_{T,k}$, 89

J (function, 22, 40
J^* (function), 47
J_k (function), 22, 40
J^*_n (function), 48
$J(i_1)$ (jump), 78
$J_k(i_1)$ (jump), 78
$J'_k(i_1)$ (jump), 78
$J_k(i_1, i_2)$ (jump), 76

K (set), 31, 65
K (function), 22, 40, 43
K^*, 47

L (direction symbol), 85, 96
L (function), 22, 40, 43
L (language), 174
L (Turing machine), 85
L^* (function), 47
L^* (Turing machine), 85
$L(\mathbf{A})$ (language), 182
Large, 129

n (numeral), 207
Next, 104, 106
NEXT (function), 100
NEXT (relation), 107
Next_M, 98

Out, 104, 106
OUT, 100
Out_M, 97
Out_T, 98, 90

p (padding function), 133
pd, 38
p_n (nth prime), 44
p_n (propositional variable), 147
P_0, P_1 (Turing machines), 85
P_i (polynomial), 126

quo, 39, 44

R (direction symbol), 85, 96
R, R^* (Turing machines), 85
R_i (r.e. set), 137
rem, 39, 44

s (function in the s-m-n theorem), 131
S (successor function), 36
s_k, 67, 70, 76, 89
Seq, 129
sg, 38
Shiftleft, Shiftright, 90
sqr, 41, 47
Step, 137
Sub_k, 89
SUC, 206

T (true), 155
T (sentences true in **N**), 212
Term, 106
TERM, 100
Term_M, 98
Test, 85
Test_k, 89
$\mathrm{Test}\{T_0, T_1\}$, 87
Th**A**, 203

W_i, 137

x_n, $x_n(a)$ (solutions of a difference equation), 119
x_n (variables of predicate logic), 174

y_n, $y_n(a)$, 119

Z (zero function), 36

$\alpha[\Phi]$, 172
γ (sequencing function), 123
δ_n (diagonal sentence), 214
μ, 25
μ', 43
π_{ni} (projection functions), 36
Φ, 29, 130
ϕ_n, 31, 130
$\phi(t/x)$, 181
χ_A, χ_{Ap}, 21

$\dot{-}$, (modified subtraction), 38
$\wedge$, 41, 147, 155
$\vee$, 41, 156
$-$, 41, 147, 155
$\rightarrow$, 147, 155
$\leftrightarrow$, 147, 157
$\perp$, 147
$\forall$, 41, 174
$\exists$, 41, 183
$\vdash$. $\vdash$, 163
$\models$, $\vdash$, 158
$\leqslant$, 116

Index

Aanderaa, S., 232
abacus, 67
abacus computable function, 71, 90, 99
abacus machine, 67
  primitive, 74
  simple, 67
  simple primitive, 75
accepted set, 107
accepting subset, 107
Ackermann's function, 53
add instruction, 76
addition, 37, 38
adequacy, 167, 170, 191
adequacy theorem
  for propositional logic, 170
  for predicate logic, 196
admissible set, 55
algorithm, 26, 230
  partial, 26
al-Khowarismi, 230
allowable function, 58
alphabet, 22, 84
Anderson, A. R., 235
argument, 145, 146
  correct, 145, 146, 158
  incorrect, 145, 146
arity, 175
associate, 96
  left-, 96
  right-, 96
atomic formula, 176
axiomatic method, 159
axioms
  for a theory, 203
  of equality, 197

base-2 exponential diophantine set, 111
base-2 exponential polynomial, 111
Bell, J. L. 235
blank, 84
Boolos, G. S., 231, 234
Börger, E., 231, 238
bounded minimisation, 43
bounded quantification, 42
bound occurrence of a variable, 180
busy beaver problem, 94

Cantor, G., 18, 230
Cantor's diagonal argument, 18, 29, 213
Carroll, Lewis, 233
characteristic function, 21
  partial, 21
Chinese Remainder Theorem, 123
Church's Thesis, 103
clearing a register, 70
closed term, 175
code, 54, 104, 105, 106
Cohen, D. E., 232
compactness theorem, 199
complete consistent set, 169, 192
completeness theorem, Gödel's, 145, 167, 190
complete theory, 204
component functions, 24
composite number, 44
composite of functions, 21, 37
computable functions, 23, 24
  informally, 24
  intuitively, 24
computation
  of a modular machine, 97
  of a register program, 76
  of a Turing machine, 86
concatenation of strings, 22
conclusion, 163, 186
configuration, 105
  of a modular machine, 96
  of a register program, 76
  of a Turing machine, 85
  terminal, 76, 85, 96
  yielded by another, 85, 96
conjunction, 41, 157
consequence, 203
  semantic, 157, 183

consistent set of formulae, 168, 192
  complete, 169, 192
  maximal, 169, 192
constant symbol, 194
contradiction, 157
coordinate functions, 40
copy of a register, 70
  destructive, 70
cosign, 38
counter-image, 64
course-of-values recursion, 45
Cutland, N. J., 234

Davis, M., 119, 232, 233
decidable set, 25
decidable theory, 203
definable set, 215
defining sequence, 37
definition by cases, 39
depth, 68
derivation, 158, 161, 185
description, tape, 84
  corresponding to ξ, 88
destructive copy, 70
diagonal argument, Cantor's, 18, 29, 213
diophantine
  function, 112
  indexing, 130, 139
  property, 112
  set, 111, 209
    base-2 exponential, 111
    exponential, 111
    main theorem on, 113
    unary exponential, 111
disjunction, 41, 157
disjunctive normal form, 171, 202
divides, 44
domain of a function, 20
double induction, proof by, 58
double recursion, 47
dovetailing, 27
downward Lowenheim–Skolem theorem, 200

elimination of quantifiers, 223, 226
elimination
  of $\wedge$, 161, 185
  of $\rightarrow$, 162, 185
  of $\forall$, 186
empty function, 20
empty register, 70
Enderton, H. D., 235
entry in register, 70
Epimenides, 17, 230
equivalent formulae, 170, 201
equivalent machines and programs, 76
Euler's identity, 127
excess over a square, 47
existential quantification, 41
existential quantifier, 183
exponent, 44
exponential diophantine set, 111
exponential polynomial, 111
exponentiation, 38
expressible set, 208
  strongly, 216
extensions of a function, 21

false, 155
false sentence, 183
first-order predicate logic, 174
formation sequence, 154
formula
  atomic, 176
  Henkin, 194
  of predicate logic, 175, 176
  of propositional logic, 148
  satisfiable, 157
formulae, equivalent, 170
four squares theorem, 127
free for x in ϕ, 181
free occurrence of a variable, 180
function
  computed by an abacus machine, 70
  computed by a modular machine, 97
  computed by a register program, 76
  computed by a Turing machine, 89
  partial, 20
  total, 20
function symbol, 174

general undecidability theorem, 212
Gödel, K., 18, 230
Gödel number
  of a configuration of a register program, 104
  of a modular machine, 99
Gödel numbering, 100, 130, 176
Gödel's Completeness Theorem, 145, 167, 190
Gödel's sequencing function, 123
Gödel's Incompleteness Theorem, 18
  original form, 220
  Rosser's extension of, 221
  strong form, 213
  weak form, 213
Gödel's Second Incompleteness Theorem, 222
good set, 48
  very, 48

halting problems, 93
halting set, 101
halting state, 86
head, read-write, 84

Henkin formula, 194
Hilbert, D., 111
Hilbert's Tenth Problem, 111, 114
history of a function, 45
Hodges, W., 235
Hofstadter, D. R., 236
Hopcroft, J. E., 234
hypotheses, 163, 186

incompleteness theorem, Gödel's, *see* Gödel's Incompleteness Theorem
inconsistent set, 168, 192
index of a function, 31, 130
indexing of functions, 130
  acceptable, 130
  diophantine, 130, 139
  modular, 130
  universal, 130
indexing of r.e. sets, 137
induction, principle of
  for formulae of predicate logic, 176
  for formulae of propositional logic, 149
  for terms, 175
informally computable function, 24
initial function, 36
input function, 97, 104, 106
instruction, 76
  add, 76
  jump, 76
  stop, 76
  substract, 76
interpretation, 184
introduction
  of $\wedge$, 161, 185
  of $\rightarrow$, 162, 185
  of $\forall$, 186
  of $\vee$, left- and right-, 164
  of $\exists$, left- and right-, 187
intuitively computable function, 24
invalid sentence, 183
isomorphism, 198
isomorphism theorem, Rogers', 135
iterate of a function, 40

Jeffrey, R. C., 234
join, 22
Jones, J. P., 232
jump instruction, 76
jump, unconditional, 78

Kleene's Normal Form Theorem, 63, 100, 126
Kleene's *s-m-n* Theorem, 131

label, 76

Lagrange, 127
language
  of first-order predicate logic, 174
    pure, 197
    with equality, 197
  of number theory, 206
  of propositional logic, 147
leaf, 160
  marked, 160
  unmarked, 160
left-associate, 96
left-introduction
  of $\vee$, 164
  of $\exists$, 187
length
  of a number, 44
  of a string, 22
letters, 84
level of a function, 55
line, 76
listable set, 25
logic, 145
  first-order predicate, 174
    undecidability of, 212
  propositional, 145
logical symbols, 147, 174
Lowenheim–Skolem theorems, 200
L-structure, 182

machine
  abacus, *see* abacus machine
  modular, *see* modular machine
  random access, 80
  Turing, *see* Turing machine
  unlimited register, 76
Machover, M., 235
Machtey, M., 234
Matiyasevič, Y. V., 232, 233
maximal consistent set, 169, 192
min-computable function, 124
  regular, 124
minimisation, 25, 32
  bounded, 43
  pseudo-, 32
  regular, 60
model, 183
modular indexing, 130
modular machine, 96, 211
  corresponding to a Turing machine, 96
  function computed by a, 97
  special, 96
multiplication, 38

natural deduction, 159
natural numbers under addition, 225
negation, 41, 157
nested recursion, 53
Newman, J. R., 235

non-determinism, 107
non-deterministic Turing machine, 85
normal form theorem, Kleene's, 63, 100, 126
number, Gödel, 99, 104
numbering, Gödel, 108, 130, 176
number theory
  language of, 206
  undecidability of, 212
numeral, 207

occurrence of a variable
  bound, 180
  free, 180
output function, 89, 97, 106

padding function, 133
pairing, 129
pairing function, 40
parentheses, 67, 147, 174
parsing tree, 152
partial algorithm, 26
partial characteristic function, 21
partial computable function, uiniversal, 30
partial function, 20
partial recursive function, 60, 73, 99
partial valuation, 157
Peter, R., 230
polynomial, 111
  exponential, 111
    base-2, 111
    unary, 111
predecessor, 38
predicate, 41
predicate logic, first-order, 174
  language of, 174
predicate symbol, 174
prenex normal form, 201
prime, 44
primitive abacus machine, 74
  simple, 75
primitive recursion, 36
  simultaneous, 45
primitive recursive function, 37, 75
primitive recursively closed set, 37
principle of induction
  for formulae
    of predicate logic, 176
    of propositional logic, 149
  for terms, 175
problem, busy beaver, 894
problems, halting, 93
program, 23, 24, 30
  register, 76, 104, 115
projection functions, 36
proof, 145, 158
property, 41
  diophantine, 112
  regular min-computable, 124
propositional logic, 145
  language of, 147
propositional variables, 147
pseudo-minimisation, 32
pure language, 197
Putnam, H., 232

quantification
  bounded, 42
  existential, 41
  universal, 41
quantifier
  existential, 183
  universal, 174
quantifiers, elimination of, 223, 226
quotient, 39

random access machine, 80
read-write head, 84
recursion
  course-of-values, 45
  double, 47
  nested, 53
  primitive, 36
recursion theorem, 132
recursive function, 61
  partial, 60
recursively axiomatisable theory, 203
recursively enumerable set, 63
recursive set, 63
register, 70
  used by an abacus machine, 71
register program, 76, 104, 115
regular min-computable
  function, 124
  property, 124
regular minimisation, 60
remainder, 39
replacement, 181
representable function, 208
  strongly, 216
r.e. set, 63
restriction of a function, 21
Rice–Shapiro Theorem, 34, 65, 136
Rice's Theorem, 32, 65
right-associate, 96
right-introduction
  of ∨, 164
  of ∃, 187
Robinson, J., 232, 233
Rödding, D., 231
Rogers' Isomorphism Theorem, 135
root, 160
Rosser's extension of Gödel's Incompleteness Theorem, 221
rule of ⊥, 162, 186
rule RAA, 162, 186

Russell, B., 18, 230, 233

satisfiable formula, 157
satisfaction, 157
scanned square, 84
segment, 22
  initial, 22
  proper, 22
self-reference, 243
semantic consequence, 157, 180
sentence, 180
  false, 183
  invalid, 183
  true, 183
  valid, 183
sequencing function, Gödel's, 123
Shepherdson, J. C., 231
sign, 30
simple abacus machine, 67
  primitive, 75
simulation, 88
simultaneous primitive recursion, 45
s-m-n theorem, Kleene's, 131
Smorynski, C., 233
Smullyan, R., 236
soundness, 167
soundness theorem
  for predicate logic, 190
  for propositional logic, 167
special modular machine, 96
square, scanned, 84
starting state, 86
state, 85
  halting, 86
  starting, 86
stop instruction, 76
string, 22
strongly expressible set, 216
strongly representable function, 216
structure, 182
Sturgis, H. E., 231
subformula, 153, 180
substitution, 172, 181
subtract instruction, 76
sucessor function, 36
symbols
  constant, 174
  function, 174
  logical, 147, 174
  predicate, 174

tape, 84
  description, 84
    corresponding to ξ, 88
tautology, 157
term, 175
  closed, 175
  free for x in ϕ, 182
terminal configuration, 76, 85, 96
theorem, 163, 186
theory, 203
  complete, 204
  decidable, 203
  recursively axiomatisable, 203
total function, 20
tree, 159
  parsing, 152
true sentence, 183
truth, 145, 155
  undecidability of, 213
  under an interpretation, 184
  undefinability of, 215
truth-table, 155
Turing, A. M., 80, 232
Turing computable function, 90, 99
Turing machine, 95
  by quadruples, 85
  by quintuples, 87, 95
  non-deterministic, 85
  universal, 95

Ullman, J. O., 234
unary exponential
  diophantine set, 111
  polynomial, 111
unconditional jump, 78
undecidability
  of logic, 212
  of number theory, 212
  of truth, 213
undecidability theorem, general, 212
undefinability of truth, 215
unique reading lemma
  for formulae of predicate logic, 179
  for formulae of propositional logic, 151
  for terms, 179
universal diophantine predicate, 125
universal partial computable function, 30
universal quantification, 41
universal quantifier, 174
universal Turing machine, 93
unlimited register machine, 76
upward Lowenheim–Skolem theorem, 200

valid sentence, 183
valuation, 155
  partial, 157
van Dalen, D., 235
variable, 174
  propositional, 147
vertex, 160

witness, 194

Young, P., 234

zero function, 36

ω-consistent set, 220

Faux, I.D. & Pratt, M.J.	**Computational Geometry for Design and Manufacture**
Firby, P.A. & Gardiner, C.F.	**Surface Topology**
Gardiner, C.F.	**Modern Algebra**
Gardiner, C.F.	**Algebraic Structures: with Applications**
Gasson, P.C.	**Geometry of Spatial Forms**
Goodbody, A.M.	**Cartesian Tensors**
Goult, R.J.	**Applied Linear Algebra**
Graham, A.	**Kronecker Products and Matrix Calculus: with Applications**
Graham, A.	**Matrix Theory and Applications for Engineers and Mathematicians**
Graham, A.	**Nonnegative Matrices and Other Topics in Linear Algebra**
Griffel, D.H.	**Applied Functional Analysis**
Griffel, D.H.	**Linear Algebra**
Hanyga, A.	**Mathematical Theory of Non-linear Elasticity**
Harris, D.J.	**Mathematics for Business, Management and Economics**
Hoskins, R.F.	**Generalised Functions**
Hoskins, R.F.	**Standard and Non-standard Analysis**
Hunter, S.C.	**Mechanics of Continuous Media, 2nd (Revised) Edition**
Huntley, I. & Johnson, R.M.	**Linear and Nonlinear Differential Equations**
Jaswon, M.A. & Rose, M.A.	**Crystal Symmetry: The Theory of Colour Crystallography**
Johnson, R.M.	**Theory and Applications of Linear Differential and Difference Equations**
Kim, K.H. & Roush, F.W.	**Applied Abstract Algebra**
Kim, K.H. & Roush, F.W.	**Team Theory**
Kosinski, W.	**Field Singularities and Wave Analysis in Continuum Mechanics**
Krishnamurthy, V.	**Combinatorics: Theory and Applications**
Lindfield, G. & Penny, J.E.T.	**Microcomputers in Numerical Analysis**
Lord, E.A. & Wilson, C.B.	**The Mathematical Description of Shape and Form**
Marichev, O.I.	**Integral Transforms of Higher Transcendental Functions**
Massey, B.S.	**Measures in Science and Engineering**
Meek, B.L. & Fairthorne, S.	**Using Computers**
Mikolas, M.	**Real Functions and Orthogonal Series**
Moore, R.	**Computational Functional Analysis**
Müller-Pfeiffer, E.	**Spectral Theory of Ordinary Differential Operators**
Murphy, J.A. & McShane, B.	**Computation in Numerical Analysis**
Nonweiler, T.R.F.	**Computational Mathematics: An Introduction to Numerical Approximation**
Ogden, R.W.	**Non-linear Elastic Deformations**
Oldknow, A.	**Microcomputers in Geometry**
Oldknow, A. & Smith, D.	**Learning Mathematics with Micros**
O'Neill, M.E. & Chorlton, F.	**Ideal and Incompressible Fluid Dynamics**
O'Neill, M.E. & Chorlton, F.	**Viscous and Compressible Fluid Dynamics**
Page, S. G.	**Mathematics: A Second Start**
Rankin, R.A.	**Modular Forms**
Ratschek, H. & Rokne, J.	**Computer Methods for the Range of Functions**
Scorer, R.S.	**Environmental Aerodynamics**
Smith, D.K.	**Network Optimisation Practice: A Computational Guide**
Srivastava, H.M. & Karlsson, P.W.	**Multiple Gaussian Hypergeometric Series**
Srivastava, H.M. & Manocha, H.L.	**A Treatise on Generating Functions**
Shivamoggi, B.K.	**Stability of Parallel Gas Flows**
Stirling, D.S.G.	**Mathematical Analysis**
Sweet, M.V.	**Algebra, Geometry and Trigonometry in Science, Engineering and Mathematics**
Temperley, H.N.V. & Trevena, D.H.	**Liquids and Their Properties**
Temperley, H.N.V.	**Graph Theory and Applications**
Thom, R.	**Mathematical Models of Morphogenesis**
Toth, G.	**Harmonic and Minimal Maps**
Townend, M. S.	**Mathematics in Sport**
Twizell, E.H.	**Computational Methods for Partial Differential Equations**
Wheeler, R.F.	**Rethinking Mathematical Concepts**
Willmore, T.J.	**Total Curvature in Riemannian Geometry**
Willmore, T.J. & Hitchin, N.	**Global Riemannian Geometry**
Wojtynski, W.	**Lie Groups and Lie Algebras**

Statistics and Operational Research

Editor: B. W. CONOLLY, Professor of Operational Research, Queen Mary College, University of London

Beaumont, G.P.	**Introductory Applied Probability**
Beaumont, G.P.	**Probability and Random Variables**
Conolly, B.W.	**Techniques in Operational Research: Vol. 1, Queueing Systems**

Conolly, B.W.	**Lecture Notes in Queueing Systems**
Conolly, B.W.	**Techniques in Operational Research: Vol. 2, Models, Search, Randomization**
French, S.	**Sequencing and Scheduling: Mathematics of the Job Shop**
French, S.	**Decision Theory: An Introduction to the Mathematics of Rationality**
Griffiths, P. & Hill, I.D.	**Applied Statistics Algorithms**
Hartley, R.	**Linear and Non-linear Programming**
Jolliffe, F.R.	**Survey Design and Analysis**
Jones, A.J.	**Game Theory**
Kemp, K.W.	**Dice, Data and Decisions: Introductory Statistics**
Oliveira-Pinto, F.	**Simulation Concepts in Mathematical Modelling**
Oliveira-Pinto, F. & Conolly, B.W.	**Applicable Mathematics of Non-physical Phenomena**
Schendel, U.	**Introduction to Numerical Methods for Parallel Computers**
Stoodley, K.D.C.	**Applied and Computational Statistics: A First Course**
Stoodley, K.D.C., Lewis, T. & Stainton, C.L.S.	**Applied Statistical Techniques**
Thomas, L.C.	**Games, Theory and Applications**
Whitehead, J.R.	**The Design and Analysis of Sequential Clinical Trials**